Linear Algebra With Matrix Theory

線性代數與矩陣理論

陳長城

東華書局

國家圖書館出版品預行編目資料

線性代數與矩陣理論 ＝ Linear algebra with matrix theory ／陳長城著． -- 初版． --
臺北市 ：臺灣東華, 民 94
　　面； 　　公分
參考書目 ：面
ISBN 957-483-316-X（平裝）

1.線性代數　2.矩陣

313.3　　　　　　　　　　　　94007397

版權所有・翻印必究

中華民國九十四年六月初版

線性代數與矩陣理論

定價　新臺幣肆佰貳拾元整
（外埠酌加運費匯費）

著　者　陳　　長　　城
發行人　卓　　鑫　　淼
出版者　臺灣東華書局股份有限公司
　　　　臺北市重慶南路一段一四七號三樓
　　　　電話：（02）2311-4027
　　　　傳真：（02）2311-6615
　　　　郵撥：0 0 0 6 4 8 1 3
　　　　網址：http://www.bookcake.com.tw
印刷者　昶　順　印　刷　廠

行政院新聞局登記證　局版臺業字第零柒貳伍號

Dedicated to
Alicia, Hope and **Molly**

自　　序

　　幾十年來，臺灣的各大學理工科系用書一直都選擇以原文書籍為主要授課用書。尤其是，近年來各大學在追求卓越、講究與國際接軌計畫的驅使之下，除了採用原文本書籍之外，更有號稱上課時全程以英文講授者。倘若前述做法能夠順利進行，學生學習成效又都能令人滿意的話，那真是可喜可賀之事。可是，根據私下進行調查發現，採用原文書籍並且全程以英文講授課程的情形之下，學生除了在語文方面有所精進之外，一般而言，學生對於專業的理解有逐漸鈍化或退步的現象。特別是理工科目的傳授，同學們課堂上普遍的反映是，不知老師所言者何物。此一現象說明著，當追求卓越與國際化之同時，我們似乎忽略了本國同學們的感受和學習成效。

　　以數學系為例，眾所皆知，要研讀一門專業課程，對於該課程之定理、定義甚至於例題的背景說明一定要有所能夠體會，才能融會貫通、掌握其之要義。然而，對於原文閱讀不甚熟稔之本國同學而言，要了解定理或定義之背景說明，實在有其難處。此時，老師上課時若又都以英文講解的話，對於學生的學習結果無疑是雪上加霜，難上加難。

　　有鑑於此，筆者在編纂此書之當時，考慮到同學們的吸收程度，特別以中、英文混合使用的方式進行編寫。除了定理、定義、例題以及習題皆以英文敘述外，在每個定理、定義之前後，或者是部分定理之證明中，筆者特別使用中文說明其背景之意義和證明。如此一來，無論講台上的老師是以英文授課也好、中文講解也罷，同學們都能夠利用課餘時間，自行閱讀本書；不但能充分的了解課程內容，也不會有與國際脫軌的現象發生。這種編寫方式實際上是符合國內學術界生態習性的，它適合以中文為母語的讀者研讀，它將會逐漸形成國內理工科目編纂的趨勢。

　　本書共分為九個 Chapters，從線性空間開始，在認識矩陣的基本運算以及相關領域之後，第四、五章將介紹矩陣在線型映射以及雙線型映射上所扮演的角色；第六章是行列式的介紹和 Cramer's rule 的應用；最後，第七、

八、九章將是矩陣理論與線性代數之精華所在；尤其是，Spectral decomposition 之"組構冪等矩陣的線性組合"是這門科目中的上乘武功。特別是在 8.4 節中，作者以引用 interpolatory polynomials 的方式，解構一個可對角化矩陣的做法更是精彩萬分。這個做法一般坊間書籍不易多見，它無需經由求出矩陣的 eigenvectors，即可求出該矩陣的組構冪等矩陣，真是方便至極。此外，本書的另一個特色是，除了每一個 section 之後，都列有習題之外，文章內容中隨時都穿插著給同學們練習的一些關鍵性的概念題目。

自二十世紀中期以降，矩陣理論已經逐漸的由線性代數的附屬地位脫離，而自成一門獨立的學問。尤其是，資訊、電機等科系興旺的近期年代，一切講求以數值方法計算的年代，矩陣理論在所有應用科學計算的領域裡扮演著舉足輕重的要角。因此之故，筆者便以線性代數與矩陣理論相伴隨的角度來編寫此書。它極為適合一學年的線性代數用書，尤其特別適合一些工程科系或是應用科學科系之同學們的閱讀。以一學年上下學期各三學分的線性代數課程講授而言，筆者建議可以從第一章介紹認識向量空間開始，接著介紹 2.1、2.2、3.1 節之後，再來就是第四章以及第五章的 5.1、5.2、5.3 節；下學期則按照第六、七、八、九章的順序進行教學，最後當可給本科目畫下完美的句點。那麼，至於矩陣理論的課程而言，可按授課時數之多寡，從介紹第二章開始，一直到最後，章章都是值得和同學們共同研讀的內容。

此書之編纂著實不易，除了豐富的教學經驗之外，還必須有客觀的教學環境配合。在此，筆者要誠摯的感謝逢甲大學所提供的教學環境、資源以及完善的教學設備，使得本書得以順利付梓。當然，筆者也一併要感謝系上學術界的前輩和同仁們的指導和相互砥礪，使得本著作內容能夠更加嚴謹，前後章節連結更為順暢。初次成稿問世，若有不盡完備之處，還盼望各界學者專家不吝指正是禱。

陳長城 謹序
西元 2005 年 6 月於
逢甲大學理學院

Contents

Chapter One　Vector Spaces　　　　　　　　　　　　　　　**1**
1.1　Definitions and some properties　　　　　　　　　　　1
1.2　Bases and dimensions　　　　　　　　　　　　　　　13
1.3　Sums and direct sums　　　　　　　　　　　　　　　25

Chapter Two　Matrices　　　　　　　　　　　　　　　　　**31**
2.1　Foundations　　　　　　　　　　　　　　　　　　　31
2.2　Column spaces, row spaces and system of linear equations　　43
2.3　Row-echelon form and Gauss-Jordan elimination　　　　50
2.4　Elementary matrices　　　　　　　　　　　　　　　61

Chapter Three　Ranks、Non-singularities、*LU*-factorizations
　　　　　　　　and Nilpotent Matrices　　　　　　　　　**67**
3.1　Null spaces, ranks, criteria for non-singularities　　　　　67
3.2　Triangular matrices and *LU*-factorizations　　　　　　78
3.3　Nilpotent matrices　　　　　　　　　　　　　　　　94

Chapter Four　Matrices as Linear Transformations　　　　**99**
4.1　Linear transformations　　　　　　　　　　　　　　99
4.2　Compositions and inverses　　　　　　　　　　　　110
4.3　Matrix representations　　　　　　　　　　　　　　114

Chapter Five　Bilinear Forms and Inner Products　　　　**131**
5.1　Bilinear forms　　　　　　　　　　　　　　　　　131
5.2　Quadratic forms　　　　　　　　　　　　　　　　144
5.3　Inner products　　　　　　　　　　　　　　　　　150
5.3*　Appendixes for orthogonal complements　　　　　　163

5.4 Gram-Schmidt and orthogonal matrices　　　　　　　　　　166

Chapter Six　Determinants　　　　　　　　　　　　　　　　175
6.1 Determinants of 2×2 matrices　　　　　　　　　　　　176
6.2 N-linear functions and the determinant　　　　　　　　184
6.3 Permutations and uniqueness of the determinant　　　　194
6.4 Computing determinants　　　　　　　　　　　　　　206
6.5 Cramer's rule and the inverse of a matrix　　　　　　　216

Chapter Seven　Eigenvalues、Diagonalizations、Idempotent
　　　　　　　Matrices and Householder Matrices　　　　223
7.1 Definitions and elementary facts　　　　　　　　　　　223
7.2 Diagonalizations and hermitian matrices　　　　　　　238
7.3 Idempotent matrices　　　　　　　　　　　　　　　　257
7.4 Householder matrices　　　　　　　　　　　　　　　267

Chapter Eight　Spectral Decomposition and Interpolatory
　　　　　　　Polynomials　　　　　　　　　　　　　　277
8.1 Upper triangulation　　　　　　　　　　　　　　　　277
8.2 Normal matrices　　　　　　　　　　　　　　　　　287
8.3 Spectral decomposition of a diagonalizable matrix　　　295
8.4 Interpolatory polynomials, constituent idempotents and
　　　Cayley-Hamilton theorem　　　　　　　　　　　　307

Chapter Nine　Canonical Forms　　　　　　　　　　　　　317
9.1 Canonical forms for nilpotent matrices　　　　　　　　317
9.2 Jordan canonical forms　　　　　　　　　　　　　　329

Index　　　　　　　　　　　　　　　　　　　　　　　　339

Bibliography　　　　　　　　　　　　　　　　　　　　　343

Chapter One

Vector Spaces

1.1 Definitions and some properties
1.2 Bases and dimensions
1.3 Sums and direct sums

　　線性代數所介紹的內容，除了線性空間結構的建立之外，線性空間與線性空間之間的線性轉換，其各種特性以及行為之研究，佔了絕大部份的篇幅。從線性映射的定義解說開始，推導出線性映射與矩陣之間的 1-1 對應關係之後，線性映射與矩陣就形成了密不可分的一體兩面之關係。所以，不管是在，特徵值、特徵向量、對角化、正規矩陣、對稱矩陣、**spectral decomposition**、以及 Jordan 正準型的領域上，矩陣理論與線性代數這兩門課程，都表現出極其親密的合作伙伴關係。

　　那麼，矩陣理論與線性代數之間，其最大的不同處乃在於，後者除了建構向量空間之外，其餘則強調空間與空間之間的線性轉換，其在抽象方面的思維與表徵，對於矩陣的運算和特有性質以及實例，則沒有給予太多的解說和實作；而矩陣理論這門課程，則以"解構矩陣"之方式來具體實踐，線性代數在線型轉換抽象思維方面之不足。因此，本課程一開始在介紹、了解矩陣之後，對於矩陣的各類運算、一些特殊矩陣的性格、以及矩陣與線性轉換之間的角色互換等，都會有詳盡的介紹和說明。

　　如此看來，在本課程開始之先，對於向量空間的架構似乎有其不可不知的必要。所以，第一章我們仍然以介紹向量空間之線性架構開始，引導大家進入矩陣理論與線性代數之間相互伴隨的領域。

§1.1　Definitions and some properties

　　向量空間的組成零件當中，有一個很重要的角色是，**set of numbers**。在這兒，我們將稱之為**體** (field)。一般常見的體有，"有理數集，Q"、"實數集，R" 以及 "複數集，C"。所以，不妨讓我們首先認識一下，"什麼叫做體？"

Definition 1.1

A subset, F, of complex numbers, C, is called a "field", if it satisfies that
1. $x+y \in F$ and $xy \in F$, $\forall\ x, y \in F$.
2. $-x \in F$, $\forall\ x \in F$ and $\frac{1}{x} \in F$, if $x \in F$, $x \neq 0$.
3. the elements 0 and 1 are in F.

譬如說，The set of real numbers, R, and the set of complex numbers, C, are fields。通常，它們被分別稱為，**real field** 和 **complex field**。當然，我們也不難看得出來，比實數體還要小的有理數集，Q 也是一個體 (why?)。然而，整數集 Z 就不是一個體了，因為，$2 \in Z$，可是 $\frac{1}{2} \notin Z$。

Remarks

1. If S is a subset of a field, F, and S is itself a field, then S is called a "subfield" of F.
2. Elements of a field are called "numbers" or "scalars".

現在我們就來看看，什麼叫**向量空間** (vector space) 或者說**線性空間** (linear space)。這個定義敘述冗長，請大家耐心思考、用心體會一下。首先弄清楚下面兩個符號的使用。

對於任意兩個集合，V 與 F，我們作如下之定義，
$$V \times V = \{(\alpha, \beta): \alpha, \beta \in V\}，F \times V = \{(a, \alpha): a \in F, \alpha \in V\}$$

Definition 1.2

Let F be a field and V a non-empty set of elements. Given a map, called "addition", which maps from $V \times V$ into V defined by
$$(\alpha, \beta) \to \alpha + \beta, \quad \forall\ (\alpha, \beta) \in V \times V,$$
and a map, called "multiplication by scalars", which maps from $F \times V$ into V defined by
$$(a, \alpha) \to a\alpha, \quad \forall\ (a, \alpha) \in F \times V.$$
Then V is called a "vector space" over F with respect to these two operations, if, $\forall\ \alpha, \beta, \gamma \in V$, and $a, b \in F$, the following properties are satisfied.

A1 $\alpha + (\beta + \gamma) = (\alpha + \beta) + \gamma$
A2 $\alpha + \beta = \beta + \alpha$
A3 $\exists\ 0 \in V$ such that $\alpha + 0 = 0 + \alpha = \alpha$
A4 If $u \in V$, then $\exists\ v \in V$ such that $u + v = 0$.
M1 $1\alpha = \alpha$
M2 $(ab)\alpha = a(b\alpha)$
M3 $(a + b)\beta = a\alpha + b\alpha$
M4 $a(\alpha + \beta) = a\alpha + a\beta$

上述之 A1 及 A2 分別叫做，加法結合律和交換律。而 A3 之 0 向量則被稱為加法單位元素。A4 之向量 v 為向量 u 之加法反元素。(這四個特性將 $(V; +)$ 形成了一個交換群) 另外要注意，我們將 V 稱為**佈於體** F **上**的向量空間，是有其必要的。否則，前述之純量 a, b 將不知為何物。只是，有時在視聽不被混淆的情形之下，我們將 V 簡單稱為一個向量空間，而省略**佈於體** F **上**罷了。

Example 1

Let $V = R^2 = \{(x,y): x, y \in R\}$, then V is a vector space over R according to the following operations;

For $a \in R$ and $X = (x_1, x_2)$, $Y = (y_1, y_2) \in V$,
$$X + Y = (x_1, x_2) + (y_1, y_2) = (x_1 + y_1, x_2 + y_2)$$
$$aX = a(x_1, x_2) = (ax_1, ax_2)$$

Proof

讀者可直接驗證，如此定義之下的"向量相加"與"常數乘法"滿足前述定義中的八個條件。下列的兩個例子也是如此。

Example 2

Let $V = C^3 = \{(x,y,z): x, y, z \in C\}$, then V is a vector space over C according to the following operations;

For $a \in C$ and $X = (x_1, x_2, x_3)$, $Y = (y_1, y_2, y_3) \in V$,
$$X + Y = (x_1, x_2, x_3) + (y_1, y_2, y_3) = (x_1 + y_1, x_2 + y_2, x_3 + y_3)$$
$$aX = a(x_1, x_2, x_3) = (ax_1, ax_2, ax_3)$$

Example 3

一般而言，若 F 是一個體，則 $V = F^n = \{(x_1, x_2, ..., x_n) : x_i, i = 1,2,...,n$ are numbers in $F\}$ 為一 (針對下列兩個運算) 佈於體 F 上的一個向量空間。

For $a \in F$ and $X = (x_1, x_2, ..., x_n)$, $Y = (y_1, y_2, ..., y_n) \in V$,
$$X + Y = (x_1, x_2, ..., x_n) + (y_1, y_2, ..., y_n) = (x_1 + y_1, x_2 + y_2, ..., x_n + y_n)$$
$$aX = a(x_1, x_2, ..., x_n) = (ax_1, ax_2, ..., ax_n)$$

The last example says that, Q^n, R^n, C^n, $n \in \{1,2,...,n\}$ 皆為分別佈於體 Q，R，C 上的向量空間。下面，再看一個比較特殊的例子。

Example 4

Let $V = \{f : f$ is a function mapping from R into $R\}$. Then V is a vector space over R with respect to the following operations.

For $c \in R$ and $f, g \in V$,
$$(f+g)(t) = f(t) + g(t), \quad \forall t \in R$$
$$(cf)(t) = cf(t), \quad \forall t \in R$$
∎

向量空間中的元素我們通常以**向量** (vectors) 稱呼之。此外，上述向量空間的八個特性中，還有一些值得弄清楚的事情，我們將其列為 exercises。希望讀者有空時，將其反覆思維，以建立起有關 vector space 的正確觀念。

Exercises

1. Show that the unit element, 0, in A3 is uniquely determined.

2. Show that $0\alpha = 0$, $\forall \alpha \in V$ and $a0 = 0$, $\forall a \in F$. (注意，這兒所出現的第一個 0 為 number zero，後面三個 0 則都為 V 中的 zero vector。)

3. Show that if α, β and γ are vectors in a vector space and if $\alpha + \beta = 0$ and $\alpha + \gamma = 0$, then $\beta = \gamma$. (i. e. the inverse element of addition is unique.) 此後，α 的加法反元素，我們將以 $-\alpha$ 表之。

4. Show that $(-1)\alpha = -\alpha$, $\forall \alpha \in V$.

Definition 1.3

> Let V be a vector space over F. A subset, W, of V is called a "subspace" of V, if W itself is a vector space over F.

譬如說，當我們把 R^2 看成，$\{(x, y, 0): x, y \in R\}$ 時，那麼，R^2 就是 R^3 的一個 subspace。當然也是所有 R^n，$n \geq 3$ 的 subspace。接著，我們看看下面一個判斷子空間的簡單方法。

Theorem 1.1

> A non-empty subset, W, of a vector space, V, is a subspace of V, if and only if for each pair of vectors $\alpha, \beta \in W$ and each scalar a, the vector, $a\alpha + \beta$, lies in W.

Proof

($\Rightarrow$) This direction follows from the definition.

($\Leftarrow$) Since W is non-empty, $\exists w \in W$. And by the assumption, $(-1)w + w \in W$.
But, $(-1)w + w = 0$, so $0 \in W$. This satisfies A3.
For $\alpha \in W$, $-\alpha = (-1)\alpha = (-1)\alpha + 0 \in W$. This satisfies A4.

All the other properties to be a vector space are quite clear as you can check.

有了這個方法之後，要判斷一個"子空間"的題目，那就太簡單了。我們一起先來試試看一些比較特殊的例子。

Example 5

If V is a vector space, then $\{0\}$ is a subspace of V.

Proof

因為，$\{0\}$只有0向量，所以，只要$\alpha, \beta \in \{0\}$，那麼$\alpha = \beta = 0$。因此，對任意純量a，$a\alpha + \beta = 0 \in \{0\}$。 ∎

Example 6

If V is a vector space, then V is itself a subspace of V.

這個當然是毫無疑問的啦。

以上這兩個子空間常被稱為，向量空間V的"trivial subspaces"。 ∎

Example 7

Let $W = \{(0, x_2, x_3, ..., x_n) : x_i \in R, \forall\ i = 2, 3, ..., n\}$, then W is a subspace of R^n.

Proof

Let $\alpha = (0, x_2, x_3, ..., x_n)$, $\beta = (0, y_2, y_3, ..., y_n)$ be any two elements in W. And let a be any scalar. Then $a\alpha + \beta = (0, ax_2 + y_2, ax_3 + y_3, ..., ax_n + y_n)$, which lies in W. So W is a subspace of R^n. ∎

Example 8

In the xy-plane, the set of points in a straight line through the origin is a subspace of xy-plane, if the "scalar multiplication" and "vector addition" are defined as that for $\forall a \in R$ and $(x_1, y_1), (x_2, y_2)$ in xy-plane,
$$a(x_1, y_1) = (ax_1, ay_1) \text{ and } (x_1, y_1) + (x_2, y_2) = (x_1 + x_2, y_1 + y_2).$$

Proof

We first note that an equation of a straight line passing through the origin is
$$y = mx \text{ for some scalar } m$$
Now let $(x_1, y_1), (x_2, y_2)$ be any two points in the line and let $a \in R$. Then,
$$a(x_1, y_1) + (x_2, y_2) = (ax_1 + x_2, ay_1 + y_2)$$
Since $m(ax_1 + x_2) = amx_1 + mx_2 = ay_1 + y_2$, $a(x_1, y_1) + (x_2, y_2)$ lies in the line. And hence the line is a subspace of xy-plane. ∎

Exercise

Show that the straight line not passing through the origin is not a subspace of xy-plane.

下面是有關子空間的一些有趣而重要的特性，讓我們慢慢的看下去。

Theorem 1.2

> If W_1 and W_2 are subspaces of a vector space, V, then the intersection, $W_1 \cap W_2$, is also a subspace of V.

Proof

利用前面的檢定方法，令 α 和 β 為 $W_1 \cap W_2$ 中的兩個向量，且 a 為一常數，則因為 W_1 和 W_2 都是 subspaces of V，所以，$a\alpha + \beta \in W_1$ and $a\alpha + \beta \in W_1$。因此，$a\alpha + \beta \in W_1 \cap W_2$，得證。

A generalization theorem 1.3

> If $\{W_i : i = 1, 2, 3, ..., n\}$ is a finite collection of subspaces of a vector space, V, then the intersection, $\bigcap_{i=1}^{n} W_i$, is also a subspace of V.

Proof

We first confirm that $\bigcap_{i=1}^{n} W_i$ is a non-empty set, since
$$0 \in W_i, \ \forall i = 1, 2, 3, ..., n \text{。}$$
The same as the last theorem, we let $\alpha, \beta \in \bigcap_{i=1}^{n} W_i$ and let a be a scalar. Since each W_i is a subspace and $\alpha, \beta \in W_i$, $\forall i = 1, 2, 3, ..., n$,
$$a\alpha + \beta \in W_i, \ \forall i = 1, 2, 3, ..., n.$$
Thus $a\alpha + \beta \in \bigcap_{i=1}^{n} W_i$. This says that $\bigcap_{i=1}^{n} W_i$ is a subspace of V.

當然，或許早有讀者已經懷疑，當 n 為無窮大時，本定理是否仍然成立？答案是肯定的，那就是下面定理。而且，它的證明與上述定理之證明相類似，我們不打算給予重複證明。

A generalization theorem 1.4

> If $\{W_i : i = 1, 2, 3, ...\}$ is an infinite collection of subspaces of a vector space, V, then the intersection, $\bigcap_{i=1}^{\infty} W_i$, is also a subspace of V.

其實，那怕 $\{W_t : t \in I\}$ 是一堆**不可數無窮多個** (uncountably infinitely many) V 的子空間，在這種情形之下，前述定理也是成立的。有關這一點，其概念當然也和前面定理一樣，我們姑且略過不提。不過，我們倒不妨反過來看看，"子空間的**聯集** (union) 是否仍為一個子空間？"這個答案是否定的。下面例題是一個簡單的反例。

Example 9

As we know in the previous example, $W_1 = \{(x, y) : y = 2x\}$ and $W_2 = \{(x, y) : y = 3x\}$ are subspaces of R^2. Show that $W_1 \cup W_2$ is not a subspace of R^2.

Proof

This proof is quite easy by considering, $\alpha = (1, 2)$, $\beta = (1, 3)$. We see that $\alpha + \beta = (2, 5)$ which is neither in W_1 nor in W_2. So, $\alpha + \beta \notin W_1 \cup W_2$. ∎

Corollary

> If S is a non-empty subset of a vector space, V, then there exists a smallest subspace of V containing S.

Proof

Let $\{W_t : t \in I\}$ be the collection of all subspaces of V containing S.

(*Note*: the collection is not empty, since V is one of them).

Then, by the previous theorems, $\bigcap_{t \in I} W_t$ is a subspace of V and $S \subseteq \bigcap_{t \in I} W_t$.

However, $\bigcap_{t \in I} W_t$ is the smallest subspace that contains S, since it is contained in every W_t, $t \in I$.

Example 10

Let V be the xy-plane and $S = \{(1, 2)\}$. Then the smallest subspace of V that contains S is the straight line passing through the point $(1, 2)$ and the origin. ∎

下一個 section 我們將介紹線性相依的、線性獨立的以及基底,這些內容都牽涉到所謂的 "linear combination" (線性組合)。所以,本節最後就以此為主題,和大家做一概略的解說。

Definition 1.4

Let V be a vector space over a field, F. Let $\alpha_1, \alpha_2, ..., \alpha_n$ be vectors in V. If $x_1, x_2, ..., x_n$ are scalars in F, then $\sum_{i=1}^{n} x_i \alpha_i = x_1 \alpha_1 + x_2 \alpha_2 + \cdots + x_n \alpha_n$ is called a linear combination of $\alpha_1, \alpha_2, ..., \alpha_n$.

舉個例子說,假設 $X = (1, 0, 0)$, $Y = (0, 1, 0)$, $Z = (0, 0, 1)$,那麼 $2X - 3Y + Z$ 就是一個 linear combination of X, Y and Z。又由於,$2X - 3Y + Z = (2, -3, 1)$,所以,我們也可以說,$(2, -3, 1)$ is a linear combination of X, Y, Z。

Definition 1.5

Let S be a non-empty subset of a vector space. The set of all linear combinations of finite number of vectors in S is called a "set spanned by S", or the "spanning set (生成集) of S" and is denoted by $SP(S)$.

或者簡單的說， $SP(S) = \{\sum_{i=1}^{n} c_i s_i : c_i \in F,\ s_i \in S\ \text{and}\ n \in Z^+\}$。

譬如說，若 $S = \{s_1, s_2, ..., s_m\}$，則

$$SP(S) = \{\sum_{i=1}^{m} x_i s_i\ :\ \forall\ x_i \in F\}$$

當然，若 $S = \{s_1, s_2, s_3 ...\}$ 為一無窮集，那麼

$$SP(S) = \{\sum_{i \in P} c_i s_i : c_i \in F,\ P \subset Z^+\ \text{and}\ P\ \text{is a finite set}\}$$

Theorem 1.5

If S is a non-empty subset of a vector space, V, then $SP(S)$ is a subspace of V and $S \subset SP(S)$.

Proof

For $\alpha = \sum_{i=1}^{m} x_i u_i$ and $\beta = \sum_{j=1}^{n} y_j v_j$ in $SP(S)$ and a scalar a, $a\alpha + \beta = \sum_{i=1}^{m}(ax_i)u_i + \sum_{j=1}^{n} y_j v_j$ is also a linear combination of finite number of vectors in $SP(S)$. So, $a\alpha + \beta \in SP(S)$. Hence, $SP(S)$ is a subspace of V. For the proof of $S \subset SP(S)$ is quite trivial, and we leave it to the readers.

這個定理告訴我們，the spanning set of S is a subspace of V which contains S。其實，更進一步的觀察，我們可以發現，$SP(S)$ 也是包含 S 的最小的子空間。那就是下列習題 2。

Exercises

1. Find $SP(S)$, given that $S = \{(1, 0, 0),\ (0, 1, 0),\ (0, 0, 1)\}$ is a subset of R^3.

2. Show that $SP(S)$ is the smallest subspace of V that contains S.

3. Let V be the xy-plane and $S = \{(0,0)\}$. What is the smallest space of V containing S?

4. Let a be a nonzero number and α a vector in a vector space, V. Show that if $av = 0$, then $v = 0$.

5. Let $W = \{(x,y,z) \in R^3 : x+y+z = 0\}$ and $V = R^3$. Show that W is a subspace of V.

6. Let $K = \{a + b\sqrt{2} : a, b \in Q \}$. Show that K is a field.

§1.2 Bases and dimensions

常聽到一般人喜歡談論，這個東西在第幾度空間？或者說這個空間有幾個維度？…等。這類幾度空間的問題，我們要從空間的**基底** (bases) 開始談起，或者更確切的說，我們應該從**線性獨立** (linear independence) 或**線性相依** (linear dependence) 著手開始介紹。

Definition 1.6

> Let V be a vector space over a field F. A set of vectors, $\alpha_1, \alpha_2, ..., \alpha_m$ in V are called "linearly dependent", if there exist scalars $a_1, a_2, ..., a_m$ in F not all equal to zero such that
> $$a_1\alpha_1 + a_2\alpha_2 + ... + a_m\alpha_m = 0.$$

e.g.

If $X = (1,1,0)$, $Y = (2,2,0)$ are vectors in R^3, then X and Y are linearly dependent, because $-2X + Y = 0$.

However, $X = (1,0,0)$, $Y = (0,1,0)$ and $Z = (0,0,1)$ are not linearly dependent. Because, if $aX + bY + cZ = 0$, then $a = b = c = 0$, a, b and c are all equal to 0.

如此，**不為線性相依** (not linearly dependent) 的向量，我們將其稱為線性獨立的。或者明確的說，

vectors, $\alpha_1, \alpha_2, ..., \alpha_m$ in V are said to be linearly independent, if $a_1\alpha_1 + a_2\alpha_2 + ... + a_m\alpha_m = 0$ implies $a_1 = a_2 = ... = a_m = 0$.

Example 1

Let $v_1, v_2, ..., v_l$ be vectors in a vector space, V. If $v_i = 0$, for some i, then $v_1, v_2, ..., v_l$ are linearly dependent.

Proof

這是一個非常重要的基本概念，其證明也很簡單。我們只要取 $a_i \neq 0$，

然後 $a_j = 0$, $\forall\, j \neq i$，就可使得 $a_1v_1 + a_2v_2 + ... + a_lv_l = 0$。下面例子也是顯而易懂的、重要的基本道理。

Example 2

Let $V = R^n$. The vectors, $E_1 = (1,0,0,...,0)$, $E_2 = (0,1,0,...,0)$,..., $E_n = (0,0,...,1)$ are linearly independent.

Example 3

Vectors, $(0,1,1)$, $(0,2,1)$, $(1,5,3)$ in R^3 are linearly independent.

Proof

Let $a(0,1,1) + b(0,2,1) + c(1,5,3) = 0$. Then $(c, a+2b+5c, a+b+3c) = 0$. This implies that
$$\begin{cases} c = 0 \\ a + 2b + 5c = 0 \\ a + b = 3c = 0 \end{cases}$$
Solve the simultaneous equations, we obtain that $a = b = c = 0$.

我們來點不一樣的看看。

Example 4

Let $V = \{f : f$ is a function mapping from R into $R\}$. (As we knew in the last section, it is a vector space over R.) If $f(t) = t$, $g(t) = t^2$, $\forall\, t \in R$, then f and g are linearly independent.

Proof

Let a and b be real numbers such that $af(t) + bg(t) = 0$. Then $at + bt^2 = 0$. This says that $a = b = 0$.

了解線性相依、線性獨立之後，接著是向量空間的基底。首先，我們回頭看看 Example 2。

$E_1 = (1,0,0,...,0)$, $E_2 = (0,1,0,...,0)$,…, $E_n = (0,0,...,1)$ 是 R^n 中的 n 個線性獨立的向量。再往下看看，R^n 中的任意一個向量，$X = (x_1, x_2,...,x_n)$ 都可以被表示為，$X = x_1E_1 + x_2E_2 +...+ x_nE_n$。這個事實暗示著下面這個定義。

Definition 1.7

Let $S = \{\alpha_1, \alpha_2,...,\alpha_n\}$ be a linearly independent set of vectors in a vector space, V. If for every $v \in V$, there are scalars $x_1, x_2,...,x_n$ such that
$$v = x_1\alpha_1 + x_2\alpha_2 +...+ x_n\alpha_n$$
then S is called a "basis" for V.

定義中的這句話，If for every $v \in V$, there are scalars $x_1, x_2,...,x_n$ such that $v = x_1\alpha_1 + x_2\alpha_2 +...+ x_n\alpha_n$，成立的話。我們稱集合 S 生成 V (S spans V or V is spanned by S)，明確的說，也就是，$V = SP(S)$ 的意思。也因此，我們可以把前述定義改口為

$S = \{\alpha_1, \alpha_2,...,\alpha_n\}$ is a basis for V if and only if $\alpha_1, \alpha_2,...,\alpha_n$ are linearly independent and S spans V.

譬如說，前面所提過的向量集合，$S = \{E_1, E_2,...,E_n\}$ is a basis for R^n。而這個基底通常被稱為**標準基底** (standard basis)。

Example 5

Show that $\{\alpha_1 = (0,1,1), \alpha_2 = (0,2,1), \alpha_3 = (1,5,3)\}$ is basis for R^3.

Proof

前面我們已經證明過，這三個向量為線性獨立的。所以，我們只消證明每一個 R^3 中的向量，$X = (x,y,z)$ 都可以被表示為 $\alpha_1 = (0,1,1)$, $\alpha_2 = (0,2,1)$ 以及 $\alpha_3 = (1,5,3)$ 的線性組合。也就是說，找得到純量，a, b, c，使得，$X = a\alpha_1 + b\alpha_2 + c\alpha_3$。

今令，$X = a\alpha_1 + b\alpha_2 + c\alpha_3$，則

$$\begin{cases} x = c \\ y = a + 2b + 5c \\ z = a + b + 3c \end{cases}$$

解之得　$a = 2z - y - x$, $b = y - z - 2x$, $c = x$。得證。
我們以 $X = (2, -3, 4)$ 為例，則 $a = 9, b = -11, c = 2$。

Example　6

Let $V = \{f : f$ is a real-valued polynomial defined on R with $\deg(f) \leq 3\}$. Show that $\{1, t, t^2, t^3\}$ is a basis for V.

Proof

前面我們提過相類似的。所以，毫無疑問的，$1, t, t^2, t^3$ are linearly independent。至於，$\{1, t, t^2, t^3\}$ 生成 V，那就更不用說了。因為任何一個次數小於或等於 3 的多項式，$f(t)$，都可以被寫為，
$$f(t) = a_0 + a_1 t + a_2 t^2 + a_3 t^3 \text{。}$$

Definition　1.8

> Let V be a vector space with a basis $\{\alpha_1, \alpha_2, ..., \alpha_n\}$. If for every $v \in V$ and $v = x_1\alpha_1 + x_2\alpha_2 + ... + x_n\alpha_n$, the n-tuple $(x_1, x_2, ..., x_n)$ is called the coordinates of the vector, v, with respect to the basis $\{\alpha_1, \alpha_2, ..., \alpha_n\}$.

以標準基底，$\{E_1, E_2, ..., E_n\}$ 而言，R^n 中的任意向量，$X = (x_1, x_2, ..., x_n)$ 皆可被寫成 $X = x_1 E_1 + x_2 E_2 + ... + x_n E_n$。所以，傳統而言，我們稱 $x_1, x_2, ..., x_n$ 為向量 X 的座標。

注意，一旦基底被選定之後，空間中的每一個向量的 coordinates 是唯一被決定的。因為，假若 $v = x_1\alpha_1 + x_2\alpha_2 + ... + x_n\alpha_n$，而且 v 又等於 $y_1\alpha_1 + y_2\alpha_2 + ... + y_n\alpha_n$ 的話，那麼
$$x_1\alpha_1 + x_2\alpha_2 + ... + x_n\alpha_n = y_1\alpha_1 + y_2\alpha_2 + ... + y_n\alpha_n \text{，}$$
或者說

$$(x_1 - y_1)\alpha_1 + (x_2 - y_2)\alpha_2 + ... + (x_n - y_n)\alpha_n = 0 \text{ 。}$$

但由於，$\alpha_1, \alpha_2, ..., \alpha_n$ 是線性獨立的，所以 $x_1 = y_1, x_2 = y_2, ..., x_n = y_n$ 。

Example 7

Show that $\{\alpha_1 = (1,1), \alpha_2 = (-1,2)\}$ is a basis for R^2. And find the coordinates of $(3,4)$ with respect to this basis.

Solution

Let $c_1(1,1) + c_2(-1,2) = 0$. Then
$$\begin{cases} c_1 - c_2 = 0 \\ c_1 + 2c_2 = 0 \end{cases}$$
This implies that $c_1 = c_2 = 0$. So, $(1,1)$ and $(-1,2)$ are linearly independent.

Now, for any (x, y) in R^2, letting $(x, y) = a(1,1) + b(-1,2)$, then
$$\begin{cases} x = a - b \\ y = a + 2b \end{cases} \text{ and hence } a = \frac{2x + y}{3}, b = \frac{y - x}{3}.$$

Therefore, $\{\alpha_1 = (1,1), \alpha_2 = (-1,2)\}$ spans R^2.

To find the coordinates of $(3,4)$, we let $(x, y) = (3,4)$, and obtain that they are $a = \frac{2x + y}{3} = \frac{10}{3}$, $b = \frac{y - x}{3} = \frac{1}{3}$. ∎

Example 8

Given a basis, $\{(1,0,0), (1,1,1), (-1,1,0)\}$, for R^3. Find the coordinates of $(5, 3, -1)$ with respect to the basis.

Solution

Letting $(5, 3, -1) = a(1,0,0) + b(1,1,1) + c(-1,1,0)$, then
$$\begin{cases} a + b - c = 5 \\ b + c = 3 \\ b = -1 \end{cases}$$
Solve the simultaneous equations, we obtain, $a = 2, b = -1, c = -4$. ∎

談到線性獨立，還有一個重要而容易令人搞亂的概念是：假若 u，v 以及 w 為向量空間 V 中的三個向量，而且它們兩兩之間是線性獨立的。那麼，請問 u，v 以及 w 三個向量擺在一起，是否一定為線性獨立的？答案是否定的。同學們帶回家想想看，就當作家庭作業吧。(參考本節最後，exercise 6)

Theorem 1.6

Let V be a vector space and $\{\alpha_1, \alpha_2, ..., \alpha_m\}$ a set of vectors in V such that $V = SP(\{\alpha_1, \alpha_2, ..., \alpha_m\})$. If $\{\beta_1, \beta_2, ..., \beta_n\}$ is a set of vectors of V and $n > m$, then $\beta_1, \beta_2, ..., \beta_n$ are linearly dependent.

Proof

我們用反證法。假設 $\beta_1, \beta_2, ..., \beta_n$ 為線性獨立的。那麼，由於 $V = SP(\{\alpha_1, \alpha_2, ..., \alpha_m\})$，所以存在 $c_1, c_2, ..., c_m$ 使得 $\beta_1 = \sum_{i=1}^{m} c_i \alpha_i$。在不失大原則的情況之下，我們假設 $c_1 \neq 0$ (注意，$\beta_1 \neq 0$)，那麼 $\alpha_1 = \frac{1}{c_1}(\beta_1 - c_2\alpha_2 - ... - c_m\alpha_m)$。接著，我們以 β_1 替代 α_1，發現 $\{\beta_1, \alpha_2, ..., \alpha_m\}$ 生成 V。再來，對 β_2 而言，存在 $d_1, d_2, ..., d_m$ 使得 $\beta_2 = d_1\beta_1 + \sum_{j=2}^{m} d_j\alpha_j$。同樣的，在大於等於 2 的 j 裡面，存在一個 $d_j \neq 0$，否則 $\beta_2 = d_1\beta_1$，這將有違前題假設。在不失大原則的情況之下，我們同樣假設，$d_2 \neq 0$，此時，我們得出 $\alpha_2 = \frac{1}{d_2}(\beta_2 - d_1\beta_1 - d_3\alpha_3 - d_4\alpha_4 - ... - d_m\alpha_m)$。接著，再以 β_2 替代 α_2，我們得知 $\{\beta_1, \beta_2, \alpha_3, ..., \alpha_m\}$ 生成 V。

按此要領，最後我們將歸納出，$\{\beta_1, \beta_2, ..., \beta_m\}$ 生成 V。既然，$\{\beta_1, \beta_2, ..., \beta_m\}$ 生成 V，那麼每一個 β_j，$j > m$ 的向量，都可以被表示為 $\beta_1, \beta_2, ..., \beta_m$ 的 linear combination。這說明了，$\beta_1, \beta_2, ..., \beta_n$ 是線性相依的，而導致矛盾。

Corollary

> Let V be a vector space. If $\{\alpha_1, \alpha_2, ..., \alpha_m\}$ and $\{\beta_1, \beta_2, ..., \beta_n\}$ are bases of V, then $n = m$.

Proof

Since $\{\alpha_1, \alpha_2, ..., \alpha_m\}$ spans V and $\beta_1, \beta_2, ..., \beta_n$ are linearly independent, n must be less than or equal to m. The other hand, $\{\beta_1, \beta_2, ..., \beta_n\}$ also spans V and $\alpha_1, \alpha_2, ..., \alpha_m$ are also linearly independent, so m must be less than or equal to n. By these, we conclude that $n = m$.

這是一個非常重要的結果。有了這個結果，我們得以定義本章節的另一個主題，所謂向量空間的**維度** (dimension)。

Definition 1.9

> Let V be a vector space and $\{\alpha_1, \alpha_2, ..., \alpha_n\}$ a basis of V. Then the dimension of V is defined to be $\dim V = n$.

舉個例子說，$S = \{E_1, E_2, ..., E_n\}$ is a basis for R^n，所以，$\dim R^n = n$。同樣，在前面我們也提過，$\{1, t, t^2, t^3\}$ is a basis for $V = \{f : f$ is a real valued polynomial defined on R with $\deg(f) \leq 3\ \}$，所以三次多項式集合，其 dimension 為 4。

一個向量空間，V，常被稱為 finite dimensional (有限維度的)，假若 $\dim V < \infty$。對於特殊的向量空間，$\{0\}$，我們則共同約定它的維度為 0。這個約定或可以從下面這個角度予以思考理解。按理說，0 向量它本身是線性相依的，所以，向量空間 $\{0\}$ 沒有線性獨立向量。也就是說，$\{0\}$ 真有基底的話，它的基底元素個數是 0。

Example 9

Let F be a field. Then F is itself a vector space having dimension 1.
Proof

第一節的時候，我們提過，F^n 是一個向量空間。所以當，$n=1$ 的時候，F 本身也是一個向量空間。由於，$F = \{x \cdot 1 : x \in F\}$，所以，$\{1\}$ 可以說是 F 的基底，因此 $\dim F = 1$。∎

Example 10

Let $V = \{f : f$ is a real valued polynomial defined on R with $\deg(f) \leq n\}$. One can easily show that $\{1, t, t^2, ..., t^n\}$ is a basis for V, so V has dimension $n+1$. ∎

下面我們從不同的角度來查看向量空間的基底。首先，介紹所謂的**最大線性獨立集** (maximal set of linear independence)。

Definition 1.10

Let V be a vector space. A set of vectors, $\{v_1, v_2, ..., v_n\}$, in V is called a "maximal set of linear independence", if they are linearly independent and for each $w \in V$, $\{v_1, v_2, ..., v_n, w\}$ is a linearly dependent set.

e.g.

Any basis of a vector space is a maximal set of linear independence. For if $\{v_1, v_2, ..., v_n\}$ is a basis of V and $w \in V$, then w is a linear combination of $v_1, v_2, ..., v_n$, and thus $v_1, v_2, ..., v_n, w$ are linearly dependent.

Theorem 1.7

Let V be a vector space. Then $\{v_1, v_2, ..., v_n\}$ is a basis for V if and only if $\{v_1, v_2, ..., v_n\}$ is a maximal set of linear independence.

Proof

($\Rightarrow$) This direction has been explained previously.

($\Leftarrow$) To show that a maximal set of linear independence, $\{v_1,v_2,...,v_n\}$, is a basis for V, it is sufficient to show that $\{v_1,v_2,...,v_n\}$ spans V.

Let $w \in V$. Since $\{v_1,v_2,...,v_n\}$ is a maximal set of linear independence, $v_1,v_2,...,v_n,w$ are linearly dependent. So there exist numbers $x_0, x_1,..., x_n$ not all equal to zero such that

$$x_0 w + x_1 v_1 + x_2 v_2 + ... + x_n v_n = 0$$

(注意，x_0 決不可能為 0，否則所有其他的 x_i 都要等於 0。)

Therefore, $w = \dfrac{-1}{x_0}(x_1 v_1 + x_2 v_2 + ... + x_n v_n)$. That is, $w \in SP\{v_1,v_2,...,v_n\}$.

這個定理說明，假若 $\dim V = n$ 而且 $\{v_1,v_2,...,v_n\}$ 是 linearly independent set 的話，$\{v_1,v_2,...,v_n\}$ 就是 V 的一個 basis。此外，假若 W 是 V 的一個 subspace 而且 $\dim V = \dim W$ 的話，W and V are equal。

前面介紹了向量空間的基底，那麼我們不禁要問，是不是每一個向量空間都有基底呢？又要如何建立一個向量空間的基底呢？下面看一個 lemma。

Lemma 1.8

Let V be a vector space and let $v_1, v_2,...,v_l$ be vectors in V which are linearly independent. If $w \in V$ and w is not in the spanning set of $\{v_1,v_2,...,v_l\}$, then $v_1, v_2,...,v_l, w$ are linearly independent.

Proof

Let $x_0, x_1,..., x_l$ be scalars such that

$$x_0 w + x_1 v_1 + x_2 v_2 + ... + x_l v_l = 0$$

If $x_0 \neq 0$, then it follows that w is a linear combination of $v_1, v_2,...,v_l$. This contradicts to the hypothesis. So x_0 must be equal to 0. And hence, all

x_i, $i = 1, 2, ..., l$ are equal to 0, since $v_1, v_2, ..., v_l$ are linearly independent. Therefore, $v_1, v_2, ..., v_l, w$ are linearly independent.

Theorem 1.9

Let V be a vector space with $\dim V = n$. If $v_1, v_2, ..., v_l$ are linearly independent vectors in V and $l \leq n$, then $\{v_1, v_2, ..., v_l\}$ can be a part of a basis for V.

Proof

If $l = n$, then $\{v_1, v_2, ..., v_l\}$ is a basis for V as we mentioned before. So, we suppose that $l < n$, then $\exists\, w_1 \in V$ which is not in $SP\{v_1, v_2, ..., v_l\}$. By the last lemma, $v_1, v_2, ..., v_l, w_1$ are linearly independent. Again, if $l + 1 = n$, then $\{v_1, v_2, ..., v_l, w_1\}$ is a basis for V and hence the proof is completed. However, if $l + 1 < n$, then $\exists\, w_2 \in V$ which is not in $SP\{v_1, v_2, ..., v_l, w_1\}$. Also, the last lemma implies that $v_1, v_2, ..., v_l, w_1, w_2$ are linearly independent.

Proceed stepwise further, we consequently obtain vectors $w_1, w_2, ..., w_r$ in V such that $v_1, v_2, ..., v_l, w_1, w_2, ..., w_r$ are linearly independent and $l + r = n$. This implies that $v_1, v_2, ..., v_l, w_1, w_2, ..., w_r$ form a basis for V which contains $\{v_1, v_2, ..., v_l\}$.

上述這個定理確立了一個非常重要的結果，那就是，每一個向量空間一定有一組基底。請看下列推論。

Corollary

Every non-trivial vector space, V, has a basis.
(注意：所謂 non-trivial 者，指的是 $V \neq \{0\}$ 的意思。)

Proof

Let w_1 be a non-zero vector in V. Then w_1 is itself linearly independent. Using the procedure in the proof of the last theorem, a set of linearly independent vectors $\{w_1, w_2, ..., w_n\}$ can be found such that it spans V, and certainly it is a basis of V.

Example 11

Given a vector, $\alpha = (2,1,0)$, in R^3. Find a basis for R^3 which contains α as a part.

Solution

We first choose a vector, $\beta = (0,0,1)$, which is obviously not in $SP\{\alpha\}$. And then a third vector, $\gamma = (0,1,0)$, which is not in $SP\{\alpha, \beta\}$. Consequently, $\alpha = (2,1,0)$, $\beta = (0,0,1)$, $\gamma = (0,1,0)$ form a basis for R^3. ∎

Exercises

1. Are the following vectors linearly independent.
   a. $(-1, 1, 0)$, $(1, 0, 2)$ and $(0, 1, 2)$
   b. $(\pi, 2, 1)$, $(0, 2, 4)$ and $(\pi, 3, 3)$
   c. $(0, 1, 1)$, $(0, 2, 1)$ and $(1, 5, 3)$
   d. $(1, 1, 1)$, $(-1, 2, 0)$ and $(2, 0, -1)$

2. Find the coordinates of the vector, $X = (2,0,1)$, with respect to the following vectors,
   a. $(0, 1, 1)$, $(0, 2, 1)$ and $(1, 5, 3)$
   b. $(1, 1, 1)$, $(-1, 2, 0)$ and $(2, 0, -1)$

3. Let (a,b) and (c,d) be two vectors in the xy-plane. Show that if $ad - bc \neq 0$, then they are linearly independent.

4. Let $V = \{ f : f \text{ is a real-valued function defined on } (0, \infty) \}$. Show that the following pairs of functions are linearly independent.
   $f(t) = t$, $g(t) = 1/t$
   $f(t) = e^t$, $g(t) = t$
   $f(t) = \sin t$, $g(t) = \cos t$
   $f(t) = e^t$, $g(t) = \ln t$

5. Let $v_1, v_2, \ldots, v_n$ be vectors in a vector space, V, and $\dim V = n$. Show that if $V = SP\{v_1, v_2, \ldots, v_n\}$, then $v_1, v_2, \ldots, v_n$ are linearly independent.

6. Find 4 vectors, $\alpha_1, \alpha_2, \alpha_3, \alpha_4$, in R^3 such that they are mutually independent and $SP\{\alpha_1, \alpha_2, \alpha_3, \alpha_4\}$ has dimension 2. (*Hint*: 在 xy-平面上找出四個不共線的向量。)

§1.3 Sums and direct sums

第七章對角化的過程裡，我們將提到所謂的 spectral theorem。而 direct sum (直和) 的認識是，到時一個必備的基本概念。所以，我們現在就給大家介紹認識，空間中兩個集合的和 (sum) 與直和 (direct sum)。

Definition 1.11

Let A and B be two subsets of a vector space. The sum, $A+B$, is defined to be
$$A+B=\{\alpha+\beta \ : \ \alpha \in A, \ \beta \in B\}.$$

Example 1

Given that $U=\{(x,0) : x \in R\}$, $W=\{(0,y) : y \in R\}$. Then $U+W=\{(x,y) : x, y \in R\}$

這也就是說，x-軸加 y-軸等於 xy-平面的意思，不要懷疑。∎

回憶一下，第 1.1 節的時候。我們曾經介紹過，任意多個子空間的交集仍舊是子空間。而且子空間的聯集，$U \cup W$，則不一定是子空間。現在，我們來看看，兩個子空間的相加又會是如何呢？

Propersition 1.10

If U and W are subspaces of a vector space, V, then $U+W$ is a subspace of V.

Proof

Let α and β be any two elements in $U+W$, and c a scalar.
If $\alpha=u_1+w_1$ and $\beta=u_2+w_2$, then $c\alpha+\beta=(cu_1+u_2)+(cw_1+w_2)$ $\in U+W$.

Definition 1.12

Let U and W be subspaces of a vector space, V. If for each vector v in V, there exist a unique pair u in U and w in W such that $v = u + w$, then V is called a "direct sum" of U and W, and we denote it as $V = U \oplus W$.

譬如說，前面的例子，$U = \{(x,0) : x \in R\}$，$W = \{(0,y) : y \in R\}$ 而言。U 與 W 的直和 $U \oplus W$ 等於 R^2。因為，對任意 $(x,y) \in R^2$，唯一存在 $(x,0) \in U$, $(0,y) \in W$ 使得，$(x,y) = (x,0) + (0,y)$。然而，怎樣的和才得以構成直和呢？下面定理有清楚的說明。

Theorem 1.11

Let U and W be subspaces of a vector space, V. If $V = U + W$ and $U \cap W = \{0\}$, then $V = U \oplus W$.

Proof

For a vector v in V, by the hypothesis, there are vectors u in U and w in W such that $v = u + w$. We shall show that the expression is uniquely determined.

Suppose that $v = u_1 + w_1$ and $v = u_2 + w_2$, then $u_1 + w_1 = u_2 + w_2$. Hence, $u_1 - u_2 = w_2 - w_1$. Since $u_1 - u_2 \in U$ and $w_2 - w_1 \in W$ and $U \cap W = \{0\}$, $u_1 - u_2 = w_2 - w_1 = 0$. So, $u_1 = u_2$, $w_1 = w_2$.

其實，我們這樣說也是可以的。令 $V = U + W$，則 $V = U \oplus W$ 若且唯若 $U \cap W = \{0\}$。也就是說，若 $V = U \oplus W$，則 $U \cap W = \{0\}$，也是成立的。我們來看看，若 $\alpha \in U$ 而且 $\alpha \in W$，則 $\alpha + \alpha = 2\alpha = 2\alpha + 0$。但是，由於 $V = U \oplus W$，所以這個式子是不可能的，除非 $\alpha = 0$。那就是說，$U \cap W = \{0\}$。

Example 2

Let $V = R^2$, $W = \{x(1,0) : x \in R\}$. If $U = \{x(1,1) : x \in R\}$, then $V = W \oplus U$.

Proof

For each $(a,b) \in V$, writing $(a,b) = x(1,0) + y(1,1)$, then
$$\begin{cases} a = x+y \\ b = y \end{cases} \text{ or } \begin{cases} y = b \\ x = a-b \end{cases}$$
So, we see that $V = W + U$. But it can be seen trivially that $W \cap U = \{0\}$. Therefore, $V = W \oplus U$. ∎

Exercise

Let $V = \{f : f \text{ is a real-valued function defined on } R\}$. If $V_e = \{f \in V : f \text{ is an even function}\}$, $V_0 = \{f \in V : f \text{ is an odd function}\}$, then $V = V_e \oplus V_0$. (*Hint*: $f(x) = \dfrac{f(x)+f(-x)}{2} + \dfrac{f(x)-f(-x)}{2}$)

Theorem 1.12

Let W be a subspace of a vector space, V. Then there exists a subspace, U, of V such that $V = W \oplus U$.

Proof

Let $\{\alpha_1, \alpha_2, ..., \alpha_l\}$ be a basis for W. Then by Theorem 1.9 in the last section, $\{\alpha_1, \alpha_2, ..., \alpha_l\}$ can be extended to a basis $\{\alpha_1, \alpha_2, ..., \alpha_l, \alpha_{l+1}, ..., \alpha_n\}$ for V. Now, let U be the subspace of V spanned by $\{\alpha_{l+1}, ..., \alpha_n\}$. Then we clearly see that $V = W + U$ and $W \cap U = \{0\}$ (why? Exercise 4). Therefore, $V = W \oplus U$.

注意，前面 Example 2 當中，我們發現若，$U' = \{x(0,1) : x \in R\}$ 時，V 也是等於 W 與 U' 的直和。由這一點，我們得知前述定理中的 U 不是唯一存在的。

Theorem 1.13

> Let U and W be subspaces of a vector space, V. Then $\dim(U+W) = \dim U + \dim W - \dim(U \cap W)$.

Proof

若 $U \cap W = \{0\}$，則 $\dim(U+W) = \dim(U \oplus W) = \dim U + \dim W$。若 $U \cap W \neq \{0\}$，則令 $\{\alpha_1, \alpha_2, ..., \alpha_l\}$ 為 $U \cap W$ 之一基底。以此基底開始，將其擴張而為 U 之基底 $\{\alpha_1, \alpha_2, ..., \alpha_l, \beta_1, ..., \beta_s\}$，以及 W 之基底 $\{\alpha_1, \alpha_2, ..., \alpha_l, \gamma_1, ..., \gamma_t\}$。此時，我們若能證明 $\{\alpha_1, \alpha_2, ..., \alpha_l, \beta_1, ..., \beta_s, \gamma_1, ..., \gamma_t\}$ 為 $U+W$ 之一組基底的話，則定理將得證。

首先，毫無疑問的，
$U+W$ is spanned by $\{\alpha_1, \alpha_2, ..., \alpha_l, \beta_1, ..., \beta_s, \gamma_1, ..., \gamma_t\}$。所以，我們只剩證明 $\{\alpha_1, \alpha_2, ..., \alpha_l, \beta_1, ..., \beta_s, \gamma_1, ..., \gamma_t\}$ 為線性獨立的就可以了。茲令

$$\sum_{i=1}^{l} x_i \alpha_i + \sum_{j=1}^{s} y_j \beta_j + \sum_{k=1}^{t} z_k \gamma_k = 0 \quad (*)$$

$$\Rightarrow \quad -\sum_{k=1}^{t} z_k \gamma_k = (\sum_{i=1}^{l} x_i \alpha_i + \sum_{j=1}^{s} y_j \beta_j) \in (U \cap W)$$

此乃因為等號左邊之東西在 W 中，等號右邊之東西在 U 中的緣故。因而，存在常數 $c_1, c_2, ..., c_l$ 使得

$$-\sum_{k=1}^{t} z_k \gamma_k = \sum_{i=1}^{l} c_i \alpha_i$$

這說明了
$$\sum_{k=1}^{t} z_k \gamma_k + \sum_{i=1}^{l} c_i \alpha_i = 0$$

然而，$\alpha_1, \alpha_2, ..., \alpha_l, \gamma_1, ..., \gamma_t$ are linearly independent，所以

$$z_1 = z_2 = ... z_t = c_1 = c_2 = ... c_l = 0$$

現在，式子 (*) 變成

$$\sum_{i=1}^{l} x_i \alpha_i + \sum_{j=1}^{s} y_j \beta_j = 0$$

同樣的，$\alpha_1, \alpha_2, ..., \alpha_l, \beta_1, ..., \beta_s$ are linearly independent，所以

$$x_1 = x_2 = ... = x_l = y_1 = y_2 = ... = y_s = 0$$

故而得證 $\{\alpha_1, \alpha_2, ..., \alpha_l, \beta_1, ..., \beta_s, \gamma_1, ..., \gamma_t\}$ is a linearly independent set。當然因此，$\dim(U+W) = l+s+t = (l+s)+(l+t)-l = \dim U + \dim W - \dim(U \cap W)$。

上述定理說明，假若 $V = W \oplus U$，那麼 $\dim V = \dim U + \dim W$。此外，要是說，$V_1, V_2, ..., V_k$ are subspaces of V such that for every $v \in V$, there uniquely exist $v_i \in V$, $i = 1, 2, ..., k$ such that $v = v_1 + v_2 + ... + v_k$, then we say,

$$V = V_1 \oplus V_2 \oplus ... \oplus V_k = \bigoplus_{i=1}^{k} V_i$$

此時，$\dim V = \dim V_1 + \dim V_2 + ... + \dim V_k$。

Exercises

1. Let U and W be subspaces of V. Define the direct product of U and W by $U \times W = \{(u, w) : u \in U, w \in W\}$. Show that if for $(u_1, w_1), (u_2, w_2) \in U \times W$ and a scalar c

$$(u_1, w_1) + (u_2, w_2) = (u_1 + u_2, w_1 + w_2)$$
$$c(u_1, w_1) = (cu_1, cw_2)$$

    then $U \times W$ is a vector space.

2. Show that $\dim(U \times W) = \dim U + \dim W$.

3. Let $V = R^3$ and $U = SP\{(1,1,0), (0,1,1)\}$, $W = SP\{(1,0,0)\}$. Show that V is the direct sum of U and W.

4. Prove, $W \cap U = \{0\}$, in the Theorem 1.12.

Chapter Two

Matrices

2.1 Foundations
2.2 Column spaces, row spaces and system of linear equations
2.3 Row-echelon form and Gauss-Jordan elimination
2.4 Elementary matrices

認識了線性空間的架構之後，第二章我們將給讀者介紹認識，矩陣的基本概念，行、列向量空間與線性系統之互動關係，列運算以及一些特殊的矩陣等。接著，到第三章的時候，有關矩陣的秩，可逆矩陣之逆矩陣的數值求法，可逆之充要條件，以及 *LU*-factorization 等，一些有關矩陣的基礎工作，我們都將一一的給讀者詳實的介紹。

§2.1 Foundations

一開始我們要對矩陣做好定義，對於矩陣的"加、減、乘"之運算，也要說明清楚。逆矩陣、矩陣之轉置以及這些動作所可能衍生出的，一些重要觀念和結果，也都將在本節中一併研讀。首先，我們看看所謂的**矩陣** (matrix)。

Definition 2.1

> An $m \times n$ matrix, denoted by $A_{m \times n}$, is an array of numbers having m rows and n columns.

通常，我們將矩陣寫成

$$A_{m\times n} = \begin{pmatrix} a_{11} & a_{12} & \cdots & a_{1n} \\ a_{21} & a_{22} & \cdots & a_{2n} \\ \cdots & \cdots & \cdots & \cdots \\ a_{m1} & a_{m2} & \cdots & a_{mn} \end{pmatrix}$$

或者簡單寫爲，$A_{m\times n} = (a_{ij})_{m\times n}$，其中 a_{ij} 表示，矩陣的第 i 列第 j 行的元素。在不影響判讀情形之下，我們甚至簡單的以大寫的英文字母，$A = (a_{ij})$ 或 $B = (b_{ij})$ 或 $C = (c_{ij}) \cdots$，來表示矩陣。譬如說

$$\begin{pmatrix} 2 & -1 & 0 \\ 1 & 4 & -3 \\ -5 & 0 & 1 \end{pmatrix} \quad \begin{pmatrix} -1 & 0 & 12 \\ 0 & -5 & 7 \end{pmatrix} \quad \begin{pmatrix} -i & 0 \\ -2 & 1 \end{pmatrix}$$

are matrices of sizes, 3×3, 2×3 and 2×2, respectively.

另外一件事情，所謂 $0_{m\times n}$ 矩陣者指的是，矩陣的所有元素都是 0 的矩陣。在 size 的認知方面若沒有問題時，我們通常簡單的以 0 表示 $0_{m\times n}$ 矩陣。此外，還有一個很特殊的矩陣，那就是 1×1 矩陣。當一個純量扮演起矩陣角色的時候，它就是 1×1 矩陣。又若一般而言，當 $m = n$ 時，我們則稱矩陣 $A_{m\times n}$ 爲一個**方陣** (square matrix)。

至於，所謂**單位矩陣** (identity matrix) 者指的則是，一個對角線元素爲 1，其餘皆爲 0 之方陣，通常以 I_n 或 I 表示。譬如說

$$I_4 = \begin{pmatrix} 1 & 0 & 0 & 0 \\ 0 & 1 & 0 & 0 \\ 0 & 0 & 1 & 0 \\ 0 & 0 & 0 & 1 \end{pmatrix} \quad I_3 = \begin{pmatrix} 1 & 0 & 0 \\ 0 & 1 & 0 \\ 0 & 0 & 1 \end{pmatrix} \quad I_2 = \begin{pmatrix} 1 & 0 \\ 0 & 1 \end{pmatrix}$$

Definition 2.2 (addition and scalar multiplication)

Given two $m \times n$ matrices, $A_{m\times n} = (a_{ij})$, $B_{m\times n} = (b_{ij})$, and a scalar k.
1. The ij^{th} entry of $C = (c_{ij}) = A + B$ is defined to be $a_{ij} + b_{ij}$.
2. The ij^{th} entry of $D = (d_{ij}) = kA$ is defined to be ka_{ij}.

Example 1

Given that $A = \begin{pmatrix} -2 & 0 & 3 \\ 4 & -3 & 1 \end{pmatrix}$ and $B = \begin{pmatrix} 3 & 1 & -2 \\ -1 & 0 & 5 \end{pmatrix}$. Find $A+B$ and $-2A$.

Solution

$$A + B = \begin{pmatrix} 1 & 1 & 1 \\ 3 & -3 & 6 \end{pmatrix}$$

$$-2A = \begin{pmatrix} 4 & 0 & -6 \\ -8 & 6 & -2 \end{pmatrix}$$ ∎

對於矩陣的加法要特別小心。要是有人說，請把矩陣 $A = \begin{pmatrix} -2 & 0 & 3 \\ 4 & -3 & 1 \end{pmatrix}$ 與 $B = \begin{pmatrix} 2 & -1 & 0 \\ 1 & 4 & -3 \\ -5 & 0 & 1 \end{pmatrix}$ 加起來，這是辦不到的。因為，矩陣 A 與 B do not have the same size。

下面是一些簡單但是重要的基礎概念。我們不打算花時間一一的說明，希望同學當作家庭作業，練習一下。

Exercises

1. Show that for any matrices, A, B and C having the same size, $A + B = B + A$ (交換律，commutative law) and $(A+B)+C = A+(B+C)$ (結合律，associative law).

2. Let $V = \{all\ 3 \times 3\ matrices\ with\ real\ entries\}$. Show that V is a vector space over R with respect to the addition and scalar multiplication defined as before.

3. Find a basis for the vector space, V, in 2.

Definition 2.3 (multiplication of matrices)

Given two matrices, $A_{m \times k}$ and $B_{k \times n}$. If $C = (c_{ij}) = AB$, then the ij^{th} entry of C is defined by

$$c_{ij} = \sum_{t=1}^{k} a_{it} b_{tj}$$

矩陣 $A_{m \times k}$ 與 $B_{k \times n}$ 相乘之結果 C 是一個 $m \times n$ 矩陣。注意，在這兒矩陣 $A_{m \times k}$ 的行數與矩陣 $B_{k \times n}$ 的列數務必相同，才能執行乘法運算。下面我們看看幾個例題。

Example 2

Given $A = \begin{pmatrix} 2 & 1 & -3 \\ 1 & 0 & -2 \end{pmatrix}$ and $B = \begin{pmatrix} 1 & 3 \\ -2 & 5 \\ 0 & 1 \end{pmatrix}$. Find $C = AB$.

Solution

$c_{11} = 2 \cdot 1 + 1 \cdot (-2) + (-3) \cdot 0 = 0$，$c_{12} = 2 \cdot 3 + 1 \cdot 5 + (-3) \cdot 1 = 8$，
$c_{21} = 1 \cdot 1 + 0 \cdot (-2) + (-2) \cdot 0 = 1$，$c_{22} = 1 \cdot 3 + 0 \cdot 5 + (-2) \cdot 1 = 1$
所以

$$C = \begin{pmatrix} 0 & 8 \\ 1 & 1 \end{pmatrix}$$

Example 3

Given $A = \begin{pmatrix} 1 & -2 \\ 3 & 5 \end{pmatrix}$ and $B = \begin{pmatrix} -2 & 0 \\ 1 & 3 \end{pmatrix}$. Find AB and BA.

Solution

$$AB = \begin{pmatrix} 0 & -6 \\ -1 & 15 \end{pmatrix} \text{ and } BA = \begin{pmatrix} -2 & -4 \\ 10 & 13 \end{pmatrix}$$

前面這個例子說明，一般而言，矩陣的乘法是**不可交換的** (not commutative)。也就是說，**交換律** (commutative law) 是不成立的。至於結合律又如何呢？我們來看看。

Example 4

Given, $A = \begin{pmatrix} 1 & 0 & -2 \\ 0 & 3 & 1 \\ -2 & -1 & 1 \end{pmatrix}$, $B = \begin{pmatrix} -1 & 2 & -3 \\ 4 & 0 & 2 \\ 1 & -2 & 3 \end{pmatrix}$ and $C = \begin{pmatrix} 0 & 3 & -5 \\ 6 & 1 & 0 \\ 1 & -4 & 2 \end{pmatrix}$.

Find $(AB)C$ and $A(BC)$.

Solution

注意，$(AB)C$ 意思，為先做 AB 之乘積之後，再與矩陣 C 相乘。$A(BC)$ 意指，先做 BC 之乘積之後，再與 A 相乘。

$AB = \begin{pmatrix} -3 & 6 & -9 \\ 13 & -2 & 9 \\ -1 & -6 & 7 \end{pmatrix}$, $BC = \begin{pmatrix} 9 & 11 & -1 \\ 2 & 4 & -16 \\ -9 & -11 & 1 \end{pmatrix}$

$(AB)C = \begin{pmatrix} 27 & 33 & -3 \\ -3 & 1 & -47 \\ -29 & -37 & 19 \end{pmatrix}$ and $A(BC) = \begin{pmatrix} 27 & 33 & -3 \\ -3 & 1 & -47 \\ -29 & -37 & 19 \end{pmatrix}$ ◼

這個例題指出，$(AB)C = A(BC)$。其實，一般而言，只要 $(AB)C$ 與 $A(BC)$ 皆為可乘的矩陣 (也就是說，它們的 sizes 符合乘法的要求)，那麼 $(AB)C = A(BC)$，**結合律** (associative law) 是成立的。(參看本節最後的 Exercise 1)

Example 5

Given that $A = \begin{pmatrix} 2 & 1 & 0 & -2 \end{pmatrix}$ a 1×4 matrix, and $B = \begin{pmatrix} -1 \\ 3 \\ 1 \\ 3 \end{pmatrix}$ a 4×1

matrix. Then $AB = -2 + 3 + 0 - 6 = -5$, which is a scalar or we say, it is a 1×1 matrix. ∎

上例說明，若 A 為一 $1 \times n$ 的**列向量** (row vector) 且 B 為一 $n \times 1$ 的**行向量** (column vector)，則 AB 為一純量。

Notations

We shall always use A^i to denote the ith column of matrix, A, and A_j to denote the jth row of matrix, A.

在上述符號的情形之下，矩陣的相乘可以被表示如下。若 $C = (c_{ij}) = AB$，則

$$c_{ij} = A_i B^j, \quad \forall \ i, j$$

或者說，c_{ij} 是 row vector, A_i, 與 column vector, B^j, 的點積吧。

因此，若 C 為 A 與 B 的乘積的話，那麼 C 可以被簡單寫為

$$C = \begin{pmatrix} A_1 B^1 & A_1 B^2 & ... & A_1 B^n \\ A_2 B^1 & A_2 B^2 & ... & A_2 B^n \\ ... & ... & ... & ... \\ A_m B^1 & A_m B^2 & ... & A_m B^n \end{pmatrix}$$

由上述的表示法來看待矩陣相乘時，其運算就變得簡單而容易理解了。就以下列矩陣乘法的**分配律** (distributive law) 來說，讀者當可以輕鬆的驗證它的真實性，我們就不予詳細說明了。

Properties

If matrices A, B and C are of the sizes necessary for the following operations to make sense, then
1. $A(B + C) = AB + AC$
2. $(A + B)C = AC + BC$

此即所謂的，乘法對加法分配律。

Exercises

Let A and B be square matrices of the same size, and suppose that $AB = BA$. Show that

1. $(A+B)^2 = A^2 + 2AB + B^2$

2. $(A+B)(A-B) = A^2 - B^2$

注意：在這本書裡，我們也將用符號 A^2 表示 AA。雖然，A^2 同時也表示矩陣 A 的第二行，但是由於它們使用場合不同，讀者應該不至於將它們混為一談才對。

下面是一個簡單的補充說明。

假若，$A_{m \times k}$ 是一個 $m \times k$ 矩陣，那麼毫無疑問的

$$I_m A_{m \times k} = A_{m \times k} \quad \text{and} \quad A_{m \times k} I_k = A_{m \times k}$$

Example 6

Given, $A = \begin{pmatrix} 2 & -1 \\ 1 & 1 \end{pmatrix}$ and $B = \begin{pmatrix} 1/3 & 1/3 \\ -1/3 & 2/3 \end{pmatrix}$.

Find AB and BA.

Solution

$$AB = \begin{pmatrix} 1 & 0 \\ 0 & 1 \end{pmatrix}, \quad BA = \begin{pmatrix} 1 & 0 \\ 0 & 1 \end{pmatrix} \qquad \blacksquare$$

Definition 2.4

> A square matrix, A, is said to be "invertible" or "non-singular", if there exists a square matrix, B, such that $AB = BA = I$.

定義中所出現的矩陣，B，被稱為矩陣 A 的**逆矩陣** (inverse matrix)，我

們以符號 A^{-1} 表之。要是說，A^{-1} 不存在的話，我們則稱"A is not invertible or A is singular."另外要注意，當我們談論"是否為 invertible"時，矩陣 A 必須是一個 square matrix。

Exercises

1. Show that $B = \begin{pmatrix} 1/13 & 4/13 & -4/13 \\ 3/26 & 6/13 & 1/26 \\ 3/13 & -1/13 & 1/13 \end{pmatrix}$ is an inverse of $A = \begin{pmatrix} 1 & 0 & 4 \\ 0 & 2 & -1 \\ -3 & 2 & 0 \end{pmatrix}$.

2. Show that if B and C are inverses of square matrix, A, then $B = C$.

3. Show that by definition $\begin{pmatrix} 0 & 1 \\ 0 & 2 \end{pmatrix}$ is singular.

上述 Exercise 2 說明，矩陣的乘法反元素 (逆矩陣) 是**唯一的** (unique)。至於，如何求出一個 square matrix 的逆矩陣，在後面的章節裡，陸續會有不同的方法介紹，姑且按下不表。

Theorem 2.1

If A and B are both $n \times n$ non-singular matrices, then AB is non-singular.

Proof

Since A and B are non-singular, A^{-1} and B^{-1} exist. Now, consider that
$$(B^{-1}A^{-1})(AB) = B^{-1}(A^{-1}A)B = B^{-1}IB = B^{-1}B = I \text{ and}$$
$$(AB)(B^{-1}A^{-1}) = A(BB^{-1})A^{-1} = AIA^{-1} = AA^{-1} = I$$
we see that $(B^{-1}A^{-1})$ is the inverse of AB. So AB is non-singular.

從上述證明過程中，我們發現。假若，矩陣 A 與 B 是可逆的話，那麼 $(AB)^{-1} = (B^{-1}A^{-1})$。

Example 7

Given, $A = \begin{pmatrix} 2 & -1 \\ 1 & 1 \end{pmatrix}$ and $B = \begin{pmatrix} 1 & -1 \\ 0 & 1 \end{pmatrix}$. Find $(AB)^{-1}$.

Solution

We first note that
$$\begin{pmatrix} 2 & -1 \\ 1 & 1 \end{pmatrix}\begin{pmatrix} 1/3 & 1/3 \\ -1/3 & 2/3 \end{pmatrix} = \begin{pmatrix} 1/3 & 1/3 \\ -1/3 & 2/3 \end{pmatrix}\begin{pmatrix} 2 & -1 \\ 1 & 1 \end{pmatrix} = \begin{pmatrix} 1 & 0 \\ 0 & 1 \end{pmatrix}$$
$$\begin{pmatrix} 1 & -1 \\ 0 & 1 \end{pmatrix}\begin{pmatrix} 1 & 1 \\ 0 & 1 \end{pmatrix} = \begin{pmatrix} 1 & 1 \\ 0 & 1 \end{pmatrix}\begin{pmatrix} 1 & -1 \\ 0 & 1 \end{pmatrix} = \begin{pmatrix} 1 & 0 \\ 0 & 1 \end{pmatrix}$$

This says that A, B are invertible. So,
$$(AB)^{-1} = B^{-1}A^{-1} = \begin{pmatrix} 1 & 1 \\ 0 & 1 \end{pmatrix}\begin{pmatrix} 1/3 & 1/3 \\ -1/3 & 2/3 \end{pmatrix} = \begin{pmatrix} 0 & 1 \\ -1/3 & 2/3 \end{pmatrix}$$

Definition 2.5

Given an $m \times n$ matrix, A. The transpose of A, denoted by ${}^t A$, is defined to be an $n \times m$ matrix obtained by interchanging the rows and columns of A. I.e. if $B = (b_{ij}) = {}^t A$, then, for each ij^{th} entry of B, $b_{ij} = a_{ji}$.

e. g.

If $A = \begin{pmatrix} 2 & 1 & -3 \\ 1 & 0 & -2 \end{pmatrix}$, then ${}^t A = \begin{pmatrix} 2 & 1 \\ 1 & 0 \\ -3 & -2 \end{pmatrix}$

一個很特殊的現象，那就是單位矩陣，I_n，的轉置等於它本身。也就是說，$I_n = {}^t(I_n)$。談到轉置，下面有一些基本的特性。

Theorem 2.2

If matrices A and B are of the sizes necessary for the following operations to make sense, then
1. ${}^t({}^tA) = A$
2. ${}^t(A+B) = {}^tA + {}^tB$
3. ${}^t(AB) = {}^tB \, {}^tA$

Proof

本定理之 1. 和 2. 顯然成立,我們來證明 3.。

Suppose that A and B are of sizes $m \times s$ and $s \times n$, and let $C = AB$, $D = {}^tC$, $E = {}^tB$ and $F = {}^tA$, $G = EF$. Then it is sufficient to show, for each ij, $d_{ij} = g_{ij}$. And this can be done by looking at this

$$d_{ij} = c_{ji} = \sum_{k=1}^{s} a_{jk} b_{ki} = \sum_{k=1}^{s} f_{kj} e_{ik} = \sum_{k=1}^{s} e_{ik} f_{kj} = g_{ij}.$$

Remarks

1. If A is a matrix in complex field, then we use $\overline{A}$ to denote the complex conjugate of A. That is, if $A = (a_{ij})$, then $\overline{A} = (\overline{a_{ij}})$. And then we use A^* to denote the transpose conjugate of A. That is $A^* = {}^t(\overline{A})$.

   e.g.

   If $A = \begin{pmatrix} 2-i & 1 & 0 \\ i & 2i & 1+i \end{pmatrix}$, then $A^* = \begin{pmatrix} 2+i & -i \\ 1 & -2i \\ 0 & 1-i \end{pmatrix}$

   有件事注意一下,${}^t(\overline{A})$ 與 $\overline{({}^tA)}$ 結果相同。

2. A matrix, A, is said to be "symmetric", if $A = {}^tA$. And "hermitian", if $A = A^*$ in complex case.

   e.g.

   $A = \begin{pmatrix} 2 & -3 & 2 \\ -3 & 0 & 1 \\ 2 & 1 & 1 \end{pmatrix}$ is symmetric. $A = \begin{pmatrix} 0 & i & 0 \\ -i & 3 & 1+i \\ 0 & 1-i & 2 \end{pmatrix}$ is hermitian.

3. If A is either symmetric or hermitian symmetric, then A has to be a square matrix.

　　另外,假設矩陣,A,是 hermitian matrix 的話,那麼它的主對角線元素務必是實數。

Theorem 2.3

> If A and B are symmetric matrices, and $AB = BA$, then AB is symmetric.

Proof

The proof is quite easy as follow.
$$^t(AB) = {^tB}\,{^tA} = BA = AB.$$

要是這個條件,$AB = BA$,不存在的話,前述定理是不成立的。譬如說,若 $A = \begin{pmatrix} 2 & 1 \\ 1 & 1 \end{pmatrix}$, $B = \begin{pmatrix} 3 & 2 \\ 2 & 5 \end{pmatrix}$,則 $AB = \begin{pmatrix} 8 & 9 \\ 5 & 7 \end{pmatrix}$ 不是對稱的。注意,

$$AB = \begin{pmatrix} 8 & 9 \\ 5 & 7 \end{pmatrix} \neq BA = \begin{pmatrix} 8 & 5 \\ 9 & 7 \end{pmatrix}$$

Exercises

1. Given three matrices, $A_{m \times n}$, $B_{n \times k}$ and $C_{k \times l}$. Show that $(AB)C = A(BC)$.

2. Let A be a square matrix, and let $A^2 = 0$. Show that $(I - A)$ is non-singular.

3. In general, if A is a square matrix, and $A^n = 0$, then $(I - A)$ is non-singular.

4. Show that if $A = \begin{pmatrix} 1 & a \\ 0 & 1 \end{pmatrix}$, then A is non-singular. (*Hint*: Apply Exercise 2. with $B = \begin{pmatrix} 0 & -a \\ 0 & 0 \end{pmatrix}$)

5. Show that if A is singular and B is non-singular, then AB is singular. (*Hint*:可

用反證法。但是注意，下列說法目前來講是沒有根據的。假設 AB 為可逆的，則 $(AB)^{-1} = (B^{-1}A^{-1})$。)

6. Show that if A is non-singular, then $({}^{t}A)^{-1} = {}^{t}(A^{-1})$. Thus if A is non-singular, then ${}^{t}A$ is also non-singular. And if A is symmetric, then A^{-1} is also symmetric.

7. If A is a square matrix and if $a_{ij} = 0$, for every $i \neq j$.(This will be called a diagonal matrix.) Show that if $a_{ii} \neq 0$, $\forall i$, then A is non-singular. What is A^{-1}?

8. A square matrix, A, is called a "strictly upper triangular matrix", if $a_{ij} = 0$, $\forall\ i \geq j$. Show that if A is a 4×4 strictly upper triangular matrix, then $A^4 = 0$, and hence, $(I - A)$ is non-singular.

§2.2 Column spaces, row spaces and system of linear equations

我們將再一次的用心觀察矩陣的相乘，然後以**線性組合** (linear combination) 來表示矩陣的乘法，和**線性聯立方程組** (linear system)，並且用以說明線性聯立方程組其為有解之充要條件。

Given an $m \times k$ matrix, A, and a $k \times n$ matrix, B. If $C = AB$, then the columns of C are

$$C^1 = \begin{pmatrix} A_1 B^1 \\ A_2 B^1 \\ \dots \\ \dots \\ A_m B^1 \end{pmatrix}, \quad C^2 = \begin{pmatrix} A_1 B^2 \\ A_2 B^2 \\ \dots \\ \dots \\ A_m B^2 \end{pmatrix}, \quad \dots, \quad C^n = \begin{pmatrix} A_1 B^n \\ A_2 B^n \\ \dots \\ \dots \\ A_m B^n \end{pmatrix}$$

Let's look at C^1,

$$C^1 = \begin{pmatrix} c_{11} \\ c_{21} \\ \dots \\ \dots \\ c_{m1} \end{pmatrix} = \begin{pmatrix} a_{11}b_{11} + a_{12}b_{21} + \dots + a_{1k}b_{k1} \\ a_{21}b_{11} + a_{22}b_{21} + \dots + a_{2k}b_{k1} \\ \dots \\ \dots \\ a_{m1}b_{11} + a_{m2}b_{21} + \dots + a_{mk}b_{k1} \end{pmatrix}$$

$$= b_{11} A^1 + b_{21} A^2 + \dots + b_{k1} A^k = \sum_{i=1}^{k} b_{i1} A^i$$

It is a linear combination of columns of A with coefficients consisting of the first column of B. Similarly, for the second column of C,

$$C^2 = b_{12} A^1 + b_{22} A^2 + \dots + b_{k2} A^k = \sum_{i=1}^{k} b_{i2} A^i$$

and the jth column of C,

$$C^j = b_{1j} A^1 + b_{2j} A^2 + \dots + b_{kj} A^k = \sum_{i=1}^{k} b_{ij} A^i$$

and of course, the last column,

$$C^n = b_{1n} A^1 + b_{2n} A^2 + \dots + b_{kn} A^k = \sum_{i=1}^{k} b_{in} A^i$$

they are all linear combinations of the columns $\{A^1, A^2, \dots, A^k\}$ of A. Where the coefficients of the first column, C^1, come from the first column of B, and the coefficients of the second column, C^2, come from the second column of B, the jth column come from the jth column of B,..., and so on.

這是一個非常漂亮的觀察結果，矩陣 C 的每一行都來自於，矩陣 A 的 k 個行向量所生成的空間裡 (the space generated by the columns of A, or simply say that the column space of A, and denoted by $CS(A)$)。也就是說，

$$C^j \in CS(A) = SP\{A^1, A^2, ..., A^k\}, \quad \forall\ j = 1, 2, ..., n$$

更進一步的說，$CS(C) \subseteq CS(A)$。這個結果，當我們在研究線性系統的解、矩陣的秩、或是向量空間的維度的時候，將給我們帶來很大的方便。

Example 1

Given that $A = \begin{pmatrix} 2 & 1 & -3 \\ 1 & 0 & -2 \end{pmatrix}$ and $B = \begin{pmatrix} 1 & 3 \\ -2 & 5 \\ 0 & 1 \end{pmatrix}$, then $C = AB = \begin{pmatrix} 0 & 8 \\ 1 & 1 \end{pmatrix}$.

In fact, $C^1 = \begin{pmatrix} 0 \\ 1 \end{pmatrix} = 1\begin{pmatrix} 2 \\ 1 \end{pmatrix} - 2\begin{pmatrix} 1 \\ 0 \end{pmatrix} + 0\begin{pmatrix} -3 \\ -2 \end{pmatrix} = b_{11}A^1 + b_{21}A^2 + b_{31}A^3$, and

$C^2 = \begin{pmatrix} 8 \\ 1 \end{pmatrix} = 3\begin{pmatrix} 2 \\ 1 \end{pmatrix} + 5\begin{pmatrix} 1 \\ 0 \end{pmatrix} + 1\begin{pmatrix} -3 \\ -2 \end{pmatrix} = b_{12}A^1 + b_{22}A^2 + b_{32}A^3$ ■

Example 2

Given that $A = \begin{pmatrix} 2 & 1 & -3 \\ 1 & 0 & -2 \end{pmatrix}$ and $B = \begin{pmatrix} 1 & 3 \\ -2 & 5 \\ 0 & 1 \end{pmatrix}$. Use the previous method to find the product, $D = BA$.

Solution

Noting that D would be a 3×3 matrix, so it has three columns. And they are linear combinations, at this moment, of the columns of B.

$$D^1 = 2\begin{pmatrix}1\\-2\\0\end{pmatrix} + 1\begin{pmatrix}3\\5\\1\end{pmatrix} = \begin{pmatrix}5\\1\\1\end{pmatrix}, \quad D^2 = 1\begin{pmatrix}1\\-2\\0\end{pmatrix} + 0\begin{pmatrix}3\\5\\1\end{pmatrix} = \begin{pmatrix}1\\-2\\0\end{pmatrix},$$

$$D^3 = -3\begin{pmatrix}1\\-2\\0\end{pmatrix} - 2\begin{pmatrix}3\\5\\1\end{pmatrix} = \begin{pmatrix}-9\\-4\\-2\end{pmatrix}$$

Therefore, $D = \begin{pmatrix}5 & 1 & -9\\1 & -2 & -4\\1 & 0 & -2\end{pmatrix}$

除了矩陣 $C = AB$ 的 columns 之外，不妨讓我們也來觀察一下，有關矩陣 C 的 rows，它們的結構又是如何？首先 the first row, C_1,

$(c_{11} \quad c_{12} \quad ... \quad c_{1n})$
$= (a_{11}b_{11} + a_{12}b_{21} + ... a_{1k}b_{k1} \quad a_{11}b_{12} + a_{12}b_{22} + ... a_{1k}b_{k2} \quad ... \quad a_{11}b_{1n} + a_{12}b_{2n} + ... a_{1k}b_{kn})$
$= a_{11}B_1 + a_{12}B_2 + ... + a_{1k}B_k = \sum_{i=1}^{k} a_{1i}B_i$

The second row, C_2,
$(c_{21} \quad c_{22} \quad ... \quad c_{2n})$
$= a_{21}B_1 + a_{22}B_2 + ... + a_{2k}B_k = \sum_{i=1}^{k} a_{2i}B_i$

The jth row, C_j,
$(c_{j1} \quad c_{j2} \quad ... \quad c_{jn})$
$= a_{j1}B_1 + a_{j2}B_2 + ... + a_{jk}B_k = \sum_{i=1}^{k} a_{ji}B_i$

相類似的我們又發現，矩陣 C 的每一個 row 都等於，矩陣 B 的 k 個 rows 的一個線性組合。其中，C_1 的係數是矩陣 A 的第一列，C_2 的係數是矩陣 A 的第二列，C_j 的係數是矩陣 A 的第 j 列，…等等。換句話說，each row vector, C_j, of the matrix, C, lies in the row space of B, or simply say that

$$C_j \in RS(B) = SP\{B_1, B_2, ..., B_k\}, \quad \forall\ j = 1, 2, ..., m.$$

所以，此時 $RS(C) \subseteq RS(B)$。

Example 3

Given that $A = \begin{pmatrix} 2 & 1 & -3 \\ 1 & 0 & -2 \end{pmatrix}$ and $B = \begin{pmatrix} 1 & 3 \\ -2 & 5 \\ 0 & 1 \end{pmatrix}$, then $C = AB = \begin{pmatrix} 0 & 8 \\ 1 & 1 \end{pmatrix}$.

Write each row of C as a linear combination of rows of B.

Solution

$$C_1 = a_{11}B_1 + a_{12}B_2 + a_{13}B_3 = 2(1\ \ 3) + 1(-2\ \ 5) - 3(0\ \ 1) = (0\ \ 8)$$
$$C_2 = a_{21}B_1 + a_{22}B_2 + a_{23}B_3 = 1(1\ \ 3) + 0(-2\ \ 5) - 2(0\ \ 1) = (1\ \ 1)$$ ■

稍微休息一下，讓我們喘口氣…。接著，再也沒那麼複雜了，我們要用前述所獲心得，來介紹有關的線性聯立方程組。

$$\begin{cases} a_{11}x_1 + a_{12}x_2 + ... + a_{1n}x_n = b_1 \\ a_{21}x_1 + a_{22}x_2 + ... + a_{2n}x_n = b_2 \\ \quad ... \\ a_{m1}x_1 + a_{m2}x_2 + ... + a_{mn}x_n = b_m \end{cases} \quad (2.1)$$

(2.1)式為一有 m 個線性方程式、n 個未知數的聯立方程。今將各等式之左邊的係數、未知數以及右邊的常數分別寫為

$$A = \begin{pmatrix} a_{11} & a_{12} & ... & a_{1n} \\ a_{21} & a_{22} & ... & a_{2n} \\ ... & ... & ... & ... \\ a_{m1} & a_{m2} & ... & a_{mn} \end{pmatrix}, \quad X = \begin{pmatrix} x_1 \\ x_2 \\ ... \\ x_n \end{pmatrix}, \quad B = \begin{pmatrix} b_1 \\ b_2 \\ ... \\ b_m \end{pmatrix}$$

那麼 (2.1) 式可以被簡化為

$$AX = B \quad (2.2)$$

或者按照前述 column space 的說法，我們發現

$$x_1 A^1 + x_2 A^2 + ... + x_n A^n = B \qquad (2.3)$$

這個式子太漂亮了，它很清楚的告訴我們，假設 $B \in CS(A)$，則 B 可以被表示為 $A^1, A^2, ..., A^n$ 的線性組合，也就是說，(2.1) 式的聯立方程組有解。換句話說，假設 (2.1) 式有解的話，那麼由於 $B \in CS(A)$，所以 the dimension of the space generated by $A^1, A^2, ..., A^n$ is the same as the dimension of the space generated by $A^1, A^2, ..., A^n, B$. That is

$$\dim(SP\{A^1, A^2, ..., A^n\}) = \dim(SP\{A^1, A^2, ..., A^n, B\})$$

前述，矩陣 A 的 n 個行向量，$A^1, A^2, ..., A^n$，加上常數向量 B，將其寫成，$(A^1, A^2, ..., A^n, B)$，或簡單寫成，$(A \mid B)$。這個矩陣被稱為**擴增矩陣** (the augmented matrix)。我們再清楚的重複一遍，

The system (2.1) of linear equations has a solution if and only if

$$\dim(CS(A)) = \dim(CS(A|B))$$

這雖然是一個簡單的概念，可到後面，在介紹 (2.1) 式之解的時候有直接的相關。

下面是一個比較特殊的線性系統，所謂的**齊次線性系統** (homogeneous linear system)。

$$x_1 A^1 + x_2 A^2 + ... + x_n A^n = 0 \qquad (2.4)$$

(注意，(2.4) 式右邊的 0 表示的是 $m \times 1$ **零行向量** (zero column vector))
很顯然的 $x_1 = x_2 = ... = x_n = 0$ 為 (2.4) 式之一組解，它被稱為 trivial solution。然而，一個 homogeneous linear system 是否有**非零解** (non-trivial solution)，可由下面定理窺其端倪。

Theorem 2.4

Equation (2.4) has a non-trivial solution if and only if $A^1, A^2, ..., A^n$ form a dependent set.

Proof

(⇒)

Equation (2.4) has a non-trivial solution means that there exists a set of numbers, $x_1, x_2, ..., x_n$, not all equal to zero such that

$$x_1 A^1 + x_2 A^2 + ... + x_n A^n = 0$$

This implies that $A^1, A^2, ..., A^n$ are linearly dependent.

(⇐)

Now if $A^1, A^2, ..., A^n$ are linearly dependent, then there are n numbers, $x_1, x_2, ..., x_n$, not all equal to zero such that

$$x_1 A^1 + x_2 A^2 + ... + x_n A^n = 0$$

Hence, $x_1, x_2, ..., x_n$ form a set of non-trivial solution to Equation (2.4).

譬如說，
$$\begin{cases} 2x_1 + x_2 - x_3 = 0 \\ -x_1 - \frac{1}{2}x_2 + 3x_3 = 0 \\ 4x_1 + 2x_2 + x_3 = 0 \end{cases}$$ 有一非零解，$x_1 = 1$, $x_2 = -2$, $x_3 = 0$。因為，

$A = \begin{pmatrix} 2 & 1 & -1 \\ -1 & -1/2 & 3 \\ 4 & 2 & 1 \end{pmatrix}$ 的三個 columns，A^1, A^2, A^3 are linearly dependent。

Exercises

1. Given, $A = \begin{pmatrix} 0 & 2 \\ -1 & 1 \\ 2 & -3 \end{pmatrix}$ and $B = \begin{pmatrix} 1 & 2 & 2 \\ 0 & -1 & 2 \end{pmatrix}$. Write each column of $C = AB$ as a linear combination of A^1, A^2. And write each row of $C = AB$ as a linear combination of B_1, B_2.

2. Find a solution to $\begin{cases} 2x_2 = -2 \\ -x_1 + x_2 = -3 \\ 2x_1 - 3x_2 = 7 \end{cases}$ by examining the result from exercise 1.

3. Show that if A is an $m \times n$ matrix and if $n > m$, then $x_1 A^1 + x_2 A^2 + ... + x_n A^n = 0$ has a non-trivial solution. (*Hint*: $A^1, A^2, ..., A^n$ are vectors in the *m*-dimensional space, R^m or C^m)

§2.3 Row-echelon form and Gauss-Jordan elimination

線性系統的數值解法有許多，本節將首先介紹所謂的 Gauss-Jordan elimination。在這之前，我們需要一些技術上的基本認識，特別是將擴增矩陣 $(A|B)$ 化為 row-echelon form 的過程與技巧。在這個過程當中，除了所謂的 distinguished columns 是先發要角之外，我們也得先認識，何謂 row-echelon form。

Definition 2.6

> Given a matrix, A. The distinguished columns of A are part of the non-zero columns of A such that no one of which is a linear combination of its predecessors. (from left to right)

e.g.

1. The distinguished columns of $A = \begin{pmatrix} 0 & 1 & 1 & 2 \\ 0 & 3 & 4 & 6 \end{pmatrix}$ are A^2 and A^3. Where A^1 is a zero column and A^4 is a linear combination of its predecessors, A^2, A^3. So, A^1, A^4 are not distinguished columns.

2. The distinguished columns of $A = \begin{pmatrix} 1 & 0 & 2 \\ 0 & 1 & 5 \\ 0 & 0 & 0 \end{pmatrix}$ are A^1 and A^2, since A^3 is a linear combination of its predecessors, A^1 and A^2.

Theorem 2.5

> The distinguished columns of a matrix, A, are linearly independent and they form a basis for the column space, $CS(A)$, of A.

Proof

這個證明不會太難。假設 $\alpha_1, \alpha_2, ..., \alpha_k$ 為矩陣 A 的 distinguished columns，並且假設，

$$x_1\alpha_1 + x_2\alpha_2 + ... + x_k\alpha_k = 0$$

今，若有一係數 $x_i \neq 0$，則該 x_i 絕不可能是 x_k，否則，α_k 就可以被表示為 $\alpha_1, \alpha_2, ..., \alpha_{k-1}$ 的線性組合，此與 distinguished columns 的定義相互違背。同樣道理，x_i 也絕不可能是 x_{k-1}，也絕不可能是 x_{k-2}。以此道理往前類推，得知，絕不可能有不為 0 的 x_i。因而得知，$\alpha_1, \alpha_2, ..., \alpha_k$ 為線性獨立的。

由 column space 的定義，我們知道，$CS(A)$ 是由矩陣 A 的最大線性獨立的行向量所生成。而且由前述 distinguished columns 的定義而言，$\alpha_1, \alpha_2, ..., \alpha_k$ 就是 the maximal set of linear independence of columns of matrix, A。所以

$$SP\{\alpha_1, \alpha_2, ..., \alpha_k\} = CS(A)$$

也就是說，$\{\alpha_1, \alpha_2, ..., \alpha_k\}$ is a basis for $CS(A)$。

Definition 2.7

> An $m \times n$ matrix, A, is called a row-echelon matrix, if there exists a positive integer, $r \leq m$, such that
> (i) the last $m - r$ rows of A are zero rows.
> (ii) the distinguished columns of A are the first r-columns of I_m. (in the same order from left to right).

這個定義需要認真的思考一下。尤其第 (ii) 部分所指的是，"there are r distinguished columns of A, and the jth distinguished column of A is the unit column vector, e^j such that $^t(e^j) = (0 \ ... \ 1 \ 0 \ ... \ 0)$"。在這兒，請特別注意，往後我們就將以符號 e^j (有時用 E^j) 來表示前述所言之單位行向

量。也就是說，行向量 e^j 的座標除了第 j 個為 1 之外，其餘皆為 0。

e.g.

1. $A = \begin{pmatrix} 1 & 2 & 3 & 0 & 4 \\ 0 & 0 & 0 & 1 & 5 \\ 0 & 0 & 0 & 0 & 0 \end{pmatrix}$ is a row-echelon matrix, since there are 2 non-zero rows

    and 2 distinguished columns, the 1st and the 4th columns of A. And they are e^1, e^2 in I_3.

2. $B = \begin{pmatrix} 0 & 1 & -1 & 0 \\ 0 & 0 & 0 & 1 \\ 0 & 0 & 0 & 0 \\ 0 & 0 & 0 & 0 \end{pmatrix}$ is a row-echelon matrix, since there are 2 non-zero rows

    and 2 distinguished columns, the 2nd and the 4th columns of B. And they are e^1, e^2 in I_4.

3. $C = \begin{pmatrix} 1 & -3 & 0 & 4 & 0 & 6 \\ 0 & 0 & 1 & 5 & 0 & 7 \\ 0 & 0 & 0 & 0 & 1 & 8 \end{pmatrix}$ is a row-echelon matrix, since there are 3 non-zero

    rows and 3 distinguished columns, the 1st, the 3rd and the 5th columns of C. And they are e^1, e^2 and e^3 in I_3.

4. $D = \begin{pmatrix} 1 & 0 & 0 & 0 & 4 \\ 0 & 1 & 0 & 0 & 2 \\ 0 & 0 & 1 & 0 & 4 \\ 0 & 0 & 0 & 1 & -1 \end{pmatrix}$ is a row-echelon matrix, since there are 4 non-zero

    rows and 4 distinguished columns, the 1st, the 2nd, the 3rd and the 4th columns.

However, the following two matrices are not row-echelon matrices.

$$\begin{pmatrix} 1 & 2 & 3 & 1 & 4 \\ 0 & 0 & 0 & 1 & 5 \\ 0 & 0 & 0 & 0 & 0 \end{pmatrix}, \text{ since the 2}^{nd} \text{ distinguished column, } \begin{pmatrix} 1 \\ 1 \\ 0 \end{pmatrix}, \text{ is not } e^2.$$

$$\begin{pmatrix} 0 & 1 & -1 & 0 \\ 0 & 0 & 0 & 0 \\ 1 & 0 & 0 & 0 \\ 0 & 0 & 0 & 0 \end{pmatrix}, \text{ since the 1}^{st} \text{ distinguished column, } \begin{pmatrix} 0 \\ 0 \\ 1 \\ 0 \end{pmatrix}, \text{ is not } e^1 \text{ and the}$$

2nd one is not e^2.

Properties

Let $A_{m \times n}$ be a row-echelon matrix.
1. If A_s is a zero row and if $t \geq s$, then A_t is also a zero row.
2. If A_s is a non-zero row, then the first non-zero entry in A_s is equal to 1.
3. The non-zero rows of A are linearly independent and $\dim(RS(A)) = \dim(CS(A))$.

Proof

1. Suppose that (w. l. o. g.) A_{s+1} is not a zero row, then, by the definition, there are at least $s+1$ distinguished columns of A, and the sth distinguished column must be e^s, which has 1 in the sth entry and 0 in the other entries. This leads to a contradiction, since A_s is a zero row.

2. If A_s is a non-zero row, and assume that the jth entry is the first non-zero entry, then, by the definition, A^j is a distinguished column. And hence, A^j must be one of $\{e^1, e^2, ...\}$, and clearly $A^j = e^j$. So the first non-zero entry in A_s is equal to 1.

3. Let $A_1, A_2, ..., A_r$ be non-zero rows. Then, by part 2, for each $i = 1, 2, ..., r$, the first non-zero entry of A_i is 1. Now, let $c_1 A_1 + c_2 A_2 + ... + c_r A_r = 0$, then we may immediately find out that $c_1 = 0$, and then $c_2 = 0$, and then $c_3 = c_4 = ... = c_r = 0$. Therefore, $A_1, A_2, ..., A_r$ are linearly independent.

Henceforth, $\dim(RS(A)) = r = \dim(CS(A))$.

了解 row-echelon matrix 之後,我們開始要來介紹 Gauss-Jordan elimination。我們的作法是,利用**列運算** (row operations) 將 linear system $AX = B$ 的 augmented matrix $(A \mid B)$ 化簡成為 row-echelon matrix,然後從中判讀出聯立方程組之解。然而,列運算為何?又如何判讀聯立方程組之解呢?讓我們看看下列定義。

Definition 2.8

Let A be an $m \times n$ matrix. There are 3 basic types of row operations defined as follow.
1. Interchanging rows i and j of A, and is denoted by $r(i, j)A$.
2. Replacing the ith row of A by kA_i, where k is a non-zero scalar. And is denoted by $r_i(k)A$.
3. Replacing the ith row of A by $A_i + kA_j$, where k is a non-zero scalar. And is denoted by $r(i + kj)A$.

Example 1

Given that $A = \begin{pmatrix} -5 & 0 & -3 & 4 \\ -3 & 1 & 2 & 0 \\ 1 & 2 & -1 & 3 \end{pmatrix}$. Find $B = r(1, 3)A$, $r_1(5)B$, $C = r(3 + 5 \cdot 1)B$ and $r(2 + 3 \cdot 1)C$.

Solution

$B = r(1, 3)A = \begin{pmatrix} 1 & 2 & -1 & 3 \\ -3 & 1 & 2 & 0 \\ -5 & 0 & -3 & 4 \end{pmatrix}$ $r_1(5)B = \begin{pmatrix} 5 & 10 & -5 & 15 \\ -3 & 1 & 2 & 0 \\ -5 & 0 & -3 & 4 \end{pmatrix}$

$$C = r(3+5\cdot 1)B = \begin{pmatrix} 1 & 2 & -1 & 3 \\ -3 & 1 & 2 & 0 \\ 0 & 10 & -8 & 19 \end{pmatrix} \quad r(2+3\cdot 1)C = \begin{pmatrix} 1 & 2 & -1 & 3 \\ 0 & 7 & -1 & 9 \\ 0 & 10 & -8 & 19 \end{pmatrix}$$

做了一個例題之後，我們了解到，以純手工執行列運算時，最重要的，莫過於要特別小心每一個步驟的計算。稍一不慎，就前功盡棄。就算使用電腦來做列運算也是如此，要特別留意程式每一步驟的編寫。

Exercise

Given any two non-zero scalars, c and d. Let $r(ci+dj)A$ be a matrix obtained from matrix, A, by replacing the ith row of A by $cA_i + dA_j$. Find two basic row operations to obtain matrix, $r(ci+dj)A$, from matrix, A.

上述習題中的 $r(ci+dj)A$，也是一個列運算。甚至於也常被視為一個基本的列運算。有了列運算，就可以切入本節的主題了。可是切入主題之前，我們似乎得再釐清幾件事情。

1. 當我們以列運算，轉換擴增矩陣 $(A \mid B)$ 之後，原線性系統 $AX = B$ 的解不會改變。
2. 使用有限個基本列運算之後，可以將擴增矩陣 $(A \mid B)$，轉換成一個 row-echelon matrix。

或者更貼切的說：

1. Suppose that $(C \mid D)$ is an augmented matrix obtained from $(A \mid B)$ by applying a finite number of row operations, then $AX = B$ and $CX = D$ have the same solutions.
2. If F is any matrix, then there are finite number of basic row operations, $R_1, R_2, ..., R_l$, such that $(R_1, R_2 ... R_l)F = G$ is a row-echelon matrix.

前述兩個事實合併起來,建構了所謂的 Gauss-Jordan elimination 的方法。講的夠清楚了。現在,我們先列出兩個簡單的例子,練習判讀 $(A \mid B)$ 被轉換成 $(C \mid D)$ 之後的解。

Example 2

假設,

$$(C \mid D) = \begin{pmatrix} 1 & 0 & 0 & 0 & | & 4 \\ 0 & 1 & 0 & 0 & | & 2 \\ 0 & 0 & 1 & 0 & | & 4 \\ 0 & 0 & 0 & 1 & | & -1 \end{pmatrix}$$

那麼,線性系統 $CX = D$ 之解即為,

$$\begin{cases} x_1 = 4 \\ x_2 = 2 \\ x_3 = 4 \\ x_4 = -1 \end{cases}$$

Example 3

假設,

$$(C \mid D) = \begin{pmatrix} 1 & -3 & 0 & 4 & 0 & | & 6 \\ 0 & 0 & 1 & 5 & 0 & | & 7 \\ 0 & 0 & 0 & 0 & 1 & | & 8 \end{pmatrix}$$

那麼 $(C \mid D)$ 所代表的線性系統即為,

$$\begin{cases} x_1 - 3x_2 + 4x_4 = 6 \\ x_3 + 5x_4 = 7 \\ x_5 = 8 \end{cases}$$

其解為

$$X = \begin{pmatrix} 6+3a-4b \\ a \\ 7-5b \\ b \\ 8 \end{pmatrix}$$，其中 a 與 b 為任意二純量。

最後，我們從頭到尾正式的來解兩個線性系統的例子。

Example 4

Solve the following linear system by Gauss-Jordan elimination.

$$\begin{cases} x - y = 2 \\ x - 2y = 8/5 \\ 2x + 3y = 6 \end{cases}$$

Solution

首先將其 augmented matrix 寫下，

$$F = (A \mid B) = \begin{pmatrix} 1 & -1 & | & 2 \\ 1 & -2 & | & 8/5 \\ 2 & 3 & | & 6 \end{pmatrix}$$

然後按下列步驟將其簡化為 row-echelon matrix。

$$r(3+(-2)1)F = \begin{pmatrix} 1 & -1 & | & 2 \\ 1 & -2 & | & 8/5 \\ 0 & 5 & | & 2 \end{pmatrix}, \quad r(2+(-1)1)r(3+(-2)1)F = \begin{pmatrix} 1 & -1 & | & 2 \\ 0 & -1 & | & -2/5 \\ 0 & 5 & | & 2 \end{pmatrix}$$

$$r(1+(-1)2)r(2+(-1)1)r(3+(-2)1)F = \begin{pmatrix} 1 & 0 & | & 12/5 \\ 0 & -1 & | & -2/5 \\ 0 & 5 & | & 2 \end{pmatrix}$$

$$r(3+(5)2)r(1+(-1)2)r(2+(-1)1)r(3+(-2)1)F = \begin{pmatrix} 1 & 0 & | & 12/5 \\ 0 & -1 & | & -2/5 \\ 0 & 0 & | & 0 \end{pmatrix}$$

將最後矩陣的第二列乘以 -1，即為一個 row-echelon matrix。然後，從該矩陣中，我們清楚的發現，本題之解為

$$X = \begin{pmatrix} x \\ y \end{pmatrix} = \begin{pmatrix} 12/5 \\ 2/5 \end{pmatrix}$$ ∎

Example 5

Solve the following linear system by Gauss-Jordan elimination.

$$\begin{cases} -5x_1 - 3x_3 = 4 \\ -3x_1 + x_2 + 2x_3 = 0 \\ x_1 + 2x_2 - x_3 = 3 \end{cases}$$

Solution

首先將其 augmented matrix 寫下，

$$F = (A \mid B) = \begin{pmatrix} -5 & 0 & -3 & \mid & 4 \\ -3 & 1 & 2 & \mid & 0 \\ 1 & 2 & -1 & \mid & 3 \end{pmatrix}$$

然後按下列步驟將其簡化為 row-echelon matrix。

$$r(1,3)F = \begin{pmatrix} 1 & 2 & -1 & \mid & 3 \\ -3 & 1 & 2 & \mid & 0 \\ -5 & 0 & -3 & \mid & 4 \end{pmatrix}, \quad r(3+5\cdot 1)r(1,3)F = \begin{pmatrix} 1 & 2 & -1 & \mid & 3 \\ -3 & 1 & 2 & \mid & 0 \\ 0 & 10 & -8 & \mid & 19 \end{pmatrix}$$

$$r(2+3\cdot 1)r(3+5\cdot 1)r(1,3)F = \begin{pmatrix} 1 & 2 & -1 & \mid & 3 \\ 0 & 7 & -1 & \mid & 9 \\ 0 & 10 & -8 & \mid & 19 \end{pmatrix}$$

$$r_2(\frac{1}{7})r(2+3\cdot 1)r(3+5\cdot 1)r(1,3)F = \begin{pmatrix} 1 & 2 & -1 & \mid & 3 \\ 0 & 1 & -1/7 & \mid & 9/7 \\ 0 & 10 & -8 & \mid & 19 \end{pmatrix}$$

$$r(1+(-2)\cdot 2)r_2(\frac{1}{7})r(2+3\cdot 1)r(3+5\cdot 1)r(1,3)F = \begin{pmatrix} 1 & 0 & -5/7 & | & 3/7 \\ 0 & 1 & -1/7 & | & 9/7 \\ 0 & 10 & -8 & | & 19 \end{pmatrix}$$

$$r(3+(-10)\cdot 2)r(1+(-2)\cdot 2)r_2(\frac{1}{7})r(2+3\cdot 1)r(3+5\cdot 1)r(1,3)F$$

$$= \begin{pmatrix} 1 & 0 & -5/7 & | & 3/7 \\ 0 & 1 & -1/7 & | & 9/7 \\ 0 & 0 & -46/7 & | & 43/7 \end{pmatrix}$$

$$r_3(-\frac{7}{46})r(3+(-10)\cdot 2)r_2(\frac{1}{7})r(2+3\cdot 1)r(3+5\cdot 1)r(1,3)F$$

$$= \begin{pmatrix} 1 & 0 & -5/7 & | & 3/7 \\ 0 & 1 & -1/7 & | & 9/7 \\ 0 & 0 & 1 & | & -43/46 \end{pmatrix}$$

$$r(2+(\frac{1}{7})\cdot 3)r_3(-\frac{7}{46})r(3+(-10)\cdot 2)r_2(\frac{1}{7})r(2+3\cdot 1)r(3+5\cdot 1)r(1,3)F$$

$$= \begin{pmatrix} 1 & 0 & -5/7 & | & 3/7 \\ 0 & 1 & 0 & | & 371/322 \\ 0 & 0 & 1 & | & -43/46 \end{pmatrix}$$

$$r(1+(\frac{5}{7})\cdot 3)r(2+(\frac{1}{7})\cdot 3)r_3(-\frac{7}{46})r(3+(-10)\cdot 2)r_2(\frac{1}{7})r(2+3\cdot 1)r(3+5\cdot 1)r(1,3)F$$

$$= \begin{pmatrix} 1 & 0 & 0 & | & -77/328 \\ 0 & 1 & 0 & | & 371/322 \\ 0 & 0 & 1 & | & -43/46 \end{pmatrix}$$

最後的矩陣為一個 row-echelon matrix，從該矩陣我們清楚的發現，本題之解為

$$X = \begin{pmatrix} x_1 \\ x_2 \\ x_3 \end{pmatrix} = \begin{pmatrix} -77/328 \\ 371/322 \\ -43/46 \end{pmatrix}$$ ∎

Exercises

Use Gauss-Jordan method to solve 1~3.

1. $\begin{cases} 2x_1 + 3x_2 - x_3 = 4 \\ x_1 - 2x_2 + 3x_3 = 1 \end{cases}$

2. $\begin{cases} x_1 + 3x_2 - x_3 = 2 \\ -2x_1 + x_2 - 2x_3 = 4 \\ 3x_1 + 2x_2 - x_3 = -1 \end{cases}$

3. $\begin{cases} 4x_1 + 2x_3 - x_4 + x_5 = 1 \\ 2x_1 + 2x_2 + 3x_4 - 2x_5 = -2 \\ -x_1 + x_2 - 2x_3 + 2x_5 = 0 \\ x_2 + x_3 - 2x_4 + 2x_5 = 3 \end{cases}$

4. Suppose that the augmented matrix, $C = (A \mid B)$, of $AX = B$ is a row-echelon matrix.

   a. What is the form of C, if $AX = B$ has no solution?

   b. What is the form of C, if $AX = B$ has a unique solution?

   c. What is the from of C, if $AX = B$ has an infinite number of solutions?

§2.4　Elementary matrices

這一節，我們想要用不同的方式來了解，第 2.3 節裡所介紹的 row operations。或者說，給大家介紹新的詞令，所謂的 "基本矩陣"。對於任何一個列運算，我們可以找出一個合適的矩陣來代表它，並且可以由矩陣的相乘，得出列運算作用在一個矩陣之後的結果。這樣講不太容易理解，我們還是來看看，到底是怎麼一回事。首先，讓我們解釋一下，什麼叫 elementary matrix？

Definition 2.9

> An elementary matrix is a matrix obtained by applying a row operation to the identity matrix.

詳細來說，

1. Interchange the ith and the jth rows of I_n, $r(i,j)I_n$. We denote it by $E(i,j)$.
2. Multiply the ith row of I_n by a non-zero constant, k, $r_i(k)I_n$. We denote it by $E_i(k)$.
3. Add a scalar multiple of the jth row to the ith row of I_n, $r(i+kj)I_n$. We denote it by $E(i+kj)$.

前述之 $E(i,j)$、$E_i(k)$ 以及 $E(i+kj)$ 都是所謂的**基本矩陣** (elementary matrices)。以 $n = 3$ 而言，

$$E(1,3) = \begin{pmatrix} 0 & 0 & 1 \\ 0 & 1 & 0 \\ 1 & 0 & 0 \end{pmatrix}, \quad E_3(k) = \begin{pmatrix} 1 & 0 & 0 \\ 0 & 1 & 0 \\ 0 & 0 & k \end{pmatrix}, \quad E(1+k3) = \begin{pmatrix} 1 & 0 & k \\ 0 & 1 & 0 \\ 0 & 0 & 1 \end{pmatrix}$$

回頭用心的端詳一下，這三個基本矩陣好像都是可逆的。沒錯，確實如此，任何一個基本矩陣都是可逆的。而且

$$E(i,j)^{-1} = E(j,i), \quad E_i(k)^{-1} = E_i(1/k), \quad E(i+kj)^{-1} = E(i-kj)$$

太棒了！不僅如此，往下看還有更精彩的東西。

Theorem 2.6

> Given an $m \times n$ matrix, A.
> 1. The matrix, B, obtained from A by interchanging the rows A_i and A_j can be realized as $B = E(i, j)A$.
> 2. The matrix, B, obtained from A by multiplying the row A_i by a scalar, k, can be realized as $B = E_i(k)A$.
> 3. The matrix, B, obtained from A by adding a scalar multiple of A_j to the ith row can be realized as $B = E(i + kj)A$.

Proof

有關 1. 和 2. 的證明不難理解和想像。我們來看看，第 j 列乘上常數 k 加到第 i 列的動作。

矩陣 $E(i + kj)$ 除了第 i 列之第 i 個位置為 1 以及第 j 個位置為 k，其餘位置皆為 0 之外，其他每一列與單位矩陣 I_n 相同。所以，矩陣 $E(i + kj)A$ 除了第 i 列和矩陣 A 不同之外，其餘列與 A 相同。至於第 i 列則為 $A_i + kA_j$，得證。

Example 1

Given that $A = \begin{pmatrix} 0 & 1 & 7 \\ 2 & 1 & 3 \\ 1 & 1 & 4 \end{pmatrix}$ and $B = \begin{pmatrix} 1 & 1 & 4 \\ 0 & -1 & -5 \\ 0 & 0 & 2 \end{pmatrix}$. Find elementary matrices, D_1, D_2, D_3 such that $D_1 D_2 D_3 A = B$.

Solution

Let $D_3 = E(1, 3)$, then $D_3 A = E(1, 3)A = \begin{pmatrix} 1 & 1 & 4 \\ 2 & 1 & 3 \\ 0 & 1 & 7 \end{pmatrix}$. Let $D_2 = E(2 + (-2) \cdot 1)$, then $D_2D_3A = E(2+(-2)\cdot 1)E(1,3)A = E(2+(-2)\cdot 1)\begin{pmatrix} 1 & 1 & 4 \\ 2 & 1 & 3 \\ 0 & 1 & 7 \end{pmatrix} = \begin{pmatrix} 1 & 1 & 4 \\ 0 & -1 & -5 \\ 0 & 1 & 7 \end{pmatrix}$.

Finally, we let $D_1 = E(3+1\cdot 2)$, then

$D_1D_2D_3A = D_1E(2+(-2)\cdot 1)E(1,3)A = D_1E(2+(-2)\cdot 1)\begin{pmatrix} 1 & 1 & 4 \\ 2 & 1 & 3 \\ 0 & 1 & 7 \end{pmatrix}$

$= E(3+1\cdot 2)\begin{pmatrix} 1 & 1 & 4 \\ 0 & -1 & -5 \\ 0 & 1 & 7 \end{pmatrix} = \begin{pmatrix} 1 & 1 & 4 \\ 0 & -1 & -5 \\ 0 & 0 & 2 \end{pmatrix} = B$ ◆

Theorem 2.7

If B is an $m \times n$ matrix obtained from A by applying a finite number of row operations, then there is a non-singular matrix, P, such that $PA = B$.

Proof

This proof is quite easy. Let $E_1, E_2, ..., E_l$ be elementary matrices such that

$$E_1E_2...E_lA = B$$

Since $E_1, E_2, ..., E_l$ are all invertible, the product, $E_1E_2...E_l$ is also invertible. Putting $P = E_1E_2...E_l$, then the result follows.

前述定理說明，If A can be transformed into B by applying a finite number of row operations，then exists a non-singular, P, such that $PA = B$。然而，由於 elementary matrix 的逆矩陣也是 elementary matrix，所以反過來說，$A = P^{-1}B$，也是正確的。此時，我們稱矩陣 A 和矩陣 B 為列等價的 (row-equivalent)。

另外，在 2.3 節的時候，我們曾經提過，every matrix, A, can be transformed into a row-echelon matrix。所以，我們說，every matrix is row-equivalent to a row-echelon matrix。

Example 2

Given, $A = \begin{pmatrix} 1 & 1 & 4 \\ 2 & 1 & 3 \\ 0 & 1 & 7 \end{pmatrix}$. Find an invertible matrix, P, such that PA is a row-echelon matrix.

Solution

Writing $B_1 = E(2 + (-2)1)A = \begin{pmatrix} 1 & 1 & 4 \\ 0 & -1 & -5 \\ 0 & 1 & 7 \end{pmatrix}$,

$B_2 = E(2, 3)B_1 = \begin{pmatrix} 1 & 1 & 4 \\ 0 & 1 & 7 \\ 0 & -1 & -5 \end{pmatrix}$,

$B_3 = E(3 + 1 \cdot 2)B_2 = \begin{pmatrix} 1 & 1 & 4 \\ 0 & 1 & 7 \\ 0 & 0 & 2 \end{pmatrix}$, $B_4 = E(1 + (-1)2)B_3 = \begin{pmatrix} 1 & 0 & -3 \\ 0 & 1 & 7 \\ 0 & 0 & 2 \end{pmatrix}$,

$B_5 = E_3(\frac{1}{2})B_4 = \begin{pmatrix} 1 & 0 & -3 \\ 0 & 1 & 7 \\ 0 & 0 & 1 \end{pmatrix}$, $B_6 = E(2 + (-7)3)B_5 = \begin{pmatrix} 1 & 0 & -3 \\ 0 & 1 & 0 \\ 0 & 0 & 1 \end{pmatrix}$,

$B_7 = E(1 + 3 \cdot 3)B_6 = \begin{pmatrix} 1 & 0 & 0 \\ 0 & 1 & 0 \\ 0 & 0 & 1 \end{pmatrix}$, we then see that

$E(1 + 3 \cdot 3)E(2 + (-7)3)E_3(\frac{1}{2})E(1 + (-1)2)E(3 + 1 \cdot 2)E(2, 3)E(2 + (-2)1)A = B_7$ is a row-echelon matrix. Therefore,

$P = \begin{pmatrix} 1 & 0 & 3 \\ 0 & 1 & 0 \\ 0 & 0 & 1 \end{pmatrix}\begin{pmatrix} 1 & 0 & 0 \\ 0 & 1 & -7 \\ 0 & 0 & 1 \end{pmatrix}\begin{pmatrix} 1 & 0 & 0 \\ 0 & 1 & 0 \\ 0 & 0 & 1/2 \end{pmatrix}\begin{pmatrix} 1 & -1 & 0 \\ 0 & 1 & 0 \\ 0 & 0 & 1 \end{pmatrix}\begin{pmatrix} 1 & 0 & 0 \\ 0 & 1 & 0 \\ 0 & 1 & 1 \end{pmatrix}\begin{pmatrix} 1 & 0 & 0 \\ 0 & 0 & 1 \\ 0 & 1 & 0 \end{pmatrix}\begin{pmatrix} 1 & 0 & 0 \\ -2 & 1 & 0 \\ 0 & 0 & 1 \end{pmatrix}$

$$= \begin{pmatrix} -2 & 3/2 & 1/2 \\ 7 & -7/2 & -5/2 \\ -1 & 1/2 & 1/2 \end{pmatrix}$$ is the answer as you may check.

前述例子顯示出，在列運算的過程中，基本矩陣扮演著不可缺少的角色。這種角色在求一個可逆矩陣的逆矩陣之時，也表現的極為活躍。對於這一點，下一章我們會有詳細的解說。

Exercises

Find the invertible matrix, P, of each of the following matrices such that PA is a row-echelon matrix.

1. $A = \begin{pmatrix} 2 & 1 & 3 \\ 3 & 0 & -1 \end{pmatrix}$

2. $A = \begin{pmatrix} 0 & 2 & 4 \\ 3 & 1 & 0 \end{pmatrix}$

3. $A = \begin{pmatrix} 1 & -2 & 0 & 4 \\ -2 & 3 & 1 & -6 \\ 1 & 1 & 3 & 5 \end{pmatrix}$

4. $A = \begin{pmatrix} 0 & 2 & -3 & 1 \\ 1 & -2 & 3 & 0 \\ 0 & 2 & 1 & 4 \end{pmatrix}$

5. $A = \begin{pmatrix} 1 & -2 & 3 & 1 & 0 \\ 0 & 2 & 4 & 6 & -4 \\ 0 & -3 & -5 & -6 & 5 \\ 2 & -4 & 6 & 1 & 1 \end{pmatrix}$

Chapter Three

Ranks, Non-Singularities, *LU*-factorizations and Nilpotent Matrices

3.1　Null spaces, ranks, criteria for non-singularities
3.2　Triangular matrices and *LU*-factorizations
3.3　Nilpotent matrices

這一章，我們將從零核空間開始談起，然後介紹矩陣的秩，秩與可逆矩陣的相關性；此外，三角矩陣、冪零矩陣以及將一般矩陣化為下三角、上三角之乘積等，都將是本章的主要述說對象。

§3.1　Null spaces, ranks, criteria for non-singularities

這一節的重點工作將是介紹，矩陣的**秩** (rank) 和它所能相關的一些特性。首先想想，在 2.3 節的時候，我們曾經證明過，If A is a row-echelon matrix, then $\dim(RS(A)) = \dim(CS(A))$。這一節的主要工作之一是要看看，"上述這種情形對一般的 $m \times n$ 矩陣 A，是否也是成立？" 不過，無論如何我們得先從認識所謂的，"the null space of a matrix" 開始說起。

Definition　3.1

> Given an $m \times n$ matrix, A. The null space, $NS(A)$, of A is defined by
> $$NS(A) = \{ X \in R^n \text{ or } C^n : AX = 0 \}$$

換句話說，$NS(A)$ 也就是，**齊次線性系統**（homogeneous linear system）$AX = 0$ 的解集。定義中，所表示出來的 R^n 或 C^n 是，依我們所討論的 real field 或 complex field 而定。

Exercise

Show that $NS(A)$ is a subspace of $R^n (or \ C^n)$.

(*Hint*: Show that for any $X, Y \in NS(A)$ and a scalar, d, $dX + Y \in NS(A)$.)

Example 1

Given, $A = \begin{pmatrix} 1 & 2 & -1 \\ 0 & -3 & 2 \end{pmatrix}$. Find a non-zero vector, $X \in R^3$, such that $AX = 0$.

Solution

Let $X = \begin{pmatrix} x \\ y \\ z \end{pmatrix}$, then $AX = 0$ implies that $\begin{cases} x + 2y - z = 0 \\ -3y + 2z = 0 \end{cases}$

Hence, we see that $x = \dfrac{-1}{3}z$, $y = \dfrac{2}{3}z$, $z \in R \setminus \{0\}$ is a non-trivial solution to $AX = 0$.

Letting, $z = 1$, then $X_0 = \begin{pmatrix} -1/3 \\ 2/3 \\ 1 \end{pmatrix}$ is an answer. ∎

注意： 當 t 是任意常數的時候， tX_0 也都是 $AX = 0$ 的解。所以，以上述例子而言，$SP\{X_0\}$ is a subspace of $NS(A)$。因此，$\dim(NS(A)) \geq 1$。可是，若 $X = \begin{pmatrix} 1 \\ 0 \\ 0 \end{pmatrix}$，則 $AX = \begin{pmatrix} 1 \\ 0 \end{pmatrix} \neq \begin{pmatrix} 0 \\ 0 \end{pmatrix}$，所以 $X = \begin{pmatrix} 1 \\ 0 \\ 0 \end{pmatrix} \notin NS(A)$。因此，我們更進一步得知，$3 > \dim(NS(A)) \geq 1$。然而，實際情況到底如何呢？$NS(A)$ 與 $CS(A)$ 在維度上會有何種程度的關係呢？我們慢慢看下去。

Lemma 3.1

> Let A be an $m \times n$ matrix. Then
> $$\dim(NS(A)) + \dim(CS(A)) = n$$

Proof

令 $\{\alpha_1, \alpha_2, ..., \alpha_r\}$ 為 $NS(A)$ 的一組基底，然後將其 (依照第一章的解說) 擴建成為 R^n (or C^n) 的一組基底，$\{\alpha_1, \alpha_2, ..., \alpha_r, \beta_1, ..., \beta_{n-r}\}$。現在，我們若能證明，$\dim(CS(A)) = n - r$，那麼定理就得證。

然而，為了要得出這樣的結果，讓我們看看下列 $n-r$ 個向量
$$A\beta_1, A\beta_2, ..., A\beta_{n-r}$$
令
$$c_1 A\beta_1 + c_2 A\beta_2 + ... c_{n-r} A\beta_{n-r} = 0$$
則
$$A(c_1 \beta_1 + c_2 \beta_2 + ... c_{n-r} \beta_{n-r}) = 0$$
所以
$$c_1 \beta_1 + c_2 \beta_2 + ... c_{n-r} \beta_{n-r} \in NS(A)$$

因此，$c_1 \beta_1 + c_2 \beta_2 + ... c_{n-r} \beta_{n-r}$ 可以被表示為 $\alpha_1, \alpha_2, ..., \alpha_r$ 的線性組合。也就是說，存在 $d_1, d_2, ..., d_r$ 使得
$$c_1 \beta_1 + c_2 \beta_2 + ... c_{n-r} \beta_{n-r} = d_1 \alpha_1 + d_2 \alpha_2 + ... + d_r \alpha_r$$
或者
$$c_1 \beta_1 + c_2 \beta_2 + ... c_{n-r} \beta_{n-r} - d_1 \alpha_1 - d_2 \alpha_2 - ... - d_r \alpha_r = 0$$

由於，$\{\alpha_1, \alpha_2, ..., \alpha_r, \beta_1, ..., \beta_{n-r}\}$ 是一組基底。所以，
$$c_1 = c_2 = ... = c_{n-r} = d_1 = ... = d_r = 0$$

因此，我們得知
$$A\beta_1, A\beta_2, ..., A\beta_{n-r} \text{ 是線性獨立的。}$$

現在，令 e^j 為 R^n (or C^n) 的標準單位行向量，且令
$$e^j = \sum_{i=1}^{r} a_i \alpha_i + \sum_{i=1}^{n-r} b_i \beta_i$$

那麼
$$Ae^j = A^j = A(\sum_{i=1}^{r} a_i \alpha_i + \sum_{i=1}^{n-r} b_i \beta_i) = \sum_{i=1}^{n-r} b_i A\beta_i$$

這意思是說
$$A^j \in SP\{A\beta_1, A\beta_2, ..., A\beta_{n-r}\}$$

或者說
$$CS(A) \subseteq SP\{A\beta_1, A\beta_2, ..., A\beta_{n-r}\}$$

然而
$$\{A\beta_1, A\beta_2, ..., A\beta_{n-r}\} \subseteq CS(A)$$

所以，我們結論說
$$CS(A) = SP\{A\beta_1, A\beta_2, ..., A\beta_{n-r}\}$$
這也就是說
$$\dim(CS(A)) = n - r.$$

Lemma 3.2

> Let A be an $m \times n$ matrix and P an $m \times m$ non-singular matrix. Then
> 1. $NS(A) = NS(PA)$
> 2. $\dim(CS(PA)) = \dim(CS(A))$

Proof

1. We first note that if $AX = 0$, then $PAX = 0$. So we see that $NS(A) \subseteq NS(PA)$.

   The other hand, P is non-singular, so $NS(P) = \{0\}$ (why? Exercise 1, at the end). Thus if $PAX = 0$, then AX has to be a zero vector, that is $AX = 0$. Hence, $NS(A) \supseteq NS(PA)$. And this proved 1.

2. Look at the previous lemma, it holds for any $m \times n$ matrix. So we have that $\dim(NS(A)) + \dim(CS(A)) = n$ and $\dim(NS(PA)) + \dim(CS(PA)) = n$.

   By 1., we conclude that
   $$\dim(CS(A)) = \dim(CS(PA)).$$

Theorem 3.3

> If A is an $m \times n$ matrix, then $\dim(RS(A)) = \dim(CS(A))$.

Proof

　　在 2.4 節第 63 頁 Theorem 2.7 之後的說明曾經提到過，A is row-equivalent to a row-echelon matrix。或者更明確的說，there exists a non-singular matrix, P, such that

$PA = B$ 以及 $A = P^{-1}B$，其中 B 是一個 row-echelon matrix。

另外從第 2.2 節的說明中，前面兩個等式又告訴我們

$$RS(B) \subseteq RS(A) \text{ 以及 } RS(A) \subseteq RS(B)$$

如此情形之下得知，$RS(A) = RS(B)$。

但是，由於 B is a row-echelon matrix，所以

$$\dim(RS(B)) = \dim(CS(B))$$

將這個結果加上前面的預備定理，最後得出

$$\dim(RS(A)) = \dim(RS(B)) = \dim(CS(B)) = \dim(CS(A))$$

換一句話說，這個定理可以被重新寫為

If A is an $m \times n$ matrix, then the number of independent rows equals the number of independent columns.

萬事俱備，現在到了主角上場的時候了。那就是，**矩陣的秩** (ranks of matrices)。

Definition 3.2

For an $m \times n$ matrix, A, the rank of A, denoted by $rank(A)$, is defined by $rank(A) = \dim(CS(A))$ or $rank(A) = \dim(RS(A))$.

Remarks

1. If A is an $m \times n$ matrix, then $rank(A) \leq \min\{m, n\}$.
2. $rank(A) = 0$ if and only if A is a zero matrix.
3. $rank(A) = rank({}^t A)$.

先睹為快，我們下面練習兩個簡單的例題。

Example 2

Find the rank of $A = \begin{pmatrix} 1 & 0 & -2 & 3 \\ 0 & 2 & 1 & 4 \\ 2 & 0 & -4 & 6 \end{pmatrix}$.

Solution

按照前述理論，我們可以在 $CS(A)$ 或 $RS(A)$ 中，找出最大的線性獨立組合。因為，A 只有 3 個 rows。所以，我們從 $RS(A)$ 下手比較容易。

由於，$A_3 = 2A_1$。而且，A_1 和 A_2 為線性獨立的。所以，得知

$$rank(A) = 2$$

Example 3

Find the rank of $A = \begin{pmatrix} 1 & 2 & 0 & 3 \\ 1 & 3 & 0 & 3 \\ 1 & 1 & 1 & 3 \\ 1 & 4 & 5 & 3 \end{pmatrix}$.

Solution

我們從 $CS(A)$ 來看看。因為 $A^4 = 3A^1$，所以得知 $rank(A) < 4$。但是，A^1, A^2 以及 A^3 是線性獨立的，所以 $rank(A) = 3$。

介紹了那麼多，在感覺上似乎有點零亂。我們不妨理一理頭緒，再將一些重點給整理出來。

1. For any two matrices, A and B, $rank(AB) \leq \min\{rank(A), rank(B)\}$

   這一點，可由 $CS(AB) \subseteq CS(A)$ 以及 $RS(AB) \subseteq RS(B)$ 得知。

2. If P is a non-singular $m \times m$ matrix, then $rank(A) = rank(PA)$. And if Q is a non-singular $n \times n$ matrix, then $rank(A) = rank(AQ)$.

   只因為，$\dim(CS(PA)) = \dim(CS(A))$，所以，$rank(A) = rank(PA)$。至於另一個結果，我們令 $AQ = C$，則 $CS(AQ) \subseteq CS(A)$；另一方面，

Chapter Three Ranks, Non-singularities, *LU*-factorizations and Nilpotent Matrices

$A = CQ^{-1}$,所以 $CS(A) \subseteq CS(C)$。這說明, $CS(AQ) = CS(A)$,意即, $rank(A) = rank(AQ)$。

3. 若 A 為一 $n \times n$ 矩陣且 $rank(A) = n$,則存在一可逆矩陣 P 使得 $PA = I_n$。

   這個也不難,因為,對任意矩陣 A 而言,存在一可逆矩陣 P 使得 PA 為一 row-echelon matrix。然而, $rank(A) = n$,所以, $rank(PA) = n$。因此, PA 必定是有 n 個 distinguished columns 的 row-echelon matrix,那就是單位矩陣 I_n。

4. 又若 A 為一 $n \times n$ 矩陣且 $rank(A) = n$,則存在一可逆矩陣 Q 使得 $AQ = I_n$。

   有關這個,我們套用前面第 3 點。因為 $rank({}^tA) = rank(A) = n$,所以,存在一可逆矩陣 P 使得 $P({}^tA) = I_n$。兩邊取其轉置,得 ${}^t(P({}^tA)) = {}^tI_n$,進而得 $A({}^tP) = I_n$。(注意:P 是可逆的,所以 ${}^tP$ 也是可逆的。見 2.1 節, Exercise 6)。今取 $Q = {}^tP$,則命題可得。

   其實,我們也可以從第 3. 項來推導第 4. 項。
   $rank(A) = n \implies \exists$ a non-singular P such that $PA = I_n$
   $\qquad\qquad\quad \implies P^{-1}PA = P^{-1}$
   $\qquad\qquad\quad \implies P^{-1}PAP = P^{-1}P$
   $\qquad\qquad\quad \implies AP = I_n$

   無論如何,第 3. 與第 4. 項告訴我們
   $$rank(A) = n \implies \exists \text{ a non-singular } P \text{ such that } PA = AP = I_n$$
   也就是說,If $rank(A) = n$, then A is non-singular.

   反過來說,若 P 是可逆的而且 $PA = AP = I_n$,則 $rank(A) = rank(PA) = n$。綜觀上述所言,我們得出,一個 non-singular matrix 的充分必要條件。

Theorem 3.4

> $rank(A) = n$ if and only if A is non-singular.

換句話說，A is non-singular if and only if the n columns, $A^1, A^2, ..., A^n$ are linearly independent.

Exercise

Let A be an $n \times n$ matrix. Show that if there exists an $n \times n$ matrix, B, such that $BA = I_n$, then A is non-singular and hence $A^{-1} = B$. (注意，這裡並沒有說，B 是可逆的。) (*Hint*：證明 $rank(A) = n$。)
(題目若改為：If there exists B such that $AB = I_n$, then A is non-singular。其道理也相同。)

Exercises

1. Find 2×2 matrices, A and B, so that $rank(AB) \neq rank(BA)$.

2. Let A and B be 3×3 matrices. List all the possibilities for $rank(AB)$ by giving examples.

我們再進一步的說明，如何用基本矩陣的乘積，來表示一個可逆的矩陣。進而說明，如同 2.3 節 Gauss-Jordan elimination 相類似，以列運算的數值方法，求出一個可逆矩陣的逆矩陣。

Theorem 3.5

> A is a non-singular matrix if and only if A is a product of elementary matrices.

Chapter Three Ranks, Non-singularities, *LU*-factorizations and Nilpotent Matrices   75

Proof

這個定理證明簡單，第 2.4 節講的很清楚，對任意一矩陣 A，存在基本矩陣 $D_1, D_2, ..., D_l$ 使得 $D_1 D_2 ... D_l A$ 為一 row-echelon matrix。由於 A is non-singular，所以 the row-echelon matrix of A 是一個單位矩陣 I_n。因此，我們得出 $D_1 D_2 ... D_l A = I_n$ 或者 $A = D_l^{-1} D_{l-1}^{-1} ... D_1^{-1} I_n = D_l^{-1} D_{l-1}^{-1} ... D_1^{-1}$。

至於，若 A 為幾個基本矩陣的乘積，則 A is non-singular。此說更為顯然，因為 2.4 節告訴我們每一個基本矩陣是可逆的。

下面，就讓我們來練習如何以列運算，row-reduce a matrix to a row-echelon matrix。我們的做法是，首先擺出 $n \times 2n$ 矩陣 $(A|I_n)$。然後，對其使用列運算，直到左邊的矩陣出現單位矩陣為止，如，$(I_n|P)$。那麼，右邊的矩陣 P 就是 A^{-1} 了。因為，若 P is such that $PA = I_n$，then $(A|I_n)$ 將被轉換成 $(PA|PI_n) = (I_n|P)$。

Example 4

Given that $A = \begin{pmatrix} 2 & 1 \\ -1 & 1 \end{pmatrix}$. Find the inverse of the matrix, A, by Gauss-Jordan elimination.

Solution

Putting, $\begin{pmatrix} 2 & 1 & | & 1 & 0 \\ -1 & 1 & | & 0 & 1 \end{pmatrix}$, then perform row-operations to this augmented matrix.

$\begin{pmatrix} 1 & 1/2 & | & 1/2 & 0 \\ -1 & 1 & | & 0 & 1 \end{pmatrix} \sim \begin{pmatrix} 1 & 1/2 & | & 1/2 & 0 \\ 0 & 3/2 & | & 1/2 & 1 \end{pmatrix} \sim \begin{pmatrix} 1 & 1/2 & | & 1/2 & 0 \\ 0 & 1 & | & 1/3 & 2/3 \end{pmatrix} \sim$
$\begin{pmatrix} 1 & 0 & | & 1/3 & -1/3 \\ 0 & 1 & | & 1/3 & 2/3 \end{pmatrix}$

So, we see that $\begin{pmatrix} 1/3 & -1/3 \\ 1/3 & 2/3 \end{pmatrix}$ is the inverse of A. ∎

Example 5

Find the inverse of the matrix, $A = \begin{pmatrix} 1 & 1 & 3 \\ 2 & 0 & 3 \\ 0 & 3 & 1 \end{pmatrix}$ by Gauss-Jordan elimination.

Solution

$$\left(\begin{array}{ccc|ccc} 1 & 1 & 3 & 1 & 0 & 0 \\ 2 & 0 & 3 & 0 & 1 & 0 \\ 0 & 3 & 1 & 0 & 0 & 1 \end{array}\right) \sim \left(\begin{array}{ccc|ccc} 1 & 1 & 3 & 1 & 0 & 0 \\ 0 & -2 & -3 & -2 & 1 & 0 \\ 0 & 3 & 1 & 0 & 0 & 1 \end{array}\right) \sim \left(\begin{array}{ccc|ccc} 1 & 1 & 3 & 1 & 0 & 0 \\ 0 & 1 & 3/2 & 1 & -1/2 & 0 \\ 0 & 3 & 1 & 0 & 0 & 1 \end{array}\right) \sim$$

$$\left(\begin{array}{ccc|ccc} 1 & 0 & 3/2 & 0 & 1/2 & 0 \\ 0 & 1 & 3/2 & 1 & -1/2 & 0 \\ 0 & 0 & -7/2 & -3 & 3/2 & 1 \end{array}\right) \sim \left(\begin{array}{ccc|ccc} 1 & 0 & 3/2 & 0 & 1/2 & 0 \\ 0 & 1 & 3/2 & 1 & -1/2 & 0 \\ 0 & 0 & 1 & 6/7 & -3/7 & -2/7 \end{array}\right) \sim$$

$$\left(\begin{array}{ccc|ccc} 1 & 0 & 0 & -9/7 & 8/7 & 3/7 \\ 0 & 1 & 0 & -2/7 & 1/7 & 3/7 \\ 0 & 0 & 1 & 6/7 & -3/7 & -2/7 \end{array}\right)$$

So, we have that $A^{-1} = \begin{pmatrix} -9/7 & 8/7 & 3/7 \\ -2/7 & 1/7 & 3/7 \\ 6/7 & -3/7 & -2/7 \end{pmatrix}$. ∎

回憶一下，2.2 節的一個結果。The system $AX = B$ has a solution if and only if $\dim(CS(A)) = \dim(CS(A|B))$。本節最後，我們想以矩陣的 rank 和 null space 來重新闡釋有關線性系統之解，以為結尾。

Theorem 3.6

The system $AX = B$ has a solution if and only if $rank(A) = rank(A|B)$.

Theorem 3.7

Let X_0 be a solution to the system $AX = B$. Then Y is a solution to $AX = B$ if and only if $Y = X_0 + Z$, for some $Z \in NS(A)$.

Chapter Three Ranks, Non-singularities, *LU*-factorizations and Nilpotent Matrices

Proof

($\Rightarrow$)

這個方向的證明簡單，由於 X_0 和 Y 皆為 $AX = B$ 之解，所以 $AX_0 = B$ 且 $AY = B$。結合這兩個式子，得出 $A(Y - X_0) = 0$。這說明，$(Y - X_0) \in NS(A)$，也就是說，存在一 $Z \in NS(A)$，使得 $Y - X_0 = Z$。

($\Leftarrow$)

若存在一 $Z \in NS(A)$，使得 $Y = X_0 + Z$，則 $AY = AX_0 + AZ = B$，所以 Y is a solution to $AX = B$。

上述定理中假若 $NS(A) = \{0\}$，則我們發現 $Y = X_0$，也就是說，假若 $AX = B$ 有解的話，那麼這個解是唯一解。將這個觀察與之前的定理擺一起時，我們可以更精確的說

The system $A_{m \times n} X_{n \times 1} = B_{m \times 1}$ has a unique solution if and only if $rank(A) = rank(A|B) = n$.

那麼，對於齊次線性系統而言，當然也是如此了。

The homogeneous system $A_{m \times n} X_{n \times 1} = 0_{m \times 1}$ has a unique solution if and only if $rank(A) = n$.

或者說，

The homogeneous system $A_{m \times n} X_{n \times 1} = 0_{m \times 1}$ has no non-trivial solution if and only if $rank(A) = n$.

Exercises

1. Show that if P is non-singular, then $NS(P) = \{0\}$.

2. Show that the system $A_{n \times n} X_{n \times 1} = B_{n \times 1}$ has a unique solution if and only if A is non-singular and hence $X = A^{-1}B$.

3. Find the inverse by using Gauss-Jordan elimination.

   a. $A = \begin{pmatrix} 2 & -1 \\ 1 & 1 \end{pmatrix}$

   b. $A = \begin{pmatrix} 1 & 0 & 4 \\ 0 & 2 & -1 \\ -3 & 2 & 0 \end{pmatrix}$

§3.2 Triangular matrices and *LU*-factorizations

本節所要介紹的是，**上三角矩陣** (upper triangular matrices)、**下三角矩陣** (lower triangular matrices)、以及**對角矩陣** (diagonal matrices)。由於，三角矩陣比一般的矩陣較為容易運算和體現，所以，三角矩陣在矩陣論裡面扮演著比較重要的角色。其實，大家將會發現，在矩陣論這門學問裡，常常有很多的定理，都在介紹如何將一般的矩陣化簡成三角矩陣或是對角矩陣的工作。譬如說，前面的 Gauss-Jordan elimination，還有這一節的 *LU*-factorizations，以及未來第 8 章的 spectral decomposition theorem 裡頭，將會有很多的例子。現在，無論如何我們還是得先弄清楚，何謂上三角矩陣、下三角矩陣、對角矩陣。

Definition 3.3

Let $A = (a_{ij})$ be an $m \times n$ matrix.
1. If $a_{ij} = 0$, $\forall i > j$, then A is called an "upper triangular matrix".
2. If $a_{ij} = 0$, $\forall j > i$, then A is called a "lower triangular matrix".
3. If $a_{ij} = 0$, $\forall i \neq j$, then A is called a "diagonal matrix".

譬如說，

$$\begin{pmatrix} 1 & 0 & 0 \\ 0 & 1 & 0 \\ 0 & 0 & 1 \end{pmatrix}, \begin{pmatrix} 2 & 0 & 1 \\ 0 & 0 & 3 \\ 0 & 0 & -1 \end{pmatrix}, \begin{pmatrix} 0 & 2 & 1 & 0 & 3 \\ 0 & -3 & -6 & 4 & 2 \\ 0 & 0 & 2 & 2 & -1 \\ 0 & 0 & 0 & -1 & 1 \end{pmatrix}$$

都是上三角矩陣。那麼，

$$\begin{pmatrix} 1 & 0 & 0 \\ 0 & 1 & 0 \\ 0 & 0 & 1 \end{pmatrix}, \begin{pmatrix} 0 & 0 & 0 & 0 \\ 0 & -3 & 0 & 0 \\ 0 & 0 & 1 & 0 \\ 0 & 0 & 0 & 2 & 0 \end{pmatrix}$$

則為對角矩陣。至於 0 矩陣 ($a_{ij} = 0$, $\forall i, j$)，毫無疑問的，它是一個上三角矩陣，也是下三角矩陣，對角矩陣。

Chapter Three　Ranks, Non-singularities, *LU*-factorizations and Nilpotent Matrices

另外，對於對角線的元素都為 1 的上三角矩陣，往後我們若有碰到的話，特別給它稱呼為**上三角單位矩陣** (upper unit triangular matrix)。下三角矩陣的情形一樣，我們不再重複說明。

有一件事情還要提醒大家注意。在前述定義中，我們並沒有限定矩陣 A 為一個 square matrix。

Exercise

Show that every row-echelon matrix is an upper triangular matrix.

Proposition 3.8

If $D_{m \times n}$ and $F_{n \times k}$ are diagonal matrices, then $G = DF$ is also a diagonal matrix and $g_{ii} = d_{ii} f_{ii}$, $\forall i$.

Proof

假設 D 與 F 皆為 square matrices 的話，這個定理是很顯然的。對於一般的 case 而言，也不會太難。我們利用矩陣的乘法來看看，若 $i \neq j$，則

$$g_{ij} = \sum_{l=1}^{n} d_{il} f_{lj} = d_{ii} f_{ij} = 0$$

所以，$g_{ij} = 0$，$\forall i \neq j$。

另一方面，對任意 i 而言

$$g_{ii} = \sum_{j=1}^{n} d_{ij} f_{ji}$$

由於，當 $i \neq j$ 時，$d_{ij} f_{ji} = 0$。所以，$g_{ii} = d_{ii} f_{ii}$。

下面還有一些關於對角矩陣乘法的特性，順便一併提出。

Proposition 3.9

Let $D_{m \times m}$ be a square diagonal matrix and $S_{m \times m} = sI_{m \times m}$ a square scalar matrix.
1. If A is an $m \times n$ matrix, then $SA = sA$.
2. If B is a $k \times m$ matrix, then $BS = sB$.
3. If A is an $m \times n$ matrix and $C = DA$, then $C_i = d_{ii} A_i$.
4. If B is a $k \times m$ matrix and $C = BD$, then $C^j = d_{jj} B^j$.

Proof

這些證明，如同上一個 proposition 一樣，都可以說是簡單的，所以我們不打算一一解說。我們將只證明 3.，其餘的由同學當作業練習。

有關矩陣的乘法，在 2.2 節時，講的很清楚。若 $C = DA$，則對任意 C 的第 i 列，C_i，而言

$$C_i = d_{i1} A_1 + d_{i2} A_2 + \cdots + d_{im} A_m = d_{ii} A_i$$

Exercise

Give an example of a 2×2 diagonal matrix, D, and a 2×2 matrix, A, such that $AD \neq DA$.

Theorem 3.10

If $A_{m \times n}$ and $B_{n \times k}$ are upper triangular matrices, then AB is also an upper triangular matrix.

Proof

只要頭腦清楚，這個證明不難。

Let $C = (c_{ij}) = AB$ and consider, for each $i > j$,

Chapter Three Ranks, Non-singularities, *LU*-factorizations and Nilpotent Matrices

$$c_{ij} = \sum_{l=1}^{n} a_{il}b_{lj}$$
$$= a_{i1}b_{1j} + a_{i2}b_{2j} + \ldots + a_{i(i-1)}b_{(i-1)j} + a_{ii}b_{ij} + a_{i(i+1)}b_{(i+1)j} + \ldots + a_{in}b_{nj}$$

Since A is an upper triangular matrix, $a_{ij} = 0$, $i > j$. So, the first $(i-1)$ terms of the last sum are 0. While the rest of the terms are all 0, since matrix, B, is upper triangular.

Exercise

Let A be a square upper (lower) triangular matrix. Show that A is non-singular if and only if $a_{ii} \neq 0, \forall i$. Furthermore, if A is non-singular, then A^{-1} is also upper (lower) triangular. (*Hint*: Consider the product of elementary matrices, P, with $PA = I$.)

Example 1

Find the inverse of $A = \begin{pmatrix} 2 & 0 & 3 \\ 0 & 1 & 4 \\ 0 & 0 & 7 \end{pmatrix}$.

Solution

由前面 Exercise，得知 A is invertible，而且 A^{-1} is upper triangular。
假設

$$A^{-1} = \begin{pmatrix} b_{11} & b_{12} & b_{13} \\ 0 & b_{22} & b_{23} \\ 0 & 0 & b_{33} \end{pmatrix}$$

則由矩陣的乘法 $AA^{-1} = I_3$ 得知

$$\begin{pmatrix} 2 & 0 & 3 \\ 0 & 1 & 4 \\ 0 & 0 & 7 \end{pmatrix}\begin{pmatrix} b_{11} \\ 0 \\ 0 \end{pmatrix} = \begin{pmatrix} 1 \\ 0 \\ 0 \end{pmatrix}, \begin{pmatrix} 2 & 0 & 3 \\ 0 & 1 & 4 \\ 0 & 0 & 7 \end{pmatrix}\begin{pmatrix} b_{12} \\ b_{22} \\ 0 \end{pmatrix} = \begin{pmatrix} 0 \\ 1 \\ 0 \end{pmatrix}, \begin{pmatrix} 2 & 0 & 3 \\ 0 & 1 & 4 \\ 0 & 0 & 7 \end{pmatrix}\begin{pmatrix} b_{13} \\ b_{23} \\ b_{33} \end{pmatrix} = \begin{pmatrix} 0 \\ 0 \\ 1 \end{pmatrix}$$

第一個式子清楚的表示

$$b_{11} = \frac{1}{2}$$

第二個式子則說明

$$2b_{12} = 0, \ b_{22} = 1$$

第三個式子則為

$$b_{33} = \frac{1}{7} , \ b_{23} + \frac{4}{7} = 0 , \ 2b_{13} + \frac{3}{7} = 0$$

所以我們得出

$$b_{11} = \frac{1}{2} , \ b_{12} = 0 , \ b_{22} = 1 , \ b_{13} = -\frac{3}{14} , \ b_{23} = -\frac{4}{7} , \ b_{33} = \frac{1}{7}$$

也就是說

$$A^{-1} = \begin{pmatrix} 1/2 & 0 & -3/14 \\ 0 & 1 & -4/7 \\ 0 & 0 & 1/7 \end{pmatrix}.$$

■

從這個例子可以看出，在求 A^{-1} 的過程中，上三角矩陣的特性給了我們省掉了不少的計算。利用這個特性，我們使用**倒推代換法** (backward substitution)，可以無需太辛苦的得到 A^{-1} 的各個元素。打鐵乘熱，利用這個要領，我們接著和大家介紹，解線性系統的另一種方式，即所謂的 LU-factorization。首先，看看所謂的 upper row-echelon matrix。

Definition 3.4

An $m \times n$ matrix, A, is called an upper row-echelon matrix, if there exists a positive integer, $r \leq m$, such that
(i) the last $m-r$ rows of A are zero rows.
(ii) the distinguished columns of A are the first r-columns of an $m \times m$ upper unit-triangular matrix.

Upper row-echelon matrix 與 row-echelon matrix 所不一樣的地方是在於，distinguished columns。Row-echelon matrix 的 distinguished columns 是單位行向量 e^i，而 upper row-echelon matrix 的 distinguished columns 就有一點不一樣了。

譬如說

$$\begin{pmatrix} 1 & 2 & 4 & 6 \\ 0 & 1 & 2 & 3 \\ 0 & 0 & 1 & 1 \end{pmatrix}, \begin{pmatrix} 0 & 1 & 3 & 4 \\ 0 & 0 & 0 & 1 \end{pmatrix}, \begin{pmatrix} 1 & 2 & 4 & 6 & 4 & 5 \\ 0 & 0 & 1 & 2 & 3 & 1 \\ 0 & 0 & 0 & 1 & 2 & 7 \\ 0 & 0 & 0 & 0 & 0 & 1 \\ 0 & 0 & 0 & 0 & 0 & 0 \end{pmatrix}$$

are upper row-echelon matrices。我們可以發現，它們的 distinguished columns 得由下而上的第一個 non-zero element 為 1 之外，1 以上的 elements 可以是任何數。

Remarks

1. 很顯然的，一個 row-echelon matrix 是一個 upper row-echelon matrix；一個 upper row-echelon matrix 是一個 upper triangular matrix。
2. 如同 row-echelon matrix 一樣，若 $A_{m \times n}$ 是一個 upper row-echelon matrix 且 r 為定義中之正整數，則
   a. 若 $A_s = 0$，則 $A_t = 0$, $\forall t \geq s$
   b. 若 $A_s \neq 0$，則 A_s 第一個非 0 元素是 1。
   c. $rank(A) = r$.

這裡我們順便看看，如何由 upper row-echelon matrix 讀出一個線性系統的解。我們就以前頭第一個矩陣為例，並且假設它就是一個 linear system 的 augmented matrix。

$$\begin{pmatrix} 1 & 2 & 4 & 6 \\ 0 & 1 & 2 & 3 \\ 0 & 0 & 1 & 1 \end{pmatrix}$$

那麼，前面矩陣所代表的 linear system 為

$$\begin{cases} x+2y+4z=6 \\ y+2z=3 \\ z=1 \end{cases}$$

現在使用 backward substitution，由第三個式子告訴我們 $z=1$，代入第二式，得 $y=1$，再代入第一式得 $x=0$。所以該 linear system 的解為

$$\begin{pmatrix} x \\ y \\ z \end{pmatrix} = \begin{pmatrix} 0 \\ 1 \\ 1 \end{pmatrix}$$

這種將 augmented matrix $(A|B)$ 化為上三角矩陣，然後使用 backward substitution 得出解的方式，我們有時稱其為 upper triangulation method (上三角化方法)。一般而言，這個方式比 Gauss-Jordan elimination 節省了大約一半的列運算的步驟。不過，有關這個方法，其重點，還得要介紹大家認識所謂的，LU-factorization。

LU-factorization

如同 2.4 節一樣，對任意矩陣 A 而言，也可以找到一個可逆矩陣 P 使得

$$PA = B$$

是一個 upper row-echelon matrix。其中，矩陣 P 是幾個基本矩陣的乘積。只是，有一件事情要特別提出注意。由於，我們是將矩陣 A，化為上三角矩陣，所以在列運算的過程中，我們所將使用的這種類型 $E(i+kj)$ 的基本矩陣，都是下三角矩陣。又因為，$E_i(k)$ 也可以被看成是下三角矩陣。所以，

Proposition 3.11

If no row interchanges are required to row reduce A into an upper row-echelon matrix, then there exist a lower triangular matrix, L, and an upper row-echelon matrix U such that $A = LU$.

Proof

由題示,要把矩陣 A 化為一個 upper row-echelon matrix, B ,的話,我們只需 $E(i+kj)$ 與 $E_i(k)$ 類型的列運算。然而,前面剛說過,這兩類型的基本矩陣都是下三角矩陣。而且,下三角矩陣的乘積仍然是下三角矩陣。所以我們說,存在一個可逆的下三角矩陣, P ,使得 $PA=B$ 是一個 upper row-echelon matrix。或者說, $A=P^{-1}B$ 。現在,只要令, $L=P^{-1}$ 、$U=B$ 。那麼, $A=LU$ 即為所求。

定理中,我們之所以使用 L 與 U ,純粹是用其分別代表 lower triangular matrix 和 upper triangular matrix。我們就以下面矩陣為例,給予完整的介紹,求出矩陣 L 以及 U 的一個方法。

Example 2

Find the lower triangular matrix, L, and the upper row-echelon matrix, U, for the matrix, $A = \begin{pmatrix} 3 & 4 & 5 \\ 2 & 1 & -1 \\ -2 & 2 & 8 \end{pmatrix}$.

Solution

$$E_1(1/3)A = \begin{pmatrix} 1 & 4/3 & 5/3 \\ 2 & 1 & -1 \\ -2 & 2 & 8 \end{pmatrix}$$

$$E(3+2\cdot 1)E(2+(-2)\cdot 1)E_1(1/3)A = \begin{pmatrix} 1 & 4/3 & 5/3 \\ 0 & -5/3 & -13/3 \\ 0 & 14/3 & 34/3 \end{pmatrix}$$

$$E_2(-3/5)E(3+2\cdot 1)E(2+(-2)\cdot 1)E_1(1/3)A = \begin{pmatrix} 1 & 4/3 & 5/3 \\ 0 & 1 & 13/5 \\ 0 & 14/3 & 34/3 \end{pmatrix}$$

$$E(3+(-14/3)\cdot 2)E_2(-3/5)E(3+2\cdot 1)E(2+(-2)\cdot 1)E_1(1/3)A$$
$$= \begin{pmatrix} 1 & 4/3 & 5/3 \\ 0 & 1 & 13/5 \\ 0 & 0 & -4/5 \end{pmatrix}$$

$$E_3(-5/4)E(3+(-14/3)\cdot 2)E_2(-3/5)E(3+2\cdot 1)E(2+(-2)\cdot 1)E_1(1/3)A$$
$$=\begin{pmatrix} 1 & 4/3 & 5/3 \\ 0 & 1 & 13/5 \\ 0 & 0 & 1 \end{pmatrix}$$

最後這個式子告訴我們，

$$P = E_3(-5/4)E(3+(-14/3)\cdot 2)E_2(-3/5)E(3+2\cdot 1)E(2+(-2)\cdot 1)E_1(1/3)$$

所以，
$$P^{-1} = E_1(3)E(2+2\cdot 1)E(3+(-2)\cdot 1)E_2(-5/3)E(3+(14/3)\cdot 2)E_3(-4/5)$$
$$=\begin{pmatrix} 3 & 0 & 0 \\ 0 & 1 & 0 \\ 0 & 0 & 1 \end{pmatrix}\begin{pmatrix} 1 & 0 & 0 \\ 2 & 1 & 0 \\ 0 & 0 & 1 \end{pmatrix}\begin{pmatrix} 1 & 0 & 0 \\ 0 & 1 & 0 \\ -2 & 0 & 1 \end{pmatrix}\begin{pmatrix} 1 & 0 & 0 \\ 0 & -5/3 & 0 \\ 0 & 0 & 1 \end{pmatrix}\begin{pmatrix} 1 & 0 & 0 \\ 0 & 1 & 0 \\ 0 & 14/3 & 1 \end{pmatrix}\begin{pmatrix} 1 & 0 & 0 \\ 0 & 1 & 0 \\ 0 & 0 & -4/5 \end{pmatrix}$$
$$=\begin{pmatrix} 3 & 0 & 0 \\ 2 & -5/3 & 0 \\ -2 & 14/3 & -4/5 \end{pmatrix}$$

因此得出，
$$A=\begin{pmatrix} 3 & 4 & 5 \\ 2 & 1 & -1 \\ -2 & 2 & 8 \end{pmatrix}=\begin{pmatrix} 3 & 0 & 0 \\ 2 & -5/3 & 0 \\ -2 & 14/3 & -4/5 \end{pmatrix}\begin{pmatrix} 1 & 4/3 & 5/3 \\ 0 & 1 & 13/5 \\ 0 & 0 & 1 \end{pmatrix}$$

下面提出另一個方法，看看是否會比較輕鬆。

Another solution

We first write

$$A=\begin{pmatrix} 3 & 4 & 5 \\ 2 & 1 & -1 \\ -2 & 2 & 8 \end{pmatrix}=\begin{pmatrix} l_{11} & 0 & 0 \\ l_{21} & l_{22} & 0 \\ l_{31} & l_{32} & l_{33} \end{pmatrix}\begin{pmatrix} 1 & u_{12} & u_{13} \\ 0 & 1 & u_{23} \\ 0 & 0 & 1 \end{pmatrix}$$

We immediately have that

Chapter Three Ranks, Non-singularities, LU-factorizations and Nilpotent Matrices 87

$$L^1 = \begin{pmatrix} l_{11} \\ l_{21} \\ l_{31} \end{pmatrix} = A^1 = \begin{pmatrix} 3 \\ 2 \\ -2 \end{pmatrix} \text{ and hence that } U_1 = \begin{pmatrix} 1 & 4/3 & 5/3 \end{pmatrix}$$

Now, rewrite

$$A = \begin{pmatrix} 3 & 4 & 5 \\ 2 & 1 & -1 \\ -2 & 2 & 8 \end{pmatrix} = \begin{pmatrix} 3 & 0 & 0 \\ 2 & l_{22} & 0 \\ -2 & l_{32} & l_{33} \end{pmatrix} \begin{pmatrix} 1 & 4/3 & 5/3 \\ 0 & 1 & u_{23} \\ 0 & 0 & 1 \end{pmatrix}$$

Next, we find the 2nd column of L,

$$2 \cdot \frac{4}{3} + l_{22} = 1, \text{ so } l_{22} = -\frac{5}{3} \text{ and}$$

$$-2 \cdot \frac{4}{3} + l_{32} = 2, \text{ so } l_{32} = \frac{14}{3}$$

And then the 2nd row of U,

$$2 \cdot \frac{5}{3} + (\frac{-5}{3}) \cdot u_{23} = -1, \text{ so } u_{23} = \frac{13}{5}$$

Finally, l_{33} can be obtained by

$$-2 \cdot \frac{5}{3} + \frac{14}{3} \cdot \frac{13}{5} + l_{33} = 8, \text{ so } l_{33} = \frac{-4}{5}$$

Thus, the LU-factorization of A is

$$A = \begin{pmatrix} 3 & 4 & 5 \\ 2 & 1 & -1 \\ -2 & 2 & 8 \end{pmatrix} = \begin{pmatrix} 3 & 0 & 0 \\ 2 & -5/3 & 0 \\ -2 & 14/3 & -4/5 \end{pmatrix} \begin{pmatrix} 1 & 4/3 & 5/3 \\ 0 & 1 & 13/5 \\ 0 & 0 & 1 \end{pmatrix}$$

Exercise

同學練習一下，請求出矩陣 $A = \begin{pmatrix} 2 & 1 & -2 \\ 1 & 2 & 3 \\ 5 & -1/2 & 0 \end{pmatrix}$ 的 LU-factorization。

(Ans: $A = \begin{pmatrix} 2 & 1 & -2 \\ 1 & 2 & 3 \\ 5 & -1/2 & 0 \end{pmatrix} = \begin{pmatrix} 2 & 0 & 0 \\ 1 & 3/2 & 0 \\ 5 & -3 & 13 \end{pmatrix} \begin{pmatrix} 1 & 1/2 & -1 \\ 0 & 1 & 8/3 \\ 0 & 0 & 1 \end{pmatrix}$)

本節的主要工作是要，將一個線性系統的係數矩陣 A 化成 LU-factorization，之後，用其判讀 $AX = B$ 的解。

Solution to a linear system by triangular factorization

Given a linear system, $AX = B$. If A has an LU-factorization, $A = LU$, then the augmented matrix becomes,

$$(A|B) = (LU|B) = L(U|L^{-1}B)$$

Writing $C = U$, $D = L^{-1}B$, then the linear systems, $AX = B$ and $CX = D$, have the same solution set.

Proof

這個概念是很容易理解的，因為

$$AX = B$$
$$\Leftrightarrow \quad L^{-1}AX = L^{-1}B$$
$$\Leftrightarrow \quad L^{-1}LUX = D$$
$$\Leftrightarrow \quad UX = D$$

Example 3

Solve $\begin{cases} 3x + 4y + 5z = 2 \\ 2x + y - z = 1 \\ -2x + 2y + 8z = 0 \end{cases}$ by using LU-factorization.

Solution

前面例子的結果告訴我們

$$A = \begin{pmatrix} 3 & 4 & 5 \\ 2 & 1 & -1 \\ -2 & 2 & 8 \end{pmatrix} = LU = \begin{pmatrix} 3 & 0 & 0 \\ 2 & -5/3 & 0 \\ -2 & 14/3 & -4/5 \end{pmatrix} \begin{pmatrix} 1 & 4/3 & 5/3 \\ 0 & 1 & 13/5 \\ 0 & 0 & 1 \end{pmatrix}$$

由於，$LD = B$，所以

$$\begin{pmatrix} 3 & 0 & 0 \\ 2 & -5/3 & 0 \\ -2 & 14/3 & -4/5 \end{pmatrix} \begin{pmatrix} d_1 \\ d_2 \\ d_3 \end{pmatrix} = \begin{pmatrix} 2 \\ 1 \\ 0 \end{pmatrix}$$

進而得出

$$d_1 = 2/3, \ d_2 = 1/5, \ d_3 = -1/2$$

現在,求線性系統 $CX = D$ 的解。

$$\begin{pmatrix} 1 & 4/3 & 5/3 \\ 0 & 1 & 13/5 \\ 0 & 0 & 1 \end{pmatrix} \begin{pmatrix} x \\ y \\ z \end{pmatrix} = \begin{pmatrix} 2/3 \\ 1/5 \\ -1/2 \end{pmatrix}$$

並且由 backward substitution,得出

$$z = -1/2, \ y = 3/2, \ x = -1/2$$

Example 4

Solve $\begin{cases} 2x + y - 2z = 1 \\ x + 2y + 3z = -2 \\ 5x - \dfrac{1}{2}y = 3 \end{cases}$ by using the triangular factorization.(使用前面的 exercise)

Solution

$$A = \begin{pmatrix} 2 & 1 & -2 \\ 1 & 2 & 3 \\ 5 & -1/2 & 0 \end{pmatrix} = LU = \begin{pmatrix} 2 & 0 & 0 \\ 1 & 3/2 & 0 \\ 5 & -3 & 13 \end{pmatrix} \begin{pmatrix} 1 & 1/2 & -1 \\ 0 & 1 & 8/3 \\ 0 & 0 & 1 \end{pmatrix}$$

再由 $LD = B$,得出

$$\begin{pmatrix} 2 & 0 & 0 \\ 1 & 3/2 & 0 \\ 5 & -3 & 13 \end{pmatrix} \begin{pmatrix} d_1 \\ d_2 \\ d_3 \end{pmatrix} = \begin{pmatrix} 1 \\ -2 \\ 3 \end{pmatrix}$$

$$d_1 = 1/2, \ d_2 = -5/3, \ d_3 = -9/26$$

接著,就是解 $CX = D$

$$\begin{pmatrix} 1 & 1/2 & -1 \\ 0 & 1 & 8/3 \\ 0 & 0 & 1 \end{pmatrix} \begin{pmatrix} x \\ y \\ z \end{pmatrix} = \begin{pmatrix} 1/2 \\ -5/3 \\ -9/26 \end{pmatrix}$$

得 $\qquad z = -9/26, \ y = -58/78, \ x = 41/78$ ∎

注意,前面是在不需要 row interchanges 的情形之下,求出一個矩陣的 LU-factorization,然後,用其解線性系統的。可一般而言,在將矩陣 A 化成 upper row-echelon matrix 的時候,row interchanges 有時是需要的。也就是說,

There exists a non-singular matrix P such that

$$PA = B$$

where B is an upper row-echelon matrix and P is a product of elementary matrices, some of which are row interchanges.

當這種情況出現的時候,矩陣 A 就不一定有 LU-factorization 了。這時候應該怎麼辦呢?看看下面的例子。

Example 5

Show that $A = \begin{pmatrix} 1 & 1 & 2 \\ 1 & 1 & 3 \\ 2 & 0 & 4 \end{pmatrix}$ does not admit an LU-factorization.

Proof

Suppose that $A = \begin{pmatrix} 1 & 1 & 2 \\ 1 & 1 & 3 \\ 2 & 0 & 4 \end{pmatrix} = \begin{pmatrix} l_{11} & 0 & 0 \\ l_{21} & l_{22} & 0 \\ l_{31} & l_{32} & l_{33} \end{pmatrix} \begin{pmatrix} 1 & u_{12} & u_{13} \\ 0 & 1 & u_{23} \\ 0 & 0 & 1 \end{pmatrix}$, then

$L^1 = \begin{pmatrix} l_{11} \\ l_{21} \\ l_{31} \end{pmatrix} = \begin{pmatrix} 1 \\ 1 \\ 2 \end{pmatrix}$ and $U_1 = \begin{pmatrix} 1 & u_{12} & u_{13} \end{pmatrix} = \begin{pmatrix} 1 & 1 & 2 \end{pmatrix}$.

Rewrite that $\begin{pmatrix} 1 & 1 & 2 \\ 1 & 1 & 3 \\ 2 & 0 & 4 \end{pmatrix} = \begin{pmatrix} 1 & 0 & 0 \\ 1 & l_{22} & 0 \\ 2 & l_{32} & l_{33} \end{pmatrix} \begin{pmatrix} 1 & 1 & 2 \\ 0 & 1 & u_{23} \\ 0 & 0 & 1 \end{pmatrix}$, we then see that

$$1 + l_{22} = 1 \text{ and } 2 + l_{22} \cdot u_{23} = 3$$

The last two identities contradict. So, A does not have an LU-factorization. ∎

其實,就拿這個例子來看,並且用基本矩陣將 A 上三角化。

$$E_2(-1/2)E(2,3)E(3+(-2)\cdot 1)E(2+(-1)\cdot 1)\begin{pmatrix} 1 & 1 & 2 \\ 1 & 1 & 3 \\ 2 & 0 & 4 \end{pmatrix} = \begin{pmatrix} 1 & 1 & 2 \\ 0 & 1 & 0 \\ 0 & 0 & 1 \end{pmatrix}$$

我們發現,在這四個基本矩陣當中,第 3 個基本矩陣是,$E(2,3)$,不是一個下三角矩陣。就是它破壞了矩陣 A 之 LU-factorization 的可能性。

現在,我們的做法是把這個"列對調,$E(2,3)$"給提出來。也就是說,先將矩陣 A 給執行 2,3 列對調,並且令,$F = E(2,3)A = \begin{pmatrix} 1 & 1 & 2 \\ 2 & 0 & 4 \\ 1 & 1 & 3 \end{pmatrix}$。接著,用其餘的三個基本矩陣將 F 上三角化,求出 F 的 LU-factorization。

$$E_2(-1/2)E(2+(-2)\cdot 1)E(3+(-1)\cdot 1)F = \begin{pmatrix} 1 & 1 & 2 \\ 0 & 1 & 0 \\ 0 & 0 & 1 \end{pmatrix}$$

(注意:這三個基本矩陣中之 $E(2+(-2)\cdot 1)$,$E(3+(-1)\cdot 1)$ 是將原來兩個基本矩陣的數字之 2 與 3 對調所得。)

今令,$P = E_2(-1/2)E(2+(-2)\cdot 1)E(3+(-1)\cdot 1)$

則,$P^{-1} = E(3+1\cdot 1)E(2+2\cdot 1)E_2(-2) = \begin{pmatrix} 1 & 0 & 0 \\ 2 & -2 & 0 \\ 1 & 0 & 1 \end{pmatrix}$

得出 F 的 LU-factorization 為,

$$F = \begin{pmatrix} 1 & 1 & 2 \\ 2 & 0 & 4 \\ 1 & 1 & 3 \end{pmatrix} = \begin{pmatrix} 1 & 0 & 0 \\ 2 & -2 & 0 \\ 1 & 0 & 1 \end{pmatrix} \begin{pmatrix} 1 & 1 & 2 \\ 0 & 1 & 0 \\ 0 & 0 & 1 \end{pmatrix}$$

用這個例題的結果,我們試看看解下列的 system of linear equations。

Example 6

Solve the linear system, $\begin{cases} x+y+2z=1 \\ x+y+3z=3 \\ 2x+4z=5 \end{cases}$.

Solution

Writing $A = \begin{pmatrix} 1 & 1 & 2 \\ 1 & 1 & 3 \\ 2 & 0 & 4 \end{pmatrix}$, $B = \begin{pmatrix} 1 \\ 3 \\ 5 \end{pmatrix}$, $V = E(2,3)$, $F = VA$ and $G = VB$,

then $AX = B$ implies that $FX = G$, where $F = \begin{pmatrix} 1 & 1 & 2 \\ 2 & 0 & 4 \\ 1 & 1 & 3 \end{pmatrix}$, $G = \begin{pmatrix} 1 \\ 5 \\ 3 \end{pmatrix}$.

By the previous result, $F = \begin{pmatrix} 1 & 1 & 2 \\ 2 & 0 & 4 \\ 1 & 1 & 3 \end{pmatrix} = LU = \begin{pmatrix} 1 & 0 & 0 \\ 2 & -2 & 0 \\ 1 & 0 & 1 \end{pmatrix} \begin{pmatrix} 1 & 1 & 2 \\ 0 & 1 & 0 \\ 0 & 0 & 1 \end{pmatrix}$

Taking $LD = G$, then

$$\begin{pmatrix} 1 & 0 & 0 \\ 2 & -2 & 0 \\ 1 & 0 & 1 \end{pmatrix} \begin{pmatrix} d_1 \\ d_2 \\ d_3 \end{pmatrix} = \begin{pmatrix} 1 \\ 5 \\ 3 \end{pmatrix} \text{ implies,}$$

$d_1 = 1$, $d_2 = -3/2$ and $d_3 = 2$.

We finally solve that $UX = D$ or that

$$\begin{pmatrix} 1 & 1 & 2 \\ 0 & 1 & 0 \\ 0 & 0 & 1 \end{pmatrix} \begin{pmatrix} x \\ y \\ z \end{pmatrix} = \begin{pmatrix} 1 \\ -3/2 \\ 2 \end{pmatrix}$$

Chapter Three Ranks, Non-singularities, *LU*-factorizations and Nilpotent Matrices

And find out that $z = 2$, $y = -3/2$ and $x = -3/2$.

一般而言，假若 V 是將矩陣 A 化為 upper row-echelon matrix 所需的 elementary row interchanges 的話，那麼我們將原來的 linear system $AX = B$ 改寫成 $VAX = VB$。然後，同前面一樣，將矩陣 VA 的 *LU*-factorization 求出，進而求出 $VAX = VB$ 的解，而這個解就是原方程式的解。

Exercises

1. Find an *LU*-factorization of $A = \begin{pmatrix} 2 & 0 & 3 \\ 0 & -1 & 4 \\ 1 & 2 & 1 \end{pmatrix}$ and solve
$$\begin{cases} 2x + 3z = 2 \\ -y + 4z = -1 \\ x + 2y + z = 0 \end{cases}$$

2. Solve the linear system, $\begin{cases} 3y + 4z = 1 \\ x + 2y + z = 1 \\ x + 2y - z = 5 \end{cases}$, by using *LU*-factorization.

3. Find an *LU*-factorization of $A = \begin{pmatrix} 3 & 0 & 1 & 2 \\ 1 & 2 & -2 & 0 \\ -1 & -1 & 1 & 3 \\ 0 & 4 & 2 & 32 \end{pmatrix}$ and solve
$$\begin{cases} 3x + z + 2w = 2 \\ x + 2y - 2z = -1 \\ -x - y + z + 3w = 1 \\ 4y + 2z + 32w = 3 \end{cases}$$

(*Hint*: $A = \begin{pmatrix} 3 & 0 & 1 & 2 \\ 1 & 2 & -2 & 0 \\ -1 & -1 & 1 & 3 \\ 0 & 4 & 2 & 32 \end{pmatrix} = \begin{pmatrix} 3 & 0 & 0 & 0 \\ 1 & 2 & 0 & 0 \\ -1 & -1 & 1/6 & 0 \\ 0 & 4 & 20/3 & 100 \end{pmatrix} \begin{pmatrix} 1 & 0 & 1/3 & 2/3 \\ 0 & 1 & -7/6 & -1/3 \\ 0 & 0 & 1 & 20 \\ 0 & 0 & 0 & 1 \end{pmatrix}$)

§3.3 Nilpotent matrices

在 3.2 節,我們花了不少篇幅介紹三角矩陣。其實談到三角矩陣,還有一件值得大家認識的事情,就是所謂的**冪零矩陣** (nilpotent matrix)。這一節,我們除了給大家介紹認識冪零矩陣外,有關它的一些基本特性當然也將順便研讀。至於更進一步相關的東西,在往後的章節裡隨時都會提及。

Definition 3.5

A matrix, A, is said to be strictly upper triangular, if A is upper triangular and $a_{ii} = 0$, $\forall i$.

e.g.

$$A = \begin{pmatrix} 0 & 1 & 2 & 3 \\ 0 & 0 & 4 & 2 \\ 0 & 0 & 0 & 1 \\ 0 & 0 & 0 & 0 \end{pmatrix}, \quad A = \begin{pmatrix} 0 & 2 & -1 & 3 \\ 0 & 0 & 4 & 0 \\ 0 & 0 & 0 & -2 \end{pmatrix} \quad \text{and} \quad A = \begin{pmatrix} 0 & 1 & -1 & 2 \\ 0 & 0 & 5 & 3 \\ 0 & 0 & 0 & -5 \\ 0 & 0 & 0 & 0 \\ 0 & 0 & 0 & 0 \end{pmatrix}$$

are strictly upper triangular matrices.

Remark

If B is an upper triangular matrix, then there are a diagonal matrix, D, and a strictly upper triangular matrix, A, such that $B = D + A$.

譬如說

$$B = \begin{pmatrix} 2 & 1 & 2 & 3 \\ 0 & -1 & 4 & 2 \\ 0 & 0 & 3 & 1 \\ 0 & 0 & 0 & 5 \end{pmatrix} \text{ can be written as a sum of}$$

$$D = \begin{pmatrix} 2 & 0 & 0 & 0 \\ 0 & -1 & 0 & 0 \\ 0 & 0 & 3 & 0 \\ 0 & 0 & 0 & 5 \end{pmatrix} \quad \text{and} \quad A = \begin{pmatrix} 0 & 1 & 2 & 3 \\ 0 & 0 & 4 & 2 \\ 0 & 0 & 0 & 1 \\ 0 & 0 & 0 & 0 \end{pmatrix}$$

這個分解動作在後面介紹，upper unit triangular matrix (上三角單位矩陣) 的逆矩陣時，扮演一定份量的角色。不過，我們還是得先認識，何謂 "冪零矩陣"？首先，看看下列一些定義。

Definition 3.6

Given an $m \times n$ matrix, A.
1. The main diagonal (or the 1st upper diagonal) of A is defined to be the set of entries, $\{a_{ii} : i = 1, 2 ..., m\}$.
2. The kth upper diagonal of A is defined to be the set of entries, $\{a_{i, i+k-1} : i = 1, 2, ..., m-k+1\}$.

譬如說，the 2nd upper diagonal of A consists of entries of the form $a_{i, i+1}$, $\forall i$

the 3rd upper diagonal of A consists of entries of the form $a_{i, i+2}$, $\forall i$

等等。

$$\begin{pmatrix} a_{11} & a_{12} & a_{13} & a_{14} & a_{15} \\ a_{21} & a_{22} & a_{23} & a_{24} & a_{25} \\ a_{31} & a_{32} & a_{33} & a_{34} & a_{35} \\ a_{41} & a_{42} & a_{43} & a_{44} & a_{45} \\ a_{51} & a_{52} & a_{53} & a_{54} & a_{55} \end{pmatrix}$$

Proposition 3.12

Let $A_{m \times n}$ and $B_{n \times k}$ be upper triangular matrices. If
1. the first through the rth ($r \neq 0$) upper diagonals of $A_{m \times n}$ are 0
2. the first through the sth ($s \neq 0$) upper diagonals of $B_{n \times k}$ are 0
then the first through the $(r+s)^{th}$ upper diagonals of AB are 0.

Proof

Let $C = AB$, then we shall show that

$$c_{ij}=0, \text{ for } 1\leq j\leq r+s+i-1 \text{ and } i=1,2,3,...,m$$

And this can be seen by noting that for each i, j

$$c_{ij}=\sum_{k=1}^{n}a_{ik}b_{kj}=\sum_{k=i+r}^{n}a_{ik}b_{kj}, \text{ since } a_{ik}=0, \ 1\leq k\leq i+r-1$$

More clearly,

$$c_{ij}=a_{i,i+r}b_{i+r,j}+a_{i,i+r+1}b_{i+r+1,j}+...+a_{i,n}b_{n,j}$$

Since the first through the sth ($s\neq 0$) upper diagonals of $B_{n\times k}$ are 0,

$$b_{ij}=0 \ , \quad \forall \ j\leq i+s-1$$

Hence that

$$b_{i+r,j}=0 \ , \quad \forall \ j\leq i+r+s-1$$

And hence that

$$b_{i+r+1,j}=b_{i+r+2,j}=\cdots=b_{n,j}=0.$$

Therefore

$$c_{ij}=0 \ , \quad \forall \ 1\leq j\leq r+s+i-1.$$

Corollary

If A is an $n\times n$ strictly upper triangular matrix, then there exists a positive integer, $r\leq n$, such that $A^r=0$.

這就是本節的重點,我們說

An $n\times n$ matrix, A, is called a "nilpotent" matrix (冪零矩陣), if $A^r=0$ for some positive integer $r\leq n$.

前述推論告訴我們,若 A 是一個 $n\times n$ strictly upper triangular matrix,那麼 A 是一個冪零矩陣。

譬如說,

Chapter Three   Ranks, Non-singularities, *LU*-factorizations and Nilpotent Matrices   97

$$\begin{pmatrix} 0 & 1 & 2 \\ 0 & 0 & 3 \\ 0 & 0 & 0 \end{pmatrix}, \begin{pmatrix} 0 & 1 & 1 & 3 \\ 0 & 0 & 2 & -1 \\ 0 & 0 & 0 & 4 \\ 0 & 0 & 0 & 0 \end{pmatrix}, \begin{pmatrix} 0 & 0 & 2 & -2 \\ 0 & 0 & 0 & 1 \\ 0 & 0 & 0 & 0 \\ 0 & 0 & 0 & 0 \end{pmatrix}$$

都是冪零矩陣。下面是關於冪零矩陣非常好的一個特性。

Theorem 3.13

If A is an $n \times n$ nilpotent matrix, then $I_n + A$ is invertible.

Proof

這個證明非常簡單。首先想想看，因為 A is an $n \times n$ nilpotent matrix，所以 $A^n = 0$。因此，

$$(I_n + A)(I_n - A + A^2 - A^3 + \cdots + (-1)^{n-1} A^{n-1}) = I_n$$

由 3.1 節，第 74 頁之 Exercise 得知，$I_n + A$ 是可逆的，而且

$$(I_n + A)^{-1} = (I_n - A + A^2 - A^3 + \cdots + (-1)^{n-1} A^{n-1})$$

Example 1

Use the previous result to find the inverse of the matrix,

$$B = \begin{pmatrix} 1 & 1 & 2 & 3 \\ 0 & 1 & 4 & 2 \\ 0 & 0 & 1 & 1 \\ 0 & 0 & 0 & 1 \end{pmatrix}$$

Solution

Let $A = \begin{pmatrix} 0 & 1 & 2 & 3 \\ 0 & 0 & 4 & 2 \\ 0 & 0 & 0 & 1 \\ 0 & 0 & 0 & 0 \end{pmatrix}$, then we have that $B = I_4 + A$.

So

$$B^{-1} = (I_4 - A + A^2 - A^3)$$

$$= \begin{pmatrix} 1 & 0 & 0 & 0 \\ 0 & 1 & 0 & 0 \\ 0 & 0 & 1 & 0 \\ 0 & 0 & 0 & 1 \end{pmatrix} - \begin{pmatrix} 0 & 1 & 2 & 3 \\ 0 & 0 & 4 & 2 \\ 0 & 0 & 0 & 1 \\ 0 & 0 & 0 & 0 \end{pmatrix} + \begin{pmatrix} 0 & 0 & 4 & 4 \\ 0 & 0 & 0 & 4 \\ 0 & 0 & 0 & 0 \\ 0 & 0 & 0 & 0 \end{pmatrix} - \begin{pmatrix} 0 & 0 & 0 & 4 \\ 0 & 0 & 0 & 0 \\ 0 & 0 & 0 & 0 \\ 0 & 0 & 0 & 0 \end{pmatrix}$$

$$= \begin{pmatrix} 1 & -1 & 2 & -3 \\ 0 & 1 & -4 & 2 \\ 0 & 0 & 1 & -1 \\ 0 & 0 & 0 & 1 \end{pmatrix}$$

∎

Exercises

1. Let A and B be nilpotent matrices of the same size and let $AB = BA$. Show that AB is nilpotent.

2. Let $A = \begin{pmatrix} a_{11} & a_{12} & \dots & \dots & a_{1n} \\ 0 & a_{22} & a_{23} & \dots & a_{2n} \\ 0 & 0 & a_{33} & \dots & a_{3n} \\ \dots & \dots & \dots & \dots & \dots \\ 0 & 0 & \dots & \dots & a_{nn} \end{pmatrix}$ be an $n \times n$ upper triangular matrix with $a_{ii} \neq 0$, $\forall\, i = 1, 2, \dots, n$, and let $B = \begin{pmatrix} a_{11}^{-1} & 0 & 0 & \dots & 0 \\ 0 & a_{22}^{-1} & 0 & \dots & 0 \\ 0 & \dots & a_{33}^{-1} & \dots & 0 \\ 0 & \dots & 0 & \dots & 0 \\ 0 & 0 & \dots & 0 & a_{nn}^{-1} \end{pmatrix}$ be an $n \times n$ diagonal matrix. Show that AB and BA are triangular matrices with entries 1 on the main diagonal.

3. Use exercise 2, to find the inverse of $A = \begin{pmatrix} 2 & 1 & 0 & 0 \\ 0 & 1 & 0 & 1 \\ 0 & 0 & 7 & 2 \\ 0 & 0 & 0 & 3 \end{pmatrix}$.

(*Hint*: Find B, then $AB = I_4 + U$ for some strictly upper triangular matrix, U. $(AB)^{-1} = B^{-1}A^{-1} = I - U + U^2 - U^3$)

Chapter Four

Matrices as Linear Transformations

4.1 Linear transformations

4.2 Compositions and inverses

4.3 Matrix representations

線性空間之間的線型轉換，是矩陣所能給予分析學界，在實務上的最具體表現。所以，討論矩陣若不介紹線型轉換，會令人感覺有遺珠之憾。因此，本章除了將扼要的介紹線型轉換之外，其與矩陣之間的密切關係也將進行適度的剖析。首先看看**線型轉換** (linear transformation)。

§4.1 Linear transformations

Definition 4.1

> Let V and W be two vector spaces over a field F. A map, T, from V into W is called a "linear transformation", if it satisfies that
> $T(au+v) = aT(u) + T(v)$ for every u, v in V and $a \in F$.

e.g.

1. The map, $T : R^2 \to R^2$, defined by $T\begin{pmatrix} x \\ y \end{pmatrix} = \begin{pmatrix} 2 & 0 \\ 1 & 1 \end{pmatrix}\begin{pmatrix} x \\ y \end{pmatrix}$, $\forall \begin{pmatrix} x \\ y \end{pmatrix} \in R^2$

   is a linear transformation.

   這個例子應該沒有多大問題，因為矩陣的相乘是符合線性分配的。所以，它是一個線型轉換。下面是兩個比較特殊的線型轉換。

2. The identity map, $I : V \to V$, $I(v) = v$, $\forall v \in V$, is a linear transformation.

99

3. The zero map, $O : V \to V$, $O(v) = 0$, $\forall\ v \in V$, is a linear transformation.

4. Let $V = \{$ infinitely differentiable real-valued functions defined on $R \}$. Defining, for every f in V, $D(f) = f'$, then D is a linear transformation.

有關這個例子，我們首先得確認 V 是實數體系上的向量空間（留給讀者當 exercise）。至於，它的 linearity 則由微分的基本特性可知。

$$D(af + g) = aD(f) + D(g)，\quad \forall\ f, g \in V\ 。$$

線型映射 (linear map or linear mapping) 有時，也是一些人常常使用來替代線型轉換的詞彙。下面，我們再多看兩個題目。

Example 1

Defining, for $\forall\ X = \begin{pmatrix} x_1 \\ x_2 \\ x_3 \end{pmatrix} \in R^3$, by $T(X) = x_1 + 2x_2 - x_3$, then T is a linear map.

Proof

For $X = \begin{pmatrix} x_1 \\ x_2 \\ x_3 \end{pmatrix}$, $Y = \begin{pmatrix} y_1 \\ y_2 \\ y_3 \end{pmatrix} \in R^3$ and a scalar c,

$$\begin{aligned}
T(cX + Y) &= (cx_1 + y_1) + 2(cx_2 + y_2) - (cx_3 + y_3) \\
&= c(x_1 + 2x_2 - x_3) + (y_1 + 2y_2 - y_3) \\
&= cT(X) + T(Y)
\end{aligned}$$

其實，若寫成，$A = \begin{pmatrix} 1 \\ 2 \\ -1 \end{pmatrix}$，則 $T(X) = {}^t\!AX$。那麼它也是一個矩陣相乘的結果。所以，T 當然也滿足線性分配的特性。下面是一個比較一般化的，典型的例子。

Example 2

Let A be an $m \times n$ real matrix. Defining a map, $T : R^n \to R^m$, by

$$T(X) = AX, \quad \forall \ X = \begin{pmatrix} x_1 \\ x_2 \\ ... \\ ... \\ x_n \end{pmatrix} \in R^n, \text{ then } T \text{ is a linear map.}$$

Proof

對於任意 R^n 中的向量，X，Y，以及實數 a，

$$T(aX + Y) = A(aX + Y) = aAX + AY = aT(X) + T(Y)$$

滿足 linearity 的性質。所以，T is a linear map。∎

一件事情特別注意。因為，A 是 $m \times n$ 矩陣，X 是 $n \times 1$ 矩陣，所以，AX 為一 $m \times 1$ 矩陣。因此，T 可以被看成是一個從 R^n 映射到 R^m 的 linear map。另外，假若矩陣 A 是一個 $m \times n$ complex matrix 的話，那麼如此定義之 T，則應為一個從 C^n 映射到 C^m 的 linear map。

此時，我們似乎感受到，一個 $m \times n$ 矩陣決定了，一個從 R^n (或 C^n) 映射到 R^m (或 C^m) 的線型映射。將這一句話謹記在心，後面還有更精彩、更詳細的陸續會出現。下面我們先來兩個簡單的練習題，請同學腦力激盪一下。

Exercises

1. Show that if $T : V \to W$ is a linear map, then
   a. $T(0) = 0$
   b. $T(c_1v_1 + c_2v_2 + ... + c_lv_l) = c_1T(v_1) + c_2T(v_2) + ... + c_lT(v_l)$, for $v_i \in V$ and $c_i \in F$.

2. Denote by $L(V, W)$ the collection of all linear maps from V into W. Define, for T_1, T_2 in $L(V, W)$ and a scalar a,
   a. $(T_1 + T_2)(v) = T_1(v) + T_2(v)$, $\quad \forall \ v \in V$

b. $(aT)(v) = aT(v)$, $\forall\ v \in V$

Show that $L(V, W)$ is a vector space.

Theorem 4.1

Let V and W be vector spaces. If $\{v_1, v_2, ..., v_n\}$ is a basis for V, and if $w_1, w_2, ..., w_n$ are any vectors in W, then there exists exactly one linear map, $T: V \to W$ such that $T(v_i) = w_i$, $\forall i = 1, 2, ..., n$.

Proof

(existence)

For the existence of $T: V \to W$, we define for each $v = x_1 v_1 + x_2 v_2 + ... + x_n v_n$ in V, by $T(v) = x_1 w_1 + x_2 w_2 + ... + x_n w_n$.

We claim that such T is a desired linear map. To show this, we let $v = \sum_{i=1}^{n} x_i v_i$, $u = \sum_{i=1}^{n} y_i v_i$ be vectors in V and c a scalar. Then

$$T(cv + u) = T(\sum_{i=1}^{n}(cx_i + y_i)v_i) = \sum_{i=1}^{n}(cx_i + y_i)w_i = c\sum_{i=1}^{n} x_i w_i + \sum_{i=1}^{n} y_i w_i$$
$$= cT(v) + T(u)$$

So, T is linear. Moreover, for each $i = 1, 2, ..., n$, no doubt, $T(v_i) = w_i$, by the definition.

(uniqueness)

Suppose that $T': V \to W$ is another linear map that satisfies, $T'(v_i) = w_i$, $\forall i = 1, 2, ..., n$. Let $v = x_1 v_1 + x_2 v_2 + ... + x_n v_n$ be in V. Then

$$T'(v) = T'(\sum_{i=1}^{n} x_i v_i) = \sum_{i=1}^{n} x_i T'(v_i) = \sum_{i=1}^{n} x_i w_i = \sum_{i=1}^{n} x_i T(v_i) = T(\sum_{i=1}^{n} x_i v_i)$$
$$= T(v), \quad \forall\ v \in V.$$

This says that $T = T'$.

第二章以及第三章的時候，我們介紹過矩陣 A 的 column space，$CS(A)$

Chapter Four　Matrices as Linear Transformations　103

以及 null space，NS(A)。線型轉換 T，在這方面也有相類似的說明。那就是所謂的核 (the kernel)"KerT"與值域 (the range)"RanT"。

Definition 4.2

> Let V and W be vector spaces over a field, F, and let $T : V \to W$ be a linear map.
> 1. The "kernel" of T, denoted by $KerT$, is defined to be the subset of vectors, v, in V such that $T(v) = 0$.
> 2. The "range" of T, denoted by $RanT$, is defined to be the subset of vectors, w, in W such that $T(v) = w$ for some v, in V.

或者簡單的說，
$$KerT = \{v \in V : T(v) = 0 \}$$
$$RanT = \{w \in W : T(v) = w \text{ for some } v \in V \}$$

Example 3

Let $T : R^3 \to R$ be a linear map defined by

$$T(X) = 3x - 2y + z, \quad \forall \ X = \begin{pmatrix} x \\ y \\ z \end{pmatrix} \in R^3$$

We see that $KerT = \{ X \in R^3 : 3x - 2y + z = 0 \}$.

It is a plane through the origin. ∎

Exercises

1. Show that $KerT$ is a subspace of V. (參考 3.1 節之 null space)

2. Show that $RanT$ is a subspace of W.

Definition 4.3

Let V and W be vector spaces over a field, F, and let $T: V \to W$ be a linear map. Then T is said to be
1. "injective", if $T(v_1) \neq T(v_2)$, whenever $v_1 \neq v_2$.
2. "surjective", if $RanT = W$.
3. "bijective", if T is both injective and surjective.

這兒，我們所使用的詞彙，"injective"、"surjective" 以及 "bijective" 指的是大家所熟悉的，"一對一"、"映成" 以及 "一對一且映成" 的意思。在線性代數的架構上，我們稱一個 injective linear map 為，一個 monomorphism；一個 surjective linear map 為，一個 epimorphism；一個 bijective linear map 為，一個 isomorphism。

譬如說，
1. 一個 identity map，$I: V \to V$，is bijective，這是顯然的。
2. 前面的例子，$T: R^3 \to R$ defined by

$$T(X) = 3x - 2y + z, \quad \forall\ X = \begin{pmatrix} x \\ y \\ z \end{pmatrix} \in R^3$$

is not injective，因為，$T(\begin{pmatrix} 0 \\ 1 \\ 2 \end{pmatrix}) = T(\begin{pmatrix} -1 \\ 0 \\ 3 \end{pmatrix}) = 0$。

但是由於，$T(\begin{pmatrix} 1 \\ 2 \\ 3 \end{pmatrix}) \neq 0$，所以，$T$ is surjective from R^3 onto R。

Theorem 4.2

Let $T: V \to W$ be a linear map. Then T is injective if and only if $KerT = \{0\}$.

Proof

($\Rightarrow$)

This direction is quite clear, for if v is a vector in V such that $T(v) = 0$, then v has to be a zero vector, owing to the fact that T is injective and $T(0) = 0$. So, $KerT = \{0\}$.

($\Leftarrow$)

Assume that $v_1 \neq v_2$ in V, then $v_1 - v_2 \neq 0$, and hence, $T(v_1 - v_2) \neq 0$, because, $KerT = \{0\}$. Thus, $T(v_1) - T(v_2) \neq 0$ or $T(v_1) \neq T(v_2)$. This says, T is injective.

Example 4

The linear map, D, defined by $D(f) = f'$ as before, is not injective. Because, $KerD = \{$ all constant functions $\}$.

Theorem 4.3

If $T: V \to W$ is a monomorphism, and $v_1, v_2, ..., v_l$ are linearly independent vectors in V, then $T(v_1), T(v_2), ..., T(v_l)$ are linearly independent in W.

Proof

這個證明簡單。假設，$c_1 T(v_1) + c_2 T(v_2) + ... + c_l T(v_l) = 0$。由於，$T$ 是線性的，所以 $T(c_1 v_1 + c_2 v_2 + ... + c_l v_l) = 0$。又由於，$T$ 是一對一的，所以，$c_1 v_1 + c_2 v_2 + ... + c_l v_l = 0$。現在，因為，$v_1, v_2, ..., v_l$ 是線性獨立的，所以，$c_1 = c_2 = ... = c_l = 0$。這證明了，$T(v_1), T(v_2), ..., T(v_l)$ 是線性獨立的。

在 3.1 節有過這樣一個結果，$\dim(NS(A)) + \dim(CS(A)) = n$。現在，以線型轉換的角度來解說的話，情況變成如下定理。

Theorem 4.4

Let $T: V \to W$ be a linear map. Then $\dim V = \dim KerT + \dim RanT$.

Proof

我們將採用與前述 3.1 節相類似的證明。令 $\{\alpha_1, \alpha_2,...,\alpha_r\}$ 為 $KerT$ 的一組基底，然後將其擴建成為 V 的一組基底，$\{\alpha_1, \alpha_2,...,\alpha_r, \beta_1,...,\beta_{n-r}\}$。現在，我們若能證明，$\dim RanT = n-r$，那麼定理就得證。

為了要得出這樣的結果，讓我們看看下列 $n-r$ 個向量

$$T\beta_1, T\beta_2,...,T\beta_{n-r}$$

令 $\quad c_1 T\beta_1 + c_2 T\beta_2 +...+ c_{n-r} T\beta_{n-r} = 0$

則 $\quad T(c_1\beta_1 + c_2\beta_2 +... c_{n-r}\beta_{n-r}) = 0$

所以 $\quad c_1\beta_1 + c_2\beta_2 +... c_{n-r}\beta_{n-r} \in KerT$

因此，$c_1\beta_1 + c_2\beta_2 +... c_{n-r}\beta_{n-r}$ 可以被表示為 $\alpha_1, \alpha_2,...,\alpha_r$ 的線性組合。也就是說，存在 $d_1, d_2,...,d_r$ 使得

$$c_1\beta_1 + c_2\beta_2 +...+ c_{n-r}\beta_{n-r} = d_1\alpha_1 + d_2\alpha_2 +...+ d_r\alpha_r$$

或者 $\quad c_1\beta_1 + c_2\beta_2 +...+ c_{n-r}\beta_{n-r} - d_1\alpha_1 - d_2\alpha_2 -...- d_r\alpha_r = 0$

由於，$\{\alpha_1, \alpha_2,...,\alpha_r, \beta_1,...,\beta_{n-r}\}$ 是一組基底。所以，

$$c_1 = c_2 =...= c_{n-r} = d_1 =...= d_r = 0$$

因此，我們得知

$T\beta_1, T\beta_2...,T\beta_{n-r}$ 是線性獨立的。

現在，我們要證明 $\{T\beta_1, T\beta_2...,T\beta_{n-r}\}$ 生成 $RanT$。

首先，$SP\{T\beta_1, T\beta_2...,T\beta_{n-r}\} \subseteq RanT$，這是毫無疑問的。至於，$SP\{T\beta_1, T\beta_2..., T\beta_{n-r}\} \supseteq RanT$，我們令，$w \in RanT$。則存在一 $v \in V$，使得

$$v = x_1\alpha_1 + x_2\alpha_2 +...+ x_r\alpha_r + y_1\beta_1 +...+ y_{n-r}\beta_{n-r} \text{ 且 } T(v) = w。$$

將上式代入得

$$\begin{aligned} w = T(v) &= T(x_1\alpha_1 + x_2\alpha_2 +...+ x_r\alpha_r + y_1\beta_1 +...+ y_{n-r}\beta_{n-r}) \\ &= y_1 T(\beta_1) + y_2 T(\beta_2) +...+ y_{n-r} T(\beta_{n-r}) \end{aligned}$$

這說明，$w \in SP\{T\beta_1, T\beta_2..., T\beta_{n-r}\}$。

結果得知，$\{T\beta_1, T\beta_2, ..., T\beta_{n-r}\}$ 是 $RanT$ 的一組基底，所以，
$$\dim RanT = n - r \text{。}$$

Example 5

再看一次前面的例題。$T: R^3 \to R$ defined by
$$T(X) = 3x - 2y + z, \quad \forall\ X = \begin{pmatrix} x \\ y \\ z \end{pmatrix} \in R^3$$

我們說過，T is surjective to R。也就是說，$RanT = R$。現在，由 Theorem 4.4 得知
$$3 = \dim R^3 = \dim KerT + 1$$
所以 $\dim KerT = 2$。 ■

一般來說，假若
$T: R^n \to R$ is not a zero map, then $\dim KerT = n - 1$。在泛函分析學上，此時之 $KerT$ 被稱為是，一個 hyperplane。

Corollary 1

Let $T: V \to W$ be a linear map with $\dim V = \dim W$. Then T is bijective, if T is either injective or surjective.

Proof

If T is injective, then $KerT = \{0\}$. So, $\dim V = \dim KerT + \dim RanT$ implies that $\dim V = \dim RanT$. By the hypothesis, we have that $\dim RanT = \dim W$. This says that T is surjective. Hence, T is bijective.

If T is surjective, then $\dim RanT = \dim W$. So, $\dim V = \dim KerT + \dim RanT$ and $\dim V = \dim W$ imply that $\dim KerT = 0$. That is $KerT = \{0\}$. This says, T is injective. Hence, T is bijective.

Corollary 2

Let V be a vector space over F with a basis, $\{\alpha_1, \alpha_2, ..., \alpha_n\}$. Define a map, $T: V \to F^n$, by

$$T(v) = \begin{pmatrix} x_1 \\ x_2 \\ ... \\ ... \\ x_n \end{pmatrix}, \quad \forall \, v = \sum_{i=1}^{n} x_i \alpha_i \in V$$

then T is an isomorphism.

Proof

Since every vector, v, in V has a unique linear combination, $\sum_{i=1}^{n} x_i \alpha_i$, the map, T, is well-defined. Also, since $KerT = \{0\}$, T is injective. And hence, T is bijective by the last corollary. To complete the proof, we are left to show that T is linear. And this can be done by considering for each $u = \sum_{i=1}^{n} x_i \alpha_i$, $v = \sum_{i=1}^{n} y_i \alpha_i$ in V and a scalar c,

$$T(cu+v) = T(\sum_{i=1}^{n}(cx_i + y_i)\alpha_i) = \begin{pmatrix} cx_1 + y_1 \\ cx_2 + y_2 \\ ... \\ ... \\ cx_n + y_n \end{pmatrix} = c\begin{pmatrix} x_1 \\ x_2 \\ ... \\ ... \\ x_n \end{pmatrix} + \begin{pmatrix} y_1 \\ y_2 \\ ... \\ ... \\ y_n \end{pmatrix} = cT(u) + T(v)$$

這是一個非常重要而活躍的結果。它清楚的說明, "Every finite dimensional vector space (say dimension n) over F is isomorphic to F^n"。這句話,在 matrix representation 或是在線型映射所謂的相似矩陣的主題中,扮演著舉足輕重、不可缺少的角色。往後,一路上我們都將領悟這個道理。

Exercises

1. Determine which of the following maps are linear.

   a. $T: R^3 \to R^2$ defined by $T\begin{pmatrix} x \\ y \\ z \end{pmatrix} = \begin{pmatrix} x \\ z \end{pmatrix}$, $\forall \begin{pmatrix} x \\ y \\ z \end{pmatrix} \in R^3$.

   b. $T: R^3 \to R^3$ defined by $T\begin{pmatrix} x \\ y \\ z \end{pmatrix} = \begin{pmatrix} x \\ y \\ z \end{pmatrix} + \begin{pmatrix} 1 \\ 0 \\ 0 \end{pmatrix}$, $\forall \begin{pmatrix} x \\ y \\ z \end{pmatrix} \in R^3$.

   c. $T: R^2 \to R^2$ defined by $T\begin{pmatrix} x \\ y \end{pmatrix} = \begin{pmatrix} 2x \\ y-x \end{pmatrix}$, $\forall \begin{pmatrix} x \\ y \end{pmatrix} \in R^2$.

   d. $T: R^2 \to R$ defined by $T\begin{pmatrix} x \\ y \end{pmatrix} = xy$, $\forall \begin{pmatrix} x \\ y \end{pmatrix} \in R^2$.

2. Given a linear map, $T: R^2 \to R^2$, defined as follow. Find $T\begin{pmatrix} 1 \\ 0 \end{pmatrix}$, for each of them.

   a. $T\begin{pmatrix} 3 \\ 1 \end{pmatrix} = \begin{pmatrix} 1 \\ 2 \end{pmatrix}$ and $T\begin{pmatrix} -1 \\ 0 \end{pmatrix} = \begin{pmatrix} 1 \\ 1 \end{pmatrix}$

   b. $T\begin{pmatrix} 4 \\ 1 \end{pmatrix} = \begin{pmatrix} 1 \\ 1 \end{pmatrix}$ and $T\begin{pmatrix} 1 \\ 1 \end{pmatrix} = \begin{pmatrix} 3 \\ -2 \end{pmatrix}$

   c. $T\begin{pmatrix} 1 \\ 1 \end{pmatrix} = \begin{pmatrix} 2 \\ 1 \end{pmatrix}$ and $T\begin{pmatrix} -1 \\ 1 \end{pmatrix} = \begin{pmatrix} 6 \\ 3 \end{pmatrix}$

3. Let $T: V \to W$ be a linear map and let $w_1, w_2, ..., w_m$ be linearly independent vectors in W. Show that if $T(v_i) = w_i$, $\forall i = 1, 2, ..., m$, then $v_1, v_2, ..., v_m$ are linearly independent.

4. Let $T: V \to W$ be a linear map. If $v_0 \in V$ and $w \in W$ are such that $T(v_0) = w$, then $T(v) = w$ if and only if $\exists u \in KerT$ such that $v = v_0 + u$.

   (*Hint*: Consider that $T(v) = T(v_0)$, 或參考 Theorem 3.7)

§4.2　Compositions and inverses

在 4.1 節提過，一個矩陣可以決定一個線型映射。這節我們要更進一步的介紹，兩個矩陣的乘積可以決定，線型映射的 composition。進而，了解線型映射的"逆映射"與"逆矩陣"的關係。

Definition 4.4

Let U, V and W be vector spaces. If $T_1 : U \to V$ and $T_2 : V \to W$ are linear maps, then the composition of T_1 and T_2, denoted by $T_2 \circ T_1$, is defined to be a linear map from U into W with
$$(T_2 \circ T_1)(u) = T_2(T_1(u)), \quad \forall\ u \in U$$

如此定義之下，我們得先確認這個 composition，$T_2 \circ T_1$，是否確為一個 linear map。茲令，$u_1, u_2 \in U$ 以及一個純量 c。

$$\begin{aligned}(T_2 \circ T_1)(cu_1 + u_2) &= T_2(T_1(cu_1 + u_2)) = T_2(cT_1(u_1) + T_1(u_2))\\ &= cT_2(T_1(u_1)) + T_2(T_1(u_2))\\ &= c(T_2 \circ T_1)(u_1) + (T_2 \circ T_1)(u_2)\end{aligned}$$

沒錯，$T_2 \circ T_1$ is indeed a linear map from U into W.

Example 1

Given, $A = \begin{pmatrix} 1 & 2 \\ 3 & -1 \end{pmatrix}$ and $B = \begin{pmatrix} 2 & 0 \\ 1 & 1 \end{pmatrix}$. If T_1, $T_2 : R^2 \to R^2$ are linear maps defined by

$$T_1(X) = AX \text{ and } T_2(X) = BX, \quad \forall\ X = \begin{pmatrix} x \\ y \end{pmatrix} \in R^2, \text{ then}$$

$(T_2 \circ T_1)(X) = T_2(T_1(X)) = T_2(AX) = BAX = (BA)X$ and
$(T_1 \circ T_2)(X) = T_1(T_2(X)) = T_1(BX) = ABX = (AB)(X)$

一般而言，$AB \neq BA$。這個事實讓我們合理的臆測，線型映射的合成，其順序當然也不可前後任意對調。另外，如同矩陣的乘法一樣，我們也臆測，線型映射的合成符合結合律，乘法對加法符合分配律。也就是說，若 $T_1: U \to V$, $T_2: V \to W$ 以及 $T_3: W \to Z$ 皆為 linear maps，則 $T_3 \circ (T_2 \circ T_1)$ 與 $(T_3 \circ T_2) \circ T_1$ 皆為，從空間 U 到空間 Z 的線型映射。而且，
$$T_3 \circ (T_2 \circ T_1) = (T_3 \circ T_2) \circ T_1 \text{。}$$
又若

$T_1: U \to V$, $T_2: U \to V$, $T_3: V \to W$ 以及 $T_4: V \to W$ are linear maps，則
$T_3 \circ (T_1 + T_2) = T_3 \circ T_1 + T_3 \circ T_2$
$(T_3 + T_4) \circ T_1 = T_3 \circ T_1 + T_4 \circ T_1$

到 4.3 節，當我們介紹"線型映射的矩陣表示法"之唯一性時，我們會了解到這些臆測是正確的。此外，一個 linear map，$T: U \to U$，mapping from U into itself，我們通常稱其為一個**線性算子** (linear operator)。這時候的 composition 被寫為，

$$T^2 = T \circ T$$
$$T^3 = T \circ T \circ T$$
$$\ldots\ldots\ldots\ldots\ldots$$
$$T^n = T \circ T \circ \ldots \circ T$$

也因此，$T^r \circ T^s = T^{r+s}$

而，當我們寫 T^0 時，意思指的是 I，the identity map。

Definition 4.5

A linear map, $T_1: U \to V$, is said to "invertible", if there exists a linear map, $T_2: V \to U$ such that

$T_2 \circ T_1 = I_U$, the identity map on U

and $T_1 \circ T_2 = I_V$, the identity map on V.

此時，我們稱，T_2 為 T_1 的**逆映射** (inverse map)。當然，T_1 也被稱為是 T_2 的逆映射。通常我們以符號，$T_2 = T_1^{-1}$ 表之。

Example 2

Given, $A = \begin{pmatrix} 2 & -1 \\ 1 & 1 \end{pmatrix}$ and $B = \begin{pmatrix} 1/3 & 1/3 \\ -1/3 & 2/3 \end{pmatrix}$. Let $T_1, T_2 : R^2 \to R^2$ be defined by $T_1(X) = AX$ and $T_2(X) = BX$, $\forall\ X = \begin{pmatrix} x \\ y \end{pmatrix} \in R^2$. Then

$$(T_2 \circ T_1)(X) = T_2(AX) = B(AX) = \begin{pmatrix} 1/3 & 1/3 \\ -1/3 & 2/3 \end{pmatrix}\begin{pmatrix} 2 & -1 \\ 1 & 1 \end{pmatrix}\begin{pmatrix} x \\ y \end{pmatrix}$$

$$= \begin{pmatrix} 1/3 & 1/3 \\ -1/3 & 2/3 \end{pmatrix}\begin{pmatrix} 2x - y \\ x + y \end{pmatrix} = \begin{pmatrix} x \\ y \end{pmatrix}$$

$$(T_1 \circ T_2)(X) = T_1(BX) = A(BX) = \begin{pmatrix} 2 & -1 \\ 1 & 1 \end{pmatrix}\begin{pmatrix} 1/3 & 1/3 \\ -1/3 & 2/3 \end{pmatrix}\begin{pmatrix} x \\ y \end{pmatrix}$$

$$= \begin{pmatrix} 2 & -1 \\ 1 & 1 \end{pmatrix}\begin{pmatrix} (x+y)/3 \\ (-x+2y)/3 \end{pmatrix} = \begin{pmatrix} x \\ y \end{pmatrix}$$ ∎

這例題說明，T_1 與 T_2 互為逆映射。其實，只因為矩陣 A 與 B 互為逆矩陣，所以它們所製造出來的 linear maps 是相互為逆映射的。有關這些，到第 4.3 節的時候，會有更進一步的介紹。理論上，什麼樣子的 linear map 是可逆的呢？看看下面的定理與推論。

Theorem 4.5

If $T : U \to V$ is a bijective linear map, then T is invertible.

Proof

Define a map, $G : V \to U$, by

$G(v) = u$ if and only if $T(u) = v$, for every $v \in V$.

Since T is bijective, G is well-defined and also

$(G \circ T)(u) = u$, $(T \circ G)(v) = v$, $\forall u \in U$, $\forall v \in V$.

We are now left to show that G is linear. To show this, we let

$v_1, v_2 \in V$ and c a scalar. Then, there exist u_1, u_2 in U such that
$$G(cv_1 + v_2) = G(cT(u_1) + T(u_2)) = G(T(cu_1 + u_2)) = cu_1 + u_2$$
$$= cG(v_1) + G(v_2)$$

And this completed the proof.

Corollary

If $\dim U = \dim V$ and if $T: U \to V$ is a linear map either injective or surjective, then T is invertible.

Proof

This follows from the last theorem combined with the Corollary 1 on page 107, section 4.1.

Exercises

1. Let $T: V \to W$ be a linear map. Show that if $\dim V > \dim W$, then T is not injective. (*Hint*: Consider $KerT$)

2. Let $T: R^2 \to R^2$ be defined by $T\begin{pmatrix} x \\ y \end{pmatrix} = \begin{pmatrix} 2x + y \\ 3x - 5y \end{pmatrix}$, $\forall \begin{pmatrix} x \\ y \end{pmatrix} \in R^2$. Show that T is invertible. (*Hint*: Consider $KerT$)

3. Let $P: V \to V$ be a linear map with $P^2 = P$. Show that $V = KerP \oplus RanP$. (*Hint*: Consider, for each $v \in V$, $v = Pv + (v - Pv)$.)

4. Let V and W be two vector spaces over a field, F, with $\dim V = \dim W$. Show that V and W are isomorphic.

§4.3 Matrix representations

在 4.1、4.2 節中，我們已經建立了一個深刻的印象。那就是，一個 $m \times n$ 矩陣可以製造出，一個從 R^n (或 C^n) 映射到 R^m (或 C^m) 的線型映射。這一節，我們想要反過來思考這個題目。也就是說，我們要把焦點瞄準到，"是否每一個從 R^n (或 C^n) 映射到 R^m (或 C^m) 的線型映射都可以用矩陣表示？"或者，更一般化的來說，"是否每一個從任意向量空間 V 映射到任意向量空間 W 的線型映射都可以用矩陣表示？"答案是肯定的。我們先從一個特殊的個案看起。

Theorem 4.6

Given a field F (real or complex) and let $T: F^n \to F^m$ be a linear transformation. Then there exists an $m \times n$ matrix, A, over F such that

$$T(X) = AX, \quad \forall \ X = \begin{pmatrix} x_1 \\ x_2 \\ \dots \\ \dots \\ x_n \end{pmatrix} \in F^n$$

Proof

首先，我們令 $E = \{E_1, E_2, ..., E_n\}$ 和 $e = \{e_1, e_2, ..., e_m\}$ 分別為 F^n 與 F^m 的標準基底。接著，我們看看，對任意 $i = 1, 2, ..., n$ 而言，$T(E_i) \in F^m$。所以，$T(E_i)$ 可以被表示為 $\{e_1, e_2, ..., e_m\}$ 的一個 linear combination。將其寫為

$$T(E_i) = a_{1i}e_1 + a_{2i}e_2 + ... + a_{mi}e_m$$

觀察一下，這些係數，$\begin{pmatrix} a_{1i} \\ a_{2i} \\ \dots \\ \dots \\ a_{mi} \end{pmatrix}$。當從 $i = 1$ 到 $i = n$ 時，可以得出 n 個此類

$m \times 1$ 的行向量係數。將這 n 個 columns 組合寫成矩陣

$$A = \begin{pmatrix} a_{11} & a_{12} & ... & ... & a_{1n} \\ a_{21} & a_{22} & ... & ... & a_{2n} \\ ... & ... & ... & ... & ... \\ ... & ... & ... & ... & ... \\ a_{m1} & a_{m2} & ... & ... & a_{mn} \end{pmatrix}$$

那麼這個矩陣將是我們所期待的。下面來驗證一下。

對任意 $X = \begin{pmatrix} x_1 \\ x_2 \\ ... \\ ... \\ x_n \end{pmatrix} \in F^n$，$X = x_1 E_1 + x_2 E_2 + ... + x_n E_n$，而且

$$\begin{aligned} T(X) &= x_1 T(E_1) + x_2 T(E_2) + ... + x_n T(E_n) \\ &= x_1 A^1 + x_2 A^2 + ... + x_n A^n \\ &= AX \end{aligned}$$

確實沒錯，此定理得證。

從定理證明的過程中，我們發現有件事情必須要補充說明。也就是，當我們給 F^n 與 F^m 選用不同的基底時，顯然會得到不同的矩陣 A。這一點務必要切實注意。不過，可以確定的是，一但基底選定之後，矩陣 A 是唯一被決定的。因為，這個 linear combination，$T(E_i) = a_{1i} e_1 + a_{2i} e_2 + ... + a_{mi} e_m$ 是唯一的。

前述之 $m \times n$ 矩陣 A 被稱為是，線型映射 T 的一個**矩陣表示** (matrix representation)。通常，我們以符號，$A = mat(T, E, e)$，表之。而這個東西，$mat(T, E, e)$，我們讀做，"the matrix representation of T from E into e"。下面，我們先簡單練習兩個例題。

Example 1

Given a map, $P: R^3 \to R^3$, defined by $P\begin{pmatrix} x \\ y \\ z \end{pmatrix} = \begin{pmatrix} x \\ y \\ 0 \end{pmatrix}$, $\forall \begin{pmatrix} x \\ y \\ z \end{pmatrix} \in R^3$. Find the matrix representation of P by using the standard basis.

Solution

Consider that

$$P\begin{pmatrix} 1 \\ 0 \\ 0 \end{pmatrix} = \begin{pmatrix} 1 \\ 0 \\ 0 \end{pmatrix}, \quad P\begin{pmatrix} 0 \\ 1 \\ 0 \end{pmatrix} = \begin{pmatrix} 0 \\ 1 \\ 0 \end{pmatrix} \text{ and } P\begin{pmatrix} 0 \\ 0 \\ 1 \end{pmatrix} = \begin{pmatrix} 0 \\ 0 \\ 0 \end{pmatrix}$$

we clearly see that the matrix representation of P by the standard basis is

$$A = \begin{pmatrix} 1 & 0 & 0 \\ 0 & 1 & 0 \\ 0 & 0 & 0 \end{pmatrix}$$

which is a 3×3 matrix. ∎

Example 2

Find the matrix representation for the plane coordinate rotation, $T: R^2 \to R^2$, $T\begin{pmatrix} x \\ y \end{pmatrix} = \begin{pmatrix} x\cos\theta - y\sin\theta \\ x\sin\theta + y\cos\theta \end{pmatrix}$, $\forall \begin{pmatrix} x \\ y \end{pmatrix} \in R^2$, with respect to the standard basis.

Solution

$$T\begin{pmatrix} 1 \\ 0 \end{pmatrix} = \begin{pmatrix} \cos\theta \\ \sin\theta \end{pmatrix}, \quad T\begin{pmatrix} 0 \\ 1 \end{pmatrix} = \begin{pmatrix} -\sin\theta \\ \cos\theta \end{pmatrix}$$

We obtain that $A = \begin{pmatrix} \cos\theta & -\sin\theta \\ \sin\theta & \cos\theta \end{pmatrix}$, which is a 2×2 matrix. ∎

Chapter Four Matrices as Linear Transformations 117

Example 3

Let $T: R^3 \to R$ be a linear transformation defined by

$T\begin{pmatrix}x\\y\\z\end{pmatrix} = 2x - y + 3z$, $\forall \begin{pmatrix}x\\y\\z\end{pmatrix} \in R^3$. Consider the number, $\{1\}$, as a standard basis

for one-dimensional space R. Then, we clearly see that the matrix representation for T with respect to the standard bases is a 1×3 matrix, $A = (2 \ -1 \ 3)$. ∎

Example 4

Given that $\alpha = \{\alpha_1 = \begin{pmatrix}1\\0\\0\end{pmatrix}, \alpha_2 = \begin{pmatrix}1\\1\\1\end{pmatrix}, \alpha_3 = \begin{pmatrix}-1\\1\\0\end{pmatrix}\}$ is a basis for R^3. And let

$E = \{E_1, E_2, E_3\}$ be the standard basis for R^3.

a. Find the matrix representation of $P\begin{pmatrix}x\\y\\z\end{pmatrix} = \begin{pmatrix}x\\y\\0\end{pmatrix}$, $\forall \begin{pmatrix}x\\y\\z\end{pmatrix} \in R^3$ from E into α.

(i.e. Find $A = mat(P, E, \alpha)$)

b. Check that, for $X = \begin{pmatrix}2\\3\\1\end{pmatrix} \in R^3$, $P\begin{pmatrix}2\\3\\1\end{pmatrix} = \begin{pmatrix}2\\3\\0\end{pmatrix} = 5\alpha_1 + 3\alpha_3$, by computing AX.

Solution

a. $P\begin{pmatrix}1\\0\\0\end{pmatrix} = \begin{pmatrix}1\\0\\0\end{pmatrix} = 1\alpha_1 + 0\alpha_2 + 0\alpha_3$, $P\begin{pmatrix}0\\1\\0\end{pmatrix} = \begin{pmatrix}0\\1\\0\end{pmatrix} = 1\alpha_1 + 0\alpha_2 + 1\alpha_3$ and

$P\begin{pmatrix}0\\0\\1\end{pmatrix} = \begin{pmatrix}0\\0\\0\end{pmatrix} = 0\alpha_1 + 0\alpha_2 + 0\alpha_3$

Therefore, the matrix representation of P from E into α is

$$\mathrm{mat}(P, E, \alpha) = \begin{pmatrix} 1 & 1 & 0 \\ 0 & 0 & 0 \\ 0 & 1 & 0 \end{pmatrix}$$

b. $AX = \begin{pmatrix} 1 & 1 & 0 \\ 0 & 0 & 0 \\ 0 & 1 & 0 \end{pmatrix} \begin{pmatrix} 2 \\ 3 \\ 1 \end{pmatrix} = \begin{pmatrix} 5 \\ 0 \\ 3 \end{pmatrix}$

也就是說，$PX = \begin{pmatrix} 2 \\ 3 \\ 0 \end{pmatrix} = 5\alpha_1 + 3\alpha_3$，正確無誤。∎

接著，我們要探討，從任意一個向量空間 V 映射至另一個向量空間 W 的 linear map 的 matrix representation。我們將發現，與前述特例一樣，一旦在基底選定之後，任意線型映射的矩陣表示是唯一被決定的。

Definition 4.6

Let V and W be vector spaces over a field, F, having bases, $\alpha = \{\alpha_1, \alpha_2, ..., \alpha_n\}$ and $\beta = \{\beta_1, \beta_2, ..., \beta_m\}$, respectively. Let $T: V \to W$ be a linear map. Define an $m \times n$ matrix, A, with the ith column, $1 \le i \le n$,

$$A^i = \begin{pmatrix} a_{1i} \\ a_{2i} \\ ... \\ ... \\ a_{mi} \end{pmatrix}, \text{ if } T(\alpha_i) = \sum_{j=1}^{m} a_{ji} \beta_j$$

The matrix, A, so obtained is called the matrix representation of T from α into β. And will also be denoted by $A = mat(T, \alpha, \beta)$.

定義中，所得到的矩陣 A 是否符合我們的需求？也就是說，對任意

$v = \sum_{i=1}^{n} c_i \alpha_i \in V$ 以及 $T(v) = \sum_{j=1}^{m} d_j \beta_j \in W$ 而言，$(A)_{m \times n} \begin{pmatrix} c_1 \\ c_2 \\ ... \\ ... \\ c_n \end{pmatrix} = \begin{pmatrix} d_1 \\ d_2 \\ ... \\ ... \\ d_m \end{pmatrix}$。有關這

個看法是沒有問題的。因為

$$T(v) = T(\sum_{i=1}^{n} c_i \alpha_i) = \sum_{i=1}^{n} c_i T(\alpha_i) = \sum_{i=1}^{n} c_i \sum_{j=1}^{m} a_{ji} \beta_j = \sum_{j=1}^{m} \sum_{i=1}^{n} a_{ji} c_i \beta_j$$

$$= \sum_{j=1}^{m} (\sum_{i=1}^{n} a_{ji} c_i) \beta_j$$

那麼，只要令 $d_j = \sum_{i=1}^{n} a_{ji} c_i$，即可得證 $(A)_{m \times n} \begin{pmatrix} c_1 \\ c_2 \\ ... \\ ... \\ c_n \end{pmatrix} = \begin{pmatrix} d_1 \\ d_2 \\ ... \\ ... \\ d_m \end{pmatrix}$。

我們用下面簡單的圖形，幫助大家把初始的概念給整理建構起來，讓同學們方便認識和容易記憶。

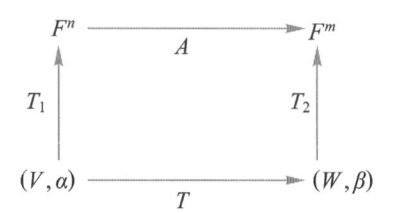

$$\begin{array}{ccc} F^n & \xrightarrow{A} & F^m \\ T_1 \uparrow & & \uparrow T_2 \\ (V, \alpha) & \xrightarrow{T} & (W, \beta) \end{array}$$

其中，T_1 以及 T_2 分別為 (V, α) 與 (W, β) 到 F^n 與 F^m 的 isomorphism。(第 108 頁 Corollary 2) 由這樣一個簡單的圖形，我們可以精確的說，$T = T_2^{-1} \circ A \circ T_1$。

由於，線性組合，$T(\alpha_i) = \sum_{j=1}^{m} a_{ji} \beta_j$ 是唯一的。所以，當 V 與 W 之基底決定之後，如此定義出來的矩陣 A 是唯一的。

現在反過來說，假設 A 是任意一個 $m \times n$ 矩陣。而且，對 V 中之

$v = \sum_{i=1}^{n} c_i \alpha_i$，我們定義，$T(v) = \sum_{j=1}^{m} d_j \beta_j$，其中，$(A)_{m \times n} \begin{pmatrix} c_1 \\ c_2 \\ ... \\ ... \\ c_n \end{pmatrix} = \begin{pmatrix} d_1 \\ d_2 \\ ... \\ ... \\ d_m \end{pmatrix}$ 的話。那麼，此時之 T 為一個從 V 映射到 W 的 linear map。

綜合上述所言，我們發現：一個 $m \times n$ 矩陣唯一決定了，一個從 n 度空間 V 映射到 m 度空間 W 的 linear map；反過來說，一個從 n 度空間 V 映射到 m 度空間 W 的 linear map，一旦它們的基底確定之後，它唯一決定了，一個 $m \times n$ 矩陣。精確的來說，假設 $L(V, W)$ 代表，所有從 n 度空間 V 映射到 m 度空間 W 的 linear maps，$M_{m \times n}$ 代表所有 $m \times n$ 矩陣。那麼，$L(V, W)$ 與 $M_{m \times n}$ 之間存在著，一個 one-to-one and onto correspondence，而且這個 correspondence 是一個 isomorphism。有關這個細節，請讀者練習下面的 Exercise。

Exercise

Let V and W be vector spaces over a field, F, with bases, $\alpha = \{\alpha_1, \alpha_2, ..., \alpha_n\}$ and $\beta = \{\beta_1, \beta_2, ..., \beta_m\}$, respectively. Show that the correspondence, $F(T) = mat(T, \alpha, \beta)$, is an isomorphism from $L(V, W)$ into $M_{m \times n}$.
(*Hint*：證明，對任意 $T_1, T_2 \in L(V, W)$，滿足 $F(cT_1 + T_2) = cF(T_1) + F(T_2)$。)

Example 5

Let $V = \{$ real-valued polynomials defined on R with degree less then or equal to 3$\}$. Define a map, $D: V \to V$, by $D(f) = f'$. Find $mat(D, \alpha, \alpha)$, given $\alpha = \{\alpha_1 = 1, \alpha_2 = x, \alpha_3 = x^2, \alpha_4 = x^3\}$ as a basis of V.

Solution

$D(1) = 0$，$D(x) = 1$，$D(x^2) = 2x$，$D(x^3) = 3x^2$

$$\therefore \quad mat(D, \alpha, \alpha) = \begin{pmatrix} 0 & 1 & 0 & 0 \\ 0 & 0 & 2 & 0 \\ 0 & 0 & 0 & 3 \\ 0 & 0 & 0 & 0 \end{pmatrix}$$

∎

Example 6

Use the result found from the last example, find the derivative of $f(x) = 4 - 6x + 7x^2 - x^3$.

Solution

Writing, $f(x) = 4\alpha_1 - 6\alpha_2 + 7\alpha_3 - \alpha_4$, then

$$\begin{pmatrix} 0 & 1 & 0 & 0 \\ 0 & 0 & 2 & 0 \\ 0 & 0 & 0 & 3 \\ 0 & 0 & 0 & 0 \end{pmatrix} \begin{pmatrix} 4 \\ -6 \\ 7 \\ -1 \end{pmatrix} = \begin{pmatrix} -6 \\ 14 \\ -3 \\ 0 \end{pmatrix}$$

Hence, $D(f) = -6 + 14x - 3x^2$.

∎

Theorem 4.7 (Composite mapping theorem)

Let U, V and W be vector spaces with bases, $\alpha = \{\alpha_1, \alpha_2, ..., \alpha_n\}$, $\beta = \{\beta_1, \beta_2, ..., \beta_p\}$ and $\gamma = \{\gamma_1, \gamma_2, ..., \gamma_m\}$, respectively. Let $T_1 : U \to V$ and $T_2 : V \to W$ be linear maps. If $A_{p \times n} = mat(T_1, \alpha, \beta)$ and $B_{m \times p} = mat(T_2, \beta, \gamma)$, then $mat(T_2 \circ T_1, \alpha, \gamma) = BA$. (參考 4.2 節第 110 頁的 Example 1)

Proof

$$\begin{array}{ccccc}
F^n & \xrightarrow{\quad A \quad} & F^p & \xrightarrow{\quad B \quad} & F^m \\
\uparrow L_1 & & \uparrow L_2 & & \uparrow L_3 \\
(U, \alpha) & \xrightarrow{\quad T_1 \quad} & (V, \beta) & \xrightarrow{\quad T_2 \quad} & (W, \gamma)
\end{array}$$

Consider for each $i = 1, 2, ..., n$,

$$(T_2 \circ T_1)(\alpha_i) = T_2(T_1(\alpha_i)) = T_2(\sum_{j=1}^{p} a_{ji}\beta_j) = \sum_{j=1}^{p} a_{ji}T_2(\beta_j)$$

$$= \sum_{j=1}^{p} a_{ji} \sum_{t=1}^{m} b_{tj}\gamma_t = \sum_{t=1}^{m} (\sum_{j=1}^{p} b_{tj}a_{ji})\gamma_t$$

We see that the ith column of $mat(T_2 \circ T_1, \alpha, \gamma)$ is $\begin{pmatrix} \sum_{j=1}^{p} b_{1j}a_{ji} \\ \sum_{j=1}^{p} b_{2j}a_{ji} \\ ... \\ \sum_{j=1}^{m} b_{mj}a_{ji} \end{pmatrix}$

and which is the ith column of BA.

越來越精彩了。前面曾經講過，一旦向量空間之基底確立之後，線型映射的矩陣表示法是唯一的。言下之意是，不同的基底將產生不同的矩陣表示。這句話是正確的。那麼問題是，不同的基底所製造出來不同的矩陣，它們相互之間會有什麼關係？或者說，它們相互之間是否有明確的式子相關連呢？答案即將揭曉。不過我們要先介紹所謂的**變換基底矩陣** (change of basis matrix)。

Definition 4.7

Let V be a vector space and let $\alpha = \{\alpha_1, \alpha_2, ..., \alpha_n\}$, $\beta = \{\beta_1, \beta_2, ..., \beta_n\}$ be two bases of V. If $I : (V, \alpha) \to (V, \beta)$ is the identity map, then the matrix representation, $mat(I, \alpha, \beta)$, is called a "change of basis matrix" from α into β.

Example 7

Given, $V = R^2$ with two bases, $\alpha = \left\{ \begin{pmatrix} 1 \\ 1 \end{pmatrix}, \begin{pmatrix} 0 \\ 1 \end{pmatrix} \right\}$, $\beta = \left\{ \begin{pmatrix} 1 \\ 2 \end{pmatrix}, \begin{pmatrix} 1 \\ 1 \end{pmatrix} \right\}$. Find the

Chapter Four　Matrices as Linear Transformations　123

change of basis matrix from α into β.

Solution

$$I(\begin{pmatrix}1\\1\end{pmatrix}) = \begin{pmatrix}1\\1\end{pmatrix} = 0\begin{pmatrix}1\\2\end{pmatrix} + 1\begin{pmatrix}1\\1\end{pmatrix} \text{ and } I(\begin{pmatrix}0\\1\end{pmatrix}) = \begin{pmatrix}0\\1\end{pmatrix} = 1\begin{pmatrix}1\\2\end{pmatrix} + (-1)\begin{pmatrix}1\\1\end{pmatrix}$$

所以，$mat(I, \alpha, \beta) = \begin{pmatrix}0 & 1\\1 & -1\end{pmatrix}$.

我們看看，上述例題 R^2 中的一個向量，$\begin{pmatrix}3\\2\end{pmatrix}$。它可以被表示為，

$$\begin{pmatrix}3\\2\end{pmatrix} = 3\begin{pmatrix}1\\1\end{pmatrix} + (-1)\begin{pmatrix}0\\1\end{pmatrix}，以及 \begin{pmatrix}3\\2\end{pmatrix} = (-1)\begin{pmatrix}1\\2\end{pmatrix} + 4\begin{pmatrix}1\\1\end{pmatrix}。$$

也就是說，在 (V,α) 中，向量 $\begin{pmatrix}3\\2\end{pmatrix}$ 的座標是 $\begin{pmatrix}3\\-1\end{pmatrix}$；在 (V,β) 中，向量 $\begin{pmatrix}3\\2\end{pmatrix}$ 的座標是 $\begin{pmatrix}-1\\4\end{pmatrix}$。而，$\begin{pmatrix}3\\-1\end{pmatrix}$ 與 $\begin{pmatrix}-1\\4\end{pmatrix}$ 之關係則為，

$$\begin{pmatrix}0 & 1\\1 & -1\end{pmatrix}\begin{pmatrix}3\\-1\end{pmatrix} = \begin{pmatrix}-1\\4\end{pmatrix}$$

因此，"變換基底矩陣" 有時被稱為**變換座標矩陣** (change of coordinates matrix)。

前述例題給了我們一個啟示，假設 $P = mat(I, \alpha, \beta)$，且任意 V 中的向量 v，若 $v = \sum_{i=1}^{n} c_i \alpha_i = \sum_{i=1}^{n} d_i \beta_i$，則 $P\begin{pmatrix}c_1\\c_2\\...\\c_n\end{pmatrix} = \begin{pmatrix}d_1\\d_2\\...\\d_n\end{pmatrix}$.

這個概念可從下列式子中體會，

$$\sum_{j=1}^{n} d_j \beta_j = v = I(v) = I(\sum_{i=1}^{n} c_i \alpha_i) = \sum_{i=1}^{n} c_i I(\alpha_i) = \sum_{i=1}^{n} c_i \sum_{j=1}^{n} p_{ji} \beta_j$$
$$= \sum_{j=1}^{n} (\sum_{i=1}^{n} p_{ji} c_i) \beta_j$$

另外，有一個特殊的 case 一併提出來，請同學們特別注意。那就是，

$mat(I, \alpha, \alpha) = I_n$ 是一個單位矩陣。也就是說，在基底沒改變的情況之下，單位映射對應單位矩陣。喘口氣吧！還沒了喔。在本節的主訴求來臨之前，我們再看 Composite mapping theorem 的兩個漂亮的推論。

Corollary 1

Let V be a vector space with two bases, $\alpha = \{\alpha_1, \alpha_2, ..., \alpha_n\}$, $\beta = \{\beta_1, \beta_2, ..., \beta_n\}$, and let $T : (V, \alpha) \to (V, \beta)$ be an invertible linear map. If $A = mat(T, \alpha, \beta)$ and $B = mat(T^{-1}, \beta, \alpha)$, then B is the inverse matrix of A.

Proof

Consider the following diagram

$$\begin{array}{ccccc} F^n & \xrightarrow{A} & F^p & \xrightarrow{B} & F^m \\ L_1 \uparrow & & L_2 \uparrow & & L_3 \uparrow \\ (V, \alpha) & \xrightarrow{T} & (V, \beta) & \xrightarrow{T^{-1}} & (V, \alpha) \end{array}$$

We clearly see, by the composite mapping theorem, that

$$mat(T^{-1} \circ T, \alpha, \alpha) = mat(I, \alpha, \alpha) = I_n = BA$$

把這個推論以及圖形稍微改一改，得到下面的特殊案例。

Corollary 2

Let V be a vector space with bases, $\alpha = \{\alpha_1, \alpha_2, ..., \alpha_n\}$ and $\beta = \{\beta_1, \beta_2, ..., \beta_n\}$. If $P_1 = mat(I, \alpha, \beta)$ and $P_2 = mat(I, \beta, \alpha)$, then $P_1 P_2 = P_2 P_1 = I_n$.
(也就是說，變換基底矩陣 P_1 是可逆的。而且，$P_1^{-1} = P_2$。)

$$\begin{array}{ccccc}
F^n & \xrightarrow{P_1} & F^p & \xrightarrow{P_2} & F^m \\
\uparrow L_1 & & \uparrow L_2 & & \uparrow L_3 \\
(V,\alpha) & \xrightarrow{I} & (V,\beta) & \xrightarrow{I} & (V,\alpha)
\end{array}$$

嚴格講起來,這個證明也不難。我們令,$P_1 = (a_{ij})_{n \times n}$,$P_2 = (b_{ij})_{n \times n}$。那麼,對任意 $j = 1, 2, \ldots, n$ 而言,由於

$$\alpha_j = I(\alpha_j) = \sum_{i=1}^{n} a_{ij} \beta_i = \sum_{i=1}^{n} a_{ij} (\sum_{t=1}^{n} b_{ti} \alpha_t) = \sum_{t=1}^{n} (\sum_{i=1}^{n} b_{ti} a_{ij}) \alpha_t$$

所以

$$\sum_{i=1}^{n} b_{ti} a_{ij} = \begin{cases} 0, & \text{if } t \neq j \\ 1, & \text{if } t = j \end{cases}$$

這說明了,$P_2 P_1 = I_n$。同理,$P_1 P_2 = I_n$ 也可得證。

Example 8

Given, $V = R^2$ with two bases, $\alpha = \left\{ \begin{pmatrix} 1 \\ 2 \end{pmatrix}, \begin{pmatrix} 1 \\ 1 \end{pmatrix} \right\}$ and $\beta = \left\{ \begin{pmatrix} 1 \\ 0 \end{pmatrix}, \begin{pmatrix} 0 \\ 1 \end{pmatrix} \right\}$. Find the change of basis matrix, $P = mat(I, \alpha, \beta)$, and its inverse, P^{-1}.

Solution

注意,$\beta = \left\{ \begin{pmatrix} 1 \\ 0 \end{pmatrix}, \begin{pmatrix} 0 \\ 1 \end{pmatrix} \right\}$ 是 $V = R^2$ 的標準基底。所以,很顯然的 $P = mat(I, \alpha, \beta) = \begin{pmatrix} 1 & 1 \\ 2 & 1 \end{pmatrix}$。由上述推論,得知,$P^{-1}$ 就是 $mat(I, \beta, \alpha)$。而,

$$I\begin{pmatrix} 1 \\ 0 \end{pmatrix} = \begin{pmatrix} 1 \\ 0 \end{pmatrix} = (-1)\begin{pmatrix} 1 \\ 2 \end{pmatrix} + 2\begin{pmatrix} 1 \\ 1 \end{pmatrix}, \quad I\begin{pmatrix} 0 \\ 1 \end{pmatrix} = \begin{pmatrix} 0 \\ 1 \end{pmatrix} = \begin{pmatrix} 1 \\ 2 \end{pmatrix} - \begin{pmatrix} 1 \\ 1 \end{pmatrix}$$

所以,$P^{-1} = \begin{pmatrix} -1 & 1 \\ 2 & -1 \end{pmatrix}$。(請同學們確認一下,$PP^{-1} = P^{-1}P = I_2$。)

準備工作做的差不多了。我們接著看上面的簡圖，再看下面的定理，把整體線型轉換的矩陣表示以及不同基底所產生不同矩陣的架構給組合起來。

Theorem 4.8

Let V be a vector space with bases, $\alpha = \{\alpha_1, \alpha_2, ..., \alpha_n\}$ and $\alpha = \{\alpha'_1, \alpha'_2, ..., \alpha'_n\}$, and let W be a vector space with bases, $\beta = \{\beta_1, \beta_2, ..., \beta_m\}$ and $\beta = \{\beta'_1, \beta'_2, ..., \beta'_m\}$. Suppose that $T: V \to W$ is a linear map, and if $A = mat(T, \alpha, \beta)$, $B = mat(T, \alpha', \beta'\}$, then there exist non-singular matrices, P_1 and P_2 such that $P_2^{-1} A P_1 = B$.

Proof

Consider the diagram as above. Writing $P_1 = mat(I, \alpha', \alpha)$ and $P_2 = mat(I, \beta', \beta)$, then by the previous theorems, we see that for each $v \in V$, if

$$v = \sum_{i=1}^{n} c_i \alpha_i = \sum c'_i \alpha'_i \text{ and } T(v) = \sum_{i=1}^{m} d_i \beta_i = \sum_{i=1}^{m} d'_i \beta'_i$$

$$P_2^{-1} A P_1 \begin{pmatrix} c'_1 \\ c'_2 \\ ... \\ c'_n \end{pmatrix} = P_2^{-1} A \begin{pmatrix} c_1 \\ c_2 \\ ... \\ c_n \end{pmatrix} = P_2^{-1} \begin{pmatrix} d_1 \\ d_2 \\ ... \\ d_m \end{pmatrix} = \begin{pmatrix} d'_1 \\ d'_2 \\ ... \\ d'_m \end{pmatrix} \text{ and } B \begin{pmatrix} c'_1 \\ c'_2 \\ ... \\ c'_n \end{pmatrix} = \begin{pmatrix} d'_1 \\ d'_2 \\ ... \\ d'_m \end{pmatrix}$$

So,

$$P_2^{-1} A P_1 \begin{pmatrix} c'_1 \\ c'_2 \\ ... \\ c'_n \end{pmatrix} = B \begin{pmatrix} c'_1 \\ c'_2 \\ ... \\ c'_n \end{pmatrix}$$

This is true for every $\begin{pmatrix} c_1' \\ c_2' \\ ... \\ c_n' \end{pmatrix} \in R^n$. Therefore, $P_2^{-1}AP_1 = B$.

看到這個式子，我們似乎想到了所謂的 "相似矩陣"。兩個 $m \times n$ 矩陣 A 與 B 被稱為**相似的** (similar)，假設存在兩個 non-singular matrices，P_1 以及 P_2 使得，$P_2^{-1}AP_1 = B$。換句話說，假設矩陣，A 與 B 代表著相同的線型映射的話，那麼 A 與 B 是相似的。

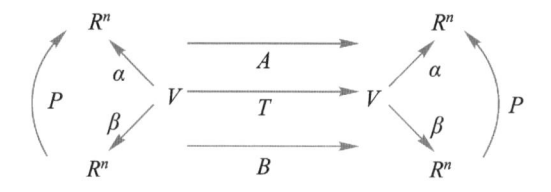

到第七章，當我們談論特徵值與特徵向量的時候，比較感興趣的是 $V = W$ 的情況。那就是，假設 $\alpha = \{\alpha_1, \alpha_2, ..., \alpha_n\}$ and $\beta = \{\beta_1, \beta_2, ..., \beta_n\}$ are bases of V and $T:V \to V$ is a linear map。若 $A = mat(T, \alpha, \alpha)$, $B = mat(T, \beta, \beta)$ 以及 $P = mat(I, \beta, \alpha)$，則 $P^{-1}AP = B$。所以，也常有些作者定義說， "Two square matrices, A and B are said to be similar, if there exists a non-singular matrix, P, such that $P^{-1}AP = B$." 。而，這個定義只不過是前述定理的一個特例罷了。

Example 9

Given a linear map, $T : R^3 \to R^3$, defined by $T\begin{pmatrix} x \\ y \\ z \end{pmatrix} = \begin{pmatrix} y + 7z \\ 2x + y + 3z \\ x + y + 4z \end{pmatrix}$. Let $\alpha = \{\begin{pmatrix} 1 \\ 0 \\ 0 \end{pmatrix}, \begin{pmatrix} 0 \\ 1 \\ 0 \end{pmatrix}, \begin{pmatrix} 0 \\ 0 \\ 1 \end{pmatrix}\}$ and $\beta = \{\begin{pmatrix} 1 \\ 0 \\ 0 \end{pmatrix}, \begin{pmatrix} 1 \\ 1 \\ 1 \end{pmatrix}, \begin{pmatrix} -1 \\ 1 \\ 0 \end{pmatrix}\}$. Find $A = mat(T, \alpha, \alpha)$, $B = mat(T, \beta, \beta)$, $P = mat(I, \beta, \alpha)$ and $P^{-1} = mat(I, \alpha, \beta)$. And check, $P^{-1}AP = B$.

Solution

$$T\begin{pmatrix}1\\0\\0\end{pmatrix}=\begin{pmatrix}0\\2\\1\end{pmatrix}=0\begin{pmatrix}1\\0\\0\end{pmatrix}+2\begin{pmatrix}0\\1\\0\end{pmatrix}+\begin{pmatrix}0\\0\\1\end{pmatrix} \qquad T\begin{pmatrix}0\\1\\0\end{pmatrix}=\begin{pmatrix}1\\1\\1\end{pmatrix}=\begin{pmatrix}1\\0\\0\end{pmatrix}+\begin{pmatrix}0\\1\\0\end{pmatrix}+\begin{pmatrix}0\\0\\1\end{pmatrix}$$

$$T\begin{pmatrix}0\\0\\1\end{pmatrix}=\begin{pmatrix}7\\3\\4\end{pmatrix}=7\begin{pmatrix}1\\0\\0\end{pmatrix}+3\begin{pmatrix}0\\1\\0\end{pmatrix}+4\begin{pmatrix}0\\0\\1\end{pmatrix}$$

所以 $\qquad A=mat(T,\alpha,\alpha)=\begin{pmatrix}0 & 1 & 7\\2 & 1 & 3\\1 & 1 & 4\end{pmatrix}$

$$T\begin{pmatrix}1\\0\\0\end{pmatrix}=\begin{pmatrix}0\\2\\1\end{pmatrix}=0\begin{pmatrix}1\\0\\0\end{pmatrix}+\begin{pmatrix}1\\1\\1\end{pmatrix}+\begin{pmatrix}-1\\1\\0\end{pmatrix} \qquad T\begin{pmatrix}1\\1\\1\end{pmatrix}=\begin{pmatrix}8\\6\\6\end{pmatrix}=2\begin{pmatrix}1\\0\\0\end{pmatrix}+6\begin{pmatrix}1\\1\\1\end{pmatrix}+0\begin{pmatrix}-1\\1\\0\end{pmatrix}$$

$$T\begin{pmatrix}-1\\1\\0\end{pmatrix}=\begin{pmatrix}1\\-1\\0\end{pmatrix}=0\begin{pmatrix}1\\0\\0\end{pmatrix}+0\begin{pmatrix}1\\1\\1\end{pmatrix}-\begin{pmatrix}-1\\1\\0\end{pmatrix}$$

所以 $\qquad B=mat(T,\beta,\beta)=\begin{pmatrix}0 & 2 & 0\\1 & 6 & 0\\1 & 0 & -1\end{pmatrix}$

$$\begin{pmatrix}1\\0\\0\end{pmatrix}=1\begin{pmatrix}1\\0\\0\end{pmatrix}+0\begin{pmatrix}1\\1\\0\end{pmatrix}+0\begin{pmatrix}0\\0\\1\end{pmatrix} \qquad \begin{pmatrix}1\\1\\1\end{pmatrix}=\begin{pmatrix}1\\0\\0\end{pmatrix}+\begin{pmatrix}0\\1\\0\end{pmatrix}+\begin{pmatrix}0\\0\\1\end{pmatrix} \qquad \begin{pmatrix}-1\\1\\0\end{pmatrix}=-1\begin{pmatrix}1\\0\\0\end{pmatrix}+\begin{pmatrix}0\\1\\0\end{pmatrix}+0\begin{pmatrix}0\\0\\1\end{pmatrix}$$

所以 $\qquad P=mat(I,\beta,\alpha)=\begin{pmatrix}1 & 1 & -1\\0 & 1 & 1\\0 & 1 & 0\end{pmatrix}$

$$\begin{pmatrix}1\\0\\0\end{pmatrix}=\begin{pmatrix}1\\0\\0\end{pmatrix}+0\begin{pmatrix}1\\1\\1\end{pmatrix}+0\begin{pmatrix}-1\\1\\0\end{pmatrix} \qquad \begin{pmatrix}0\\1\\0\end{pmatrix}=\begin{pmatrix}1\\0\\0\end{pmatrix}+0\begin{pmatrix}1\\1\\1\end{pmatrix}+\begin{pmatrix}-1\\1\\0\end{pmatrix} \qquad \begin{pmatrix}0\\0\\1\end{pmatrix}=-2\begin{pmatrix}1\\0\\0\end{pmatrix}+\begin{pmatrix}1\\1\\1\end{pmatrix}-\begin{pmatrix}-1\\1\\0\end{pmatrix}$$

所以 $\qquad P^{-1}=mat(I,\alpha,\beta)=\begin{pmatrix}1 & 1 & -2\\0 & 0 & 1\\0 & 1 & -1\end{pmatrix}$

Chapter Four Matrices as Linear Transformations   129

$$\text{Check}: P^{-1}AP = \begin{pmatrix} 1 & 1 & -2 \\ 0 & 0 & 1 \\ 0 & 1 & -1 \end{pmatrix} \begin{pmatrix} 0 & 1 & 7 \\ 2 & 1 & 3 \\ 1 & 1 & 4 \end{pmatrix} \begin{pmatrix} 1 & 1 & -1 \\ 0 & 1 & 1 \\ 0 & 1 & 0 \end{pmatrix} = \begin{pmatrix} 0 & 2 & 0 \\ 1 & 6 & 0 \\ 1 & 0 & -1 \end{pmatrix} = B \quad \blacksquare$$

Exercises

1. Let $\alpha = \{v_1, v_2, v_3, v_4\}$ be a basis for vector space, V, and let $T: V \to V$ be a linear map. If $T(v_1) = v_2, T(v_2) = v_3, T(v_3) = v_1 + v_2, T(v_4) = v_1 + 3v_4$. Find $mat(T, \alpha, \alpha)$.

2. Let $T: R^n \to R^n$ be a linear map defined by $T(X) = cX$, $\forall\ X \in R^n$, where $c \in R$. Find the matrix representation, $mat(T, E, E)$, where E denotes the standard basis.

3. Let V be the vector space generated by each of the following bases, α. Let $D = \dfrac{d}{dt}$ be the derivative. Find $mat(D, \alpha, \alpha)$

   a. $\{e^t,\ e^{2t}\}$

   b. $\{1, t, e^t,\ e^{2t},\ te^{2t}\}$

   c. $\{\sin t,\ \cos t\}$

4. Let $T: R^4 \to R^4$ be a linear map defined by $T\begin{pmatrix} a \\ b \\ c \\ d \end{pmatrix} = \begin{pmatrix} -3a + 5d \\ 3b - 5c \\ -2c \\ 2d \end{pmatrix}$, $\forall \begin{pmatrix} a \\ b \\ c \\ d \end{pmatrix} \in R^4$.

   And let $\alpha = \left\{ \begin{pmatrix} 1 \\ 0 \\ 0 \\ 1 \end{pmatrix}, \begin{pmatrix} 0 \\ 1 \\ 1 \\ 0 \end{pmatrix}, \begin{pmatrix} 0 \\ 1 \\ 0 \\ 0 \end{pmatrix}, \begin{pmatrix} 1 \\ 0 \\ 0 \\ 0 \end{pmatrix} \right\}$ be a basis for R^4. Find $mat(T, \alpha, \alpha)$.

5. Let $T: R^2 \to R^2$ be a linear map defined by $T\begin{pmatrix} x \\ y \end{pmatrix} = \begin{pmatrix} x - y \\ 3y \end{pmatrix}$, $\forall \begin{pmatrix} x \\ y \end{pmatrix} \in R^2$. Let $\alpha = \left\{ \begin{pmatrix} 1 \\ 1 \end{pmatrix}, \begin{pmatrix} 3 \\ 1 \end{pmatrix} \right\}$ and $\beta = \left\{ \begin{pmatrix} 4 \\ 3 \end{pmatrix}, \begin{pmatrix} 2 \\ 3 \end{pmatrix} \right\}$ be two basis for R^2. Find $A = mat(T, \alpha, \alpha)$, $B = mat(T, \beta, \beta)$, $P = mat(I, \beta, \alpha)$ and $P^{-1}mat(I, \alpha, \beta)$. Check, $P^{-1}AP = B$.

6. Let $T: R^3 \to R^2$ be a linear map defined by $T\begin{pmatrix} x \\ y \\ z \end{pmatrix} = \begin{pmatrix} x+2y+z \\ 3x+4y \end{pmatrix}$, $\forall \begin{pmatrix} x \\ y \\ z \end{pmatrix} \in R^3$. Let

$\alpha = \{\begin{pmatrix} 1 \\ 0 \\ 0 \end{pmatrix}, \begin{pmatrix} 0 \\ 1 \\ 0 \end{pmatrix}, \begin{pmatrix} 0 \\ 0 \\ 1 \end{pmatrix}\}$, $\alpha' = \{\begin{pmatrix} 1 \\ 0 \\ 1 \end{pmatrix}, \begin{pmatrix} 0 \\ 1 \\ 0 \end{pmatrix}, \begin{pmatrix} 1 \\ 2 \\ 0 \end{pmatrix}\}$ be two bases for R^3 and

$\beta = \{\begin{pmatrix} 1 \\ 0 \end{pmatrix}, \begin{pmatrix} 0 \\ 1 \end{pmatrix}\}$, $\beta' = \{\begin{pmatrix} 1 \\ 2 \end{pmatrix}, \begin{pmatrix} 2 \\ 1 \end{pmatrix}\}$ be two bases for R^2. Find
$A = mat(T, \alpha, \beta)$, $B = mat(T, \alpha', \beta')$, $P_1 = mat(I, \alpha', \alpha)$ and
$P_2 = mat(I, \beta', \beta)$. Check, $P_2^{-1} A P_1 = B$.

Chapter Five

Bilinear Forms and Inner Products

- **5.1** Bilinear forms
- **5.2** Quadratic forms
- **5.3** Inner products
- **5.3*** Appendixes for orthogonal complements
- **5.4** Gram-Schmidt and orthogonal matrices

　　Inner product space 是十九世紀歐洲數學界的精彩創作，它在分析學領域裡，扮演著承先啓後的重要角色。它繼承了傳統的歐基里得距離模式，進一步的推展，而爲現代數學分析學界不可缺少的基石。上一章，我們認識了矩陣在線型轉換上所扮演的角色。如今在雙線型的領域上，矩陣也有其同等重要與關鍵性的演出。一個雙線型可以唯一製造出一個矩陣，一個矩陣也唯一決定了一個雙線型。而且，這種相互的對應也是 one-to-one and onto correspondence。

§5.1　Bilinear forms

　　首先我們看看，一個 real $n \times n$ matrix，A，和 R^n 中的兩個 column vectors，α and β。計算一下 $({}^t\alpha) A \beta$，我們發現它是一個 1×1 矩陣，或說是一個純量。這個東西正是大家所謂的**雙線性型態** (bilinear form)。當然，要是說 A 是一個 complex $n \times n$ matrix，而 α，β 是 C^n 中的兩個 column vectors 的話，那麼前述所稱的 bilinear form 將被改寫爲，$({}^t\alpha) A \overline{\beta}$，其中 $\overline{\beta}$ 指的是 β 的共軛數。姑且按下不表。

Definition 5.1

> Let V be a vector space over R. A "bilinear form", φ, on V is a map which maps from $V \times V$ into R such that
> 1. $\varphi(a\alpha + b\beta, \gamma) = a\varphi(\alpha, \gamma) + b\varphi(\beta, \gamma)$, $\quad \forall \alpha, \beta, \gamma \in V$ and $a, b \in R$
> 2. $\varphi(\alpha, b\beta + c\gamma) = b\varphi(\alpha, \beta) + c\varphi(\alpha, \gamma)$, $\quad \forall \alpha, \beta, \gamma \in V$ and $b, c \in R$

譬如說

For $X = \begin{pmatrix} x_1 \\ x_2 \\ \cdots \\ x_n \end{pmatrix}, Y = \begin{pmatrix} y_1 \\ y_2 \\ \cdots \\ y_n \end{pmatrix} \in R^n$, if we define $\varphi(X,Y) = \sum_{i=1}^{n} x_i y_i$, then φ is a bilinear form on R^n。因為，若

$$X = \begin{pmatrix} x_1 \\ x_2 \\ \cdots \\ x_n \end{pmatrix}, Y = \begin{pmatrix} y_1 \\ y_2 \\ \cdots \\ y_n \end{pmatrix}, Z = \begin{pmatrix} z_1 \\ z_2 \\ \cdots \\ z_n \end{pmatrix} \in R^n，且\ a,b,c \in R，則$$

$$\varphi(aX + bY, Z) = \sum_{i=1}^{n}(ax_i + by_i)z_i = a\sum_{i=1}^{n} x_i z_i + b\sum_{i=1}^{n} y_i z_i = a\varphi(X,Z) + b\varphi(Y,Z)$$

$$\varphi(X, bY + cZ) = \sum_{i=1}^{n} x_i(by_i + cz_i) = b\sum_{i=1}^{n} x_i y_i + c\sum_{i=1}^{n} x_i z_i = b\varphi(X,Y) + c\varphi(X,Z)$$

注意，假若 V 是一個 vector space over a complex field 的時候，上述定義要稍微修正為，φ is a map which maps from $V \times V$ into C such that
1. $\varphi(a\alpha + b\beta, \gamma) = a\varphi(\alpha, \gamma) + b\varphi(\beta, \gamma)$, $\quad \forall \alpha, \beta, \gamma \in V$ and $a, b \in C$
2. $\varphi(\alpha, b\beta + c\gamma) = \bar{b}\varphi(\alpha, \beta) + \bar{c}\varphi(\alpha, \gamma)$, $\quad \forall \alpha, \beta, \gamma \in V$ and $b, c \in C$

我們稱其為雙線型，所指的是，φ is linear to the first coordinate and conjugate linear to the second coordinate。下面，我們同樣也看一個複數型態的例子。

Example 1

For $X = \begin{pmatrix} x_1 \\ x_2 \\ \cdots \\ x_n \end{pmatrix}, Y = \begin{pmatrix} y_1 \\ y_2 \\ \cdots \\ y_n \end{pmatrix} \in C^n$, if we define $\varphi(X,Y) = \sum_{i=1}^{n} x_i \bar{y}_i$, then φ is a bilinear form on C^n。

Proof

若，$X = \begin{pmatrix} x_1 \\ x_2 \\ \cdots \\ x_n \end{pmatrix}, Y = \begin{pmatrix} y_1 \\ y_2 \\ \cdots \\ y_n \end{pmatrix}, Z = \begin{pmatrix} z_1 \\ z_2 \\ \cdots \\ z_n \end{pmatrix} \in C^n$，且 $a,b,c \in C$，則

$$\varphi(aX+bY,Z) = \sum_{i=1}^{n}(ax_i+by_i)\bar{z}_i = a\sum_{i=1}^{n}x_i\bar{z}_i + b\sum_{i=1}^{n}y_i\bar{z}_i = a\varphi(X,Z) + b\varphi(Y,Z)$$

$$\varphi(X,bY+cZ) = \sum_{i=1}^{n}x_i\overline{(by_i+cz_i)} = \bar{b}\sum_{i=1}^{n}x_i\bar{y}_i + \bar{c}\sum_{i=1}^{n}x_i\bar{z}_i = \bar{b}\,\varphi(X,Y) + \bar{c}\,\varphi(X,Z)$$ ∎

很顯然的，假設 φ 是一個 bilinear form 的話，那麼對於 V 中的任意向量，X 以及 Y，

$$-\varphi(X,Y) = \varphi(-X,Y) = \varphi(X,-Y)$$

此外，由於，$\varphi(X,0) = \varphi(X,0+0) = 2\varphi(X,0)$

所以，$\varphi(X,0) = 0, \forall X \in V$

同理，$\varphi(0,Y) = 0, \forall Y \in V$

另外，在 complex case 當中，若 $X,Y \in V$ 且 $a \in C$，則

$$\varphi(aX,aY) = a\bar{a}\,\varphi(X,Y) = |a|^2 \varphi(X,Y)$$

Example 2

Let A be a real $n \times n$ matrix. Defining for $X = \begin{pmatrix} x_1 \\ \vdots \\ x_n \end{pmatrix}, Y = \begin{pmatrix} y_1 \\ \vdots \\ y_n \end{pmatrix} \in R^n$,

$\varphi_A(X,Y) = ({}^t X)AY$, then φ_A is a bilinear form on R^n.

Proof

For $a, b, c \in R$ and $X, Y, Z \in R^n$

$$\varphi_A(aX+bY, Z) = {}^t(aX+bY)AZ$$
$$= a({}^t X)AZ + b({}^t Y)AZ$$
$$= a\varphi(X,Z) + b\varphi(Y,Z)$$

$$\varphi_A(X, bY+cZ) = ({}^t X)A(bY+cZ)$$
$$= b({}^t X)AY + c({}^t X)AZ$$
$$= b\varphi(X,Y) + c\varphi(X,Z)$$

前述例子的結果純粹是來自於，矩陣乘法分配律的特性。我們再看看下面更一般化的例題。

Example 3

Let V be a vector space over R with a basis, $\{\alpha_1, \alpha_2, ..., \alpha_n\}$, and A a real $n \times n$ matrix. Defining for, $X = \sum_{i=1}^{n} x_i \alpha_i$, $Y = \sum_{i=1}^{n} y_i \alpha_i$,

$$\varphi_A(X,Y) = (x_1, x_2, ..., x_n) A \begin{pmatrix} y_1 \\ y_2 \\ ... \\ ... \\ y_n \end{pmatrix}$$

then φ_A is a bilinear form on V.

前面兩個例題清楚的告訴我們，一個 $n \times n$ 矩陣決定了一個雙線性型態。反過來說，一個雙線性型態是否也可以決定一個 $n \times n$ 矩陣呢？當然，我們所期待的答案就如同第四章一樣是肯定的。有關這一點，下面的定理會有說明。不過，還是請同學先來兩個 exercises。

Exercises

1. Show that if A is an $n \times n$ matrix over R and if $(^tX)AY = 0$, $\forall\ X, Y \in R^n$, then A has to be a zero matrix. (*Hint*: Consider the standard unit vectors.)

2. Show that if A and B are $n \times n$ matrices over R and if $(^tX)AY = (^tX)BY$, $\forall\ X, Y \in R^n$, then $A = B$.

Theorem 5.1

Let V be a vector space over R with a basis, $\{\alpha_1, \alpha_2, ..., \alpha_n\}$. If φ is a bilinear form on V, then there exists a real $n \times n$ matrix, A, such that for every $X = \sum_{i=1}^{n} x_i \alpha_i$, $Y = \sum_{i=1}^{n} y_i \alpha_i$,

$$\varphi(X,Y) = (x_1, x_2, ..., x_n) A \begin{pmatrix} y_1 \\ y_2 \\ \vdots \\ \vdots \\ y_n \end{pmatrix}$$

Proof

We first define for each $1 \le i, j \le n$

$$a_{ij} = \varphi(\alpha_i, \alpha_j)$$

and let

$$A = (a_{ij})$$

Then for $X = \sum_{i=1}^{n} x_i \alpha_i$, $Y = \sum_{i=1}^{n} y_i \alpha_i$ in V

$$\varphi(X,Y) = \varphi(\sum_{i=1}^{n} x_i \alpha_i, \sum_{j=1}^{n} y_j \alpha_j) = \sum_{i=1}^{n} x_i \varphi(\alpha_i, \sum_{j=1}^{n} y_j \alpha_j) = \sum_{i=1}^{n} x_i \sum_{j=1}^{n} y_j \varphi(\alpha_i, \alpha_j)$$

$$= \sum_{i=1}^{n}\sum_{j=1}^{n} x_i y_j \varphi(\alpha_i, \alpha_j) = (x_1, x_2, \ldots, x_n) A \begin{pmatrix} y_1 \\ y_2 \\ \ldots \\ \ldots \\ y_n \end{pmatrix}$$

定理中所得到的矩陣 A 被稱為 the "Gram matrix" of φ with respect to the basis, $\{\alpha_1, \alpha_2, \ldots, \alpha_n\}$。當然，同上一章的線型轉換一樣，不同的 basis 會產生出不同的 Gram matrix。有關這一點，想必讀者早已發現。不過，可以確定的是，一旦 basis 選定之後，Gram matrix 是唯一被決定的。這點可由前述 Exercise 2 得知，即

If $({}^t X)AY = ({}^t X)BY$, $\forall\ X, Y \in R^n$, then $A = B$

現在，我們令，$\{V, \alpha, \varphi\}$ 表示 the collection of all bilinear forms on V with respect to basis, α；$M_{n \times n}$ 表示 the collection of all real $n \times n$ matrices。那麼，$\{V, \alpha, \varphi\}$ 與 $M_{n \times n}$ 之間有一個 one-to-one and onto correspondence。

上面的定理以及前述的所有說法，在複數體系裡，也有完全相同的結果。只是要特別注意，在複數體系裡，我們的定義改寫為

$$\varphi(X, Y) = ({}^t X) A \overline{Y} = 0, \ \forall\ X, Y \in C^n$$

Example 4

Let $\varphi : R^n \times R^n \to R$ be defined by

$$\varphi(X, Y) = ({}^t X)Y = \sum_{i=1}^{n} x_i y_i, \quad \forall\ X = \begin{pmatrix} x_1 \\ x_2 \\ \ldots \\ x_n \end{pmatrix}, Y = \begin{pmatrix} y_1 \\ y_2 \\ \ldots \\ y_n \end{pmatrix} \in R^n$$

Find the Gram matrix of φ with respect to the standard basis.

Solution

Write $E = \{e_1, e_2, ..., e_n\}$ as the standard basis, then

$$a_{ij} = \varphi(e_i, e_j) = \begin{cases} 1, & \text{if } i = j \\ 0, & \text{if } i \neq j \end{cases}$$

So, the Gram matrix of φ with respect to the standard basis, E, is an identity matrix, $A = I_n$. ∎

Example 5

Let $\varphi : R^2 \times R^2 \to R$ be defined by

$$\varphi(X,Y) = x_1 y_1 + x_2 y_2, \quad X = \begin{pmatrix} x_1 \\ x_2 \end{pmatrix}, \quad Y = \begin{pmatrix} y_1 \\ y_2 \end{pmatrix} \in R^2$$

Let $\alpha = \{\alpha_1 = \begin{pmatrix} 2 \\ 1 \end{pmatrix}, \alpha_2 = \begin{pmatrix} 0 \\ 2 \end{pmatrix}\}$ be a basis of R^2.

1. Find the Gram matrix of φ with respect to the basis, α.

2. Use the result obtained from 1, to find $\varphi(\begin{pmatrix} 3 \\ -1 \end{pmatrix}, \begin{pmatrix} 1 \\ 2 \end{pmatrix})$.

Solution

$\varphi(\alpha_1, \alpha_1) = 5$, $\varphi(\alpha_1, \alpha_2) = 2$, $\varphi(\alpha_2, \alpha_1) = 2$, $\varphi(\alpha_2, \alpha_2) = 4$,

The Gram matrix is $\begin{pmatrix} 5 & 2 \\ 2 & 4 \end{pmatrix}$.

To find $\varphi(\begin{pmatrix} 3 \\ -1 \end{pmatrix}, \begin{pmatrix} 1 \\ 2 \end{pmatrix})$, we need to express $\begin{pmatrix} 3 \\ -1 \end{pmatrix}$ and $\begin{pmatrix} 1 \\ 2 \end{pmatrix}$ as linear combinations of the basis, α.

$$\begin{pmatrix} 3 \\ -1 \end{pmatrix} = \frac{3}{2}\begin{pmatrix} 2 \\ 1 \end{pmatrix} + (-\frac{5}{4})\begin{pmatrix} 0 \\ 2 \end{pmatrix} \quad \text{and} \quad \begin{pmatrix} 1 \\ 2 \end{pmatrix} = \frac{1}{2}\begin{pmatrix} 2 \\ 1 \end{pmatrix} + \frac{3}{4}\begin{pmatrix} 0 \\ 2 \end{pmatrix}$$

So, $\varphi(\begin{pmatrix} 3 \\ -1 \end{pmatrix}, \begin{pmatrix} 1 \\ 2 \end{pmatrix}) = (3/2 \quad -5/4) \begin{pmatrix} 5 & 2 \\ 2 & 4 \end{pmatrix} \begin{pmatrix} 1/2 \\ 3/4 \end{pmatrix} = 1$

(Check: $\varphi(\begin{pmatrix}3\\-1\end{pmatrix},\begin{pmatrix}1\\2\end{pmatrix})=3-2=1$, indeed, it is correct.) ∎

Example 6

Let $V=\{f:R\to R\,;\ f$ is a polynomial with degree $\leq 2\}$. Define, for $f,g\in V$, $\varphi(f,g)=\int_0^1 f(t)g(t)\,dt$.

a. Show that φ is a bilinear form on V.

b. Find the Gram matrix of φ with respect to the basis, $\{1,t,t^2\}$.

c. Compute $\int_0^1(2t^2-t+6)(4t^2+t-7)\,dt$ by using the Gram matrix obtained from b.

Solution

a. The proof that φ is a bilinear form on V is quite easy, and we leave it to the readers.

b. Let $\alpha_1=1$, $\alpha_2=t$, $\alpha_3=t^2$. Then

$\varphi(\alpha_1,\alpha_1)=\int_0^1 1\,dt=1,\quad \varphi(\alpha_1,\alpha_2)=\int_0^1 t\,dt=\dfrac{1}{2},\quad \varphi(\alpha_1,\alpha_3)=\int_0^1 t^2\,dt=\dfrac{1}{3}$

$\varphi(\alpha_2,\alpha_1)=\int_0^1 t\,dt=\dfrac{1}{2},\quad \varphi(\alpha_2,\alpha_2)=\int_0^1 t^2\,dt=\dfrac{1}{3},\quad \varphi(\alpha_2,\alpha_3)=\int_0^1 t^3\,dt=\dfrac{1}{4}$

$\varphi(\alpha_3,\alpha_1)=\int_0^1 t^2\,dt=\dfrac{1}{3},\quad \varphi(\alpha_3,\alpha_2)=\int_0^1 t^3\,dt=\dfrac{1}{4},\quad \varphi(\alpha_3,\alpha_3)=\int_0^1 t^4\,dt=\dfrac{1}{5}$

So the Gram matrix of φ with respect to the basis, $\{1,t,t^2\}$, is

$$\begin{pmatrix}1 & 1/2 & 1/3\\ 1/2 & 1/3 & 1/4\\ 1/3 & 1/4 & 1/5\end{pmatrix}$$

c. Writing $f(t)=6-t+2t^2$ and $g(t)=-7+t+4t^2$, then

$$\varphi(f,g)=(6,-1,2)\begin{pmatrix}1 & 1/2 & 1/3\\ 1/2 & 1/3 & 1/4\\ 1/3 & 1/4 & 1/5\end{pmatrix}\begin{pmatrix}-7\\1\\4\end{pmatrix}=157/5$$

所以，$\int_0^1 (2t^2 - t + 6)(4t^2 + t - 7)dt = 157/5$.

Definition 5.2

Let V be a vector space over R. A bilinear form, φ, on V is said to be "symmetric", if for every $X, Y \in V$, $\varphi(X, Y) = \varphi(Y, X)$.

譬如說，前面的兩個例子，

$$\varphi(X,Y) = ({}^t X)Y = \sum_{i=1}^n x_i y_i \quad \text{以及} \quad \varphi(f,g) = \int_0^1 f(t)g(t)\, dt$$

都是對稱的雙線型。

注意：

若 V 是一個 vector space over complex field 的話，我們則用 "hermitian" 這個字眼。即，φ is said to be hermitian, if

$$\varphi(X,Y) = \overline{\varphi(Y,X)}, \quad \forall\, X, Y \in V$$

譬如說，在先前我們曾經看過的例子當中

$$\varphi(X,Y) = \sum_{i=1}^n x_i \bar{y}_i, \quad \forall\, X, Y \in V$$

就是一個 hermitian bilinear form。

因為

$$\varphi(X,Y) = \sum_{i=1}^n x_i \bar{y}_i = \overline{\sum_{i=1}^n \bar{x}_i y_i} = \overline{\sum_{i=1}^n y_i \bar{x}_i} = \overline{\varphi(Y,X)}$$

Example 7

Let $V = \{f : [0, 2\pi] \to C\,;\, f \text{ is continuous}\}$. Define, for $\forall\, f, g \in V$, $\varphi(f,g) = \int_0^{2\pi} f(t)\overline{g(t)}\, dt$, then φ is hermitian.

Solution

For $f, g, h \in V$ and $a, b, c \in C$,

$$\varphi(af+bg, h) = \int_0^{2\pi} (af+bg)(t)\overline{h(t)}\,dt$$

$$= a\int_0^{2\pi} f(t)\overline{h(t)}\,dt + b\int_0^{2\pi} g(t)\overline{h(t)}\,dt$$

$$= a\varphi(f,h) + b\varphi(g,h)$$

$$\varphi(f, bg+ch) = \int_0^{2\pi} f(t)\overline{(bg+ch)(t)}\,dt$$

$$= \bar{b}\int_0^{2\pi} f(t)\overline{g(t)}\,dt + \bar{c}\int_0^{2\pi} f(t)\overline{h(t)}\,dt$$

$$= \bar{b}\varphi(f,g) + \bar{c}\varphi(f,h)$$

So, φ is a bilinear form on V. To show, φ is hermitian, we let $f, g \in V$. Then

$$\varphi(f,g) = \int_0^{2\pi} f(t)\overline{g(t)}\,dt = \overline{\int_0^{2\pi} \overline{f(t)}g(t)\,dt}$$

$$= \overline{\varphi(g,f)}$$

So, φ is hermitian. ∎

其實，無論 symmetric 也好，hermitian 也罷；一個雙線性型與它所代表的 Gram matrix 有著絕對密不可分的關係。

Theorem 5.2

> Let V be a vector space over R with basis, $\{\alpha_1, \alpha_2, ..., \alpha_n\}$. Suppose that φ is a bilinear form on V and A is the $n \times n$ "Gram matrix" of φ with respect to $\{\alpha_1, \alpha_2, ..., \alpha_n\}$, then φ is a symmetric bilinear form if and only if A is a symmetric matrix.

Proof

This theorem can be proved by considering that for $X = \sum_{i=1}^{n} x_i \alpha_i$, $Y = \sum_{i=1}^{n} y_i \alpha_i$,

$$\varphi(X,Y) = (x_1,x_2,...,x_n)A\begin{pmatrix}y_1\\y_2\\...\\y_n\end{pmatrix} = {}^t[(x_1,x_2,...,x_n)A\begin{pmatrix}y_1\\y_2\\...\\y_n\end{pmatrix}]$$

$$= (y_1,y_2,...,y_n)({}^tA)\begin{pmatrix}x_1\\x_2\\...\\x_n\end{pmatrix}$$

Now, if A is symmetric, then the last equation equals $\varphi(Y,X)$.

注意,前述證明中, $(x_1,x_2,...,x_n)A\begin{pmatrix}y_1\\y_2\\...\\y_n\end{pmatrix} = {}^t[(x_1,x_2,...,x_n)A\begin{pmatrix}y_1\\y_2\\...\\y_n\end{pmatrix}]$,只因為,

$(x_1,x_2,...,x_n)A\begin{pmatrix}y_1\\y_2\\...\\y_n\end{pmatrix}$ 是一個 scalar (一個 1×1 矩陣)。

To the contrary, we simply write

$$X = \begin{pmatrix}x_1\\x_2\\...\\x_n\end{pmatrix}, \quad Y = \begin{pmatrix}y_1\\y_2\\...\\y_n\end{pmatrix},$$

and suppose that $\varphi(X,Y) = \varphi(Y,X)$, then

$$({}^tX)AY = ({}^tY)AX = {}^t(({}^tY)AX) = ({}^tX)({}^tA)Y.$$

And, this is true for $\forall\ X,Y$. Thus, $^tA = A$. (Exercise 2, page 135)

同樣的，在 complex case 中，前述定理則應該被改寫為

A bilinear form, φ, is hermitian if and only if the corresponding $n \times n$ matrix, A, has the property, $A = {}^t\overline{A}$.

我們通常使用符號，A^* 表示 $^t\overline{A}$，並且將其稱為，the adjoint of A (A 的伴隨)。而且，當 $A^* = A$ 時，我們說，A is hermitian or self-adjoint。(有關這些，前面已經提過。) 因此，我們可以將前面敘述改寫為，

A bilinear form, φ, is hermitian if and only if the corresponding $n \times n$ Gram matrix, A, is hermitian.

Exercises

1. Which of the following maps, φ, defined on $R^2 \times R^2$ are bilinear forms, for every
$$X = \begin{pmatrix} x_1 \\ x_2 \end{pmatrix}, Y = \begin{pmatrix} y_1 \\ y_2 \end{pmatrix} \in R^2$$
   a. $\varphi(X,Y) = 1$
   b. $\varphi(X,Y) = (x_1 - y_1)^2 + x_2 y_2$
   c. $\varphi(X,Y) = (x_1 + y_1)^2 - (x_1 - y_1)^2$
   d. $\varphi(X,Y) = x_1 y_2 - x_2 y_1$

2. Define, for $\forall\ X = \begin{pmatrix} x_1 \\ x_2 \end{pmatrix}, Y = \begin{pmatrix} y_1 \\ y_2 \end{pmatrix} \in R^2$, $\varphi(X,Y) = ({}^tX)Y$.

    Find the Gram matrix of φ with respect to the basis, $\{\begin{pmatrix} 1 \\ -1 \end{pmatrix}, \begin{pmatrix} 1 \\ 1 \end{pmatrix}\}$. And use this to find $\varphi(\begin{pmatrix} 1 \\ 3 \end{pmatrix}, \begin{pmatrix} 1 \\ 2 \end{pmatrix})$.

3. Define, for $\forall\ X = \begin{pmatrix} x_1 \\ x_2 \\ x_3 \end{pmatrix}, Y = \begin{pmatrix} y_1 \\ y_2 \\ y_3 \end{pmatrix} \in R^3$, $\varphi(X,Y) = ({}^tX)Y$.

Chapter Five Bilinear Forms and Inner Products   143

Find the Gram matrix of φ with respect to the basis, $\{\begin{pmatrix}1\\0\\0\end{pmatrix}, \begin{pmatrix}1\\1\\1\end{pmatrix}, \begin{pmatrix}-1\\1\\0\end{pmatrix}\}$. And use this to

find $\varphi(\begin{pmatrix}2\\-1\\0\end{pmatrix}, \begin{pmatrix}1\\1\\2\end{pmatrix})$.

4. Define, for $\forall\ X = \begin{pmatrix}x_1\\x_2\end{pmatrix}, Y = \begin{pmatrix}y_1\\y_2\end{pmatrix} \in C^2$, $\varphi(X,Y) = ({}^tX)\overline{Y}$.

Find the Gram matrix of φ with respect to the basis, $\{\begin{pmatrix}i\\1\end{pmatrix}, \begin{pmatrix}0\\2i\end{pmatrix}\}$. And use this to find

$\varphi(\begin{pmatrix}i\\2\end{pmatrix}, \begin{pmatrix}2\\3i\end{pmatrix})$.

5. Show that if φ is a hermitian bilinear form on C^n, then

$$\varphi(X,X) \in R,\ \forall\ X \in C^n$$

6. Let $V = \{$all real $n \times n$ matrices$\}$, and define, $\forall\ A \in V$, $trace(A) = \sum_{i=1}^{n} a_{ii}$. Show

that if, for $A, B \in V$, $\varphi(A,B) = trace(AB)$, then φ is a symmetric bilinear form.
(Hint: Show, $trace(AB) = trace(BA)$)

7. Let $V = \{$all complex $n \times n$ matrices$\}$. Define, for $A, B \in V$, $\varphi(A,B) = trace(B^*A)$, then φ is a hermitian bilinear form.
(Hint: Show, $\overline{trace(B^*A)} = trace(A^*B)$)

§5.2 Quadratic forms

假設 φ 是一個在 R^2 上被定義如下的 bilinear form

$$\varphi(X,Y) = ({}^tX)\begin{pmatrix} 2 & 0 \\ 1 & -1 \end{pmatrix}Y, \quad \forall X = \begin{pmatrix} x_1 \\ x_2 \end{pmatrix}, Y = \begin{pmatrix} y_1 \\ y_2 \end{pmatrix} \in R^2$$

我們看看，當 $X = Y$ 的時候，

$$\varphi(X,X) = \begin{pmatrix} x_1 & x_2 \end{pmatrix}\begin{pmatrix} 2 & 0 \\ 1 & -1 \end{pmatrix}\begin{pmatrix} x_1 \\ x_2 \end{pmatrix} = 2x_1^2 + x_1x_2 - x_2^2$$

它是一個二次式。這個東西正是，我們所感興趣要和大家一起研讀的、所謂的 quadratic forms。

Definition 5.3

> Let V be a vector space over R and φ a bilinear form on V. A "quadratic form", q, from V into R is defined by
> $$q(\alpha) = \varphi(\alpha,\alpha), \quad \forall \alpha \in V$$

正確的來說，我們將其說成，q is a quadratic form induced by the bilinear form, φ。注意，在複數體系上而言，q is a quadratic form from V into complex field, C。而且，請讀者記得，當 φ 是一個 hermitian bilinear form 時，二次式，q，would always be real-valued (參看 5.1 節，最後的 Exercise 5)。

Example 1

Given A as below, and define $\varphi(\alpha,\beta) = ({}^t\alpha)A\overline{\beta}$, $\forall \alpha,\beta \in C^2$. Find $q(\alpha)$ for each of the following case.

1. $A = \begin{pmatrix} i & 0 \\ i & 0 \end{pmatrix}$, $\alpha = \begin{pmatrix} i \\ 1 \end{pmatrix}$

2. $A = \begin{pmatrix} 1 & -i \\ i & -1 \end{pmatrix}, \alpha = \begin{pmatrix} x_1 \\ x_2 \end{pmatrix}$

3. $A = \begin{pmatrix} 1 & 0 \\ 0 & 2 \end{pmatrix}, \alpha = \begin{pmatrix} x_1 \\ x_2 \end{pmatrix}$

Solution

1. $q(\alpha) = (i, 1) \begin{pmatrix} i & 0 \\ i & 0 \end{pmatrix} \overline{\begin{pmatrix} i \\ 1 \end{pmatrix}} = (-1+i, 0) \overline{\begin{pmatrix} i \\ 1 \end{pmatrix}} = 1+i$

2. $q(\alpha) = (x_1, x_2) \begin{pmatrix} 1 & -i \\ i & -1 \end{pmatrix} \overline{\begin{pmatrix} x_1 \\ x_2 \end{pmatrix}} = (x_1 + ix_2, -ix_1 - x_2) \overline{\begin{pmatrix} x_1 \\ x_2 \end{pmatrix}}$
$= |x_1|^2 + i\overline{x_1}x_2 - ix_1\overline{x_2} - |x_2|^2 = |x_1|^2 + 2\operatorname{Re}(i\overline{x_1}x_2) - |x_2|^2$

3. $q(\alpha) = (x_1, x_2) \begin{pmatrix} 1 & 0 \\ 0 & 2 \end{pmatrix} \overline{\begin{pmatrix} x_1 \\ x_2 \end{pmatrix}} = (x_1, 2x_2) \overline{\begin{pmatrix} x_1 \\ x_2 \end{pmatrix}} = |x_1|^2 + 2|x_2|^2$ ∎

上述例題中之 2. 和 3.，由於矩陣 A 是 hermitian，所以我們發現它們的二次式皆為實數。不僅如此，我們還發現 3. 中的二次式永遠大於或等於零。這種情形是最能吸引我們特別注意的地方。

Definition 5.4

> A quadratic form, q, from a vector space, V, into R (or C) is said to be
> 1. "positive semi-definite", if $q(\alpha) \geq 0$, $\forall\ \alpha \in V$.
> 2. "positive definite", if $q(\alpha) > 0$, $\forall\ \alpha \in V, \alpha \neq 0$.

很顯然的，假設 q 是 positive semi-definite，而且若 $q(\alpha) = 0$，則 $\alpha = 0$ 的話。那麼，q 是 positive definite。又若，$-q$ 是 positive definite 的話，我們則將 q 稱為 negative definite。至於，negative semi-definite，我們也做相同的規定。有一種情況被稱為，non-definite；那就是，既存在一個 α，使得 $q(\alpha) > 0$，又存在一個 β，使得 $q(\beta) < 0$。

Example 2

Given a bilinear form, $\varphi: R^n \times R^n \to R$, defined by

$$\varphi(X,Y) = \sum_{i=1}^{n} x_i y_i, \ \forall \ X = \begin{pmatrix} x_1 \\ x_2 \\ \cdots \\ x_n \end{pmatrix}, Y = \begin{pmatrix} y_1 \\ y_2 \\ \cdots \\ y_n \end{pmatrix} \in R^n$$

The quadratic form, q, induced by φ is positive definite.

Proof

We shall prove, $q(X) > 0, \forall X \in R^n, X \neq 0$. And this can be seen immediately, since, for $X = \begin{pmatrix} x_1 \\ x_2 \\ \cdots \\ x_n \end{pmatrix} \neq 0$,

$$q(X) = \sum_{i=1}^{n} x_i^2 > 0$$ ∎

在複數體系裡，我們也有相類同之結果。看看下一個例子。

Example 3

Given a bilinear form, $\varphi: C^n \times C^n \to C$, defined by

$$\varphi(X,Y) = \sum_{i=1}^{n} x_i \overline{y_i} \ , \quad \forall \ X = \begin{pmatrix} x_1 \\ x_2 \\ \cdots \\ x_n \end{pmatrix}, Y = \begin{pmatrix} y_1 \\ y_2 \\ \cdots \\ y_n \end{pmatrix} \in C^n$$

The quadratic form, q, induced by φ is positive definite.

Proof

同樣的，若 $X = \begin{pmatrix} x_1 \\ x_2 \\ \cdots \\ x_n \end{pmatrix} \neq 0$ in C^n，則

$$q(X) = \sum_{i=1}^{n} x_i \bar{x}_i = \sum_{i=1}^{n} |x_i|^2 > 0$$

所以，q is positive definite. ∎

前面所講的是，一個雙線性型態其二次式的 positive definite。根據前頭介紹，一個 $n \times n$ 矩陣可以導出一個雙線性型態。所以，我們也可以對一個 $n \times n$ 矩陣，做 positive definite 相類似的定義。

Definition 5.5

> An $n \times n$ real matrix, A, is said to be
> 1. "positive definite", if the quadratic form, $q: R^n \to R$, defined by
> $q(X) = (^tX)AX$, $\forall\ X \in R^n$ is positive definite.
> 2. "positive semi-definite", if the quadratic form, $q: R^n \to R$, defined by
> $q(X) = (^tX)AX$, $\forall\ X \in R^n$, is positive semi-definite.

當然，在 complex case 當中，如同前面一樣，我們將條件，$q(X) = (^tX)AX$，$\forall\ X \in R^n$，改為 $q(X) = (^tX)A\bar{X}$，$\forall\ X \in C^n$。這個 real and complex 之間的區別，相信大家已經非常熟悉了。

到第七章，當我們介紹 eigenvalues 的時候，我們將說明，判定一個矩陣是否為 positive definite 的簡單方法。那就是，An $n \times n$ symmetric (hermitian) matrix is positive definite if and only if all its eigenvalues are positive. 目前無論如何，我們還是可以用直接計算的方式，來練習判定比較小型的矩陣，是否為 positive definite。

Example 4

Show that real matrices, $\begin{pmatrix} 2 & 1 \\ 1 & 1 \end{pmatrix}$ and $\begin{pmatrix} 2 & 0 & 1 \\ 1 & 2 & 1 \\ 1 & 1 & 2 \end{pmatrix}$ are positive definite.

Proof

For $\begin{pmatrix} x \\ y \end{pmatrix} \in R^2$ and $\begin{pmatrix} x \\ y \end{pmatrix} \neq 0$,

$$\begin{pmatrix} x & y \end{pmatrix} \begin{pmatrix} 2 & 1 \\ 1 & 1 \end{pmatrix} \begin{pmatrix} x \\ y \end{pmatrix} = \begin{pmatrix} 2x+y & x+y \end{pmatrix} \begin{pmatrix} x \\ y \end{pmatrix} = 2x^2 + xy + xy + y^2$$
$$= x^2 + (x+y)^2 > 0$$

For $\begin{pmatrix} x \\ y \\ z \end{pmatrix} \in R^3$ and $\begin{pmatrix} x \\ y \\ z \end{pmatrix} \neq 0$,

$$\begin{pmatrix} x & y & z \end{pmatrix} \begin{pmatrix} 2 & 0 & 1 \\ 1 & 2 & 1 \\ 1 & 1 & 2 \end{pmatrix} \begin{pmatrix} x \\ y \\ z \end{pmatrix} = 2x^2 + xy + 2xz + 2y^2 + 2yz + 2z^2$$
$$= (x+z)^2 + (y+z)^2 + (x^2 + xy + y^2) > 0$$

注意，前面式子最後，$x^2 + y^2 + xy \geq |xy| + xy \geq 0$。∎

Exercises

1. Let A be an $n \times n$ real matrix. Show that if A is positive definite, then A is non-singular. (*Hint*: 參考第三章，證明 $NS(A) = \{0\}$)

2. Let A be an $n \times n$ real matrix. Show that there exists a symmetric matrix, B, such that $({}^tX)AX = ({}^tX)BX$, $\forall\ X \in R^n$. (*Hint*: consider, the matrix, $\dfrac{A + {}^tA}{2}$)

在多變數函數的極值問題上，矩陣正定則的應用是，一件值得我們共同欣賞的結尾。我們來看看

Functions mapping from R^2 into R 的特殊案例。至於,一般的 case,Functions mapping from R^n into R,讀者可以類推。

Let f be a real-valued function defined on R^2 such that all its 2nd partial derivatives are continuous. We all see that the 2nd degree Taylor polynomial of f at the point (a,b) is given by

$$f(x,y) = f(a,b) + \nabla f(a,b) \cdot (x-a, y-b) + (x-a, y-b) \begin{pmatrix} \frac{\partial^2 f}{\partial x^2} & \frac{\partial^2 f}{\partial x \partial y} \\ \frac{\partial^2 f}{\partial y \partial x} & \frac{\partial^2 f}{\partial y^2} \end{pmatrix}_{(a,b)} \begin{pmatrix} x-a \\ y-b \end{pmatrix} + R_2$$

where R_2 is the Taylor's remainder which is very small, provided that (x,y) is close to (a,b).

Now, if $\nabla f(a,b) = 0$, (that means (a,b) is a critical point of f) then

$$f(x,y) - f(a,b) = (x-a, y-b) \begin{pmatrix} \frac{\partial^2 f}{\partial x^2} & \frac{\partial^2 f}{\partial x \partial y} \\ \frac{\partial^2 f}{\partial y \partial x} & \frac{\partial^2 f}{\partial y^2} \end{pmatrix}_{(a,b)} \begin{pmatrix} x-a \\ y-b \end{pmatrix} + R_2$$

現在,假設 $A = \begin{pmatrix} \frac{\partial^2 f}{\partial x^2} & \frac{\partial^2 f}{\partial x \partial y} \\ \frac{\partial^2 f}{\partial y \partial x} & \frac{\partial^2 f}{\partial y^2} \end{pmatrix}_{(a,b)}$ 是 positive definite 的話,那麼 $f(x,y) - f(a,b) > 0$,for every (x,y) near (a,b)。因此,$f(a,b)$ 為一局部極小。又若 A 是 negative definite 的話,那麼 $f(x,y) - f(a,b) < 0$,for every (x,y) near (a,b)。因此,$f(a,b)$ 為一局部極大。

§5.3 Inner products

一個比較特殊的雙線性型態，就是 symmetric (or Hermitian) and positive definite 的型態。實際上，它這就是本節所將要介紹的，**內積運算** (inner product) 或稱為**純量運算** (scalar products)。

Definition 5.6

> A symmetric bilinear form, φ, on a real vector space, V, is called an "inner product" on V, if the quadratic form, $q(\alpha) = \varphi(\alpha, \alpha)$, $\forall \alpha \in V$, is positive definite.

譬如說，
若 $V = R^n$，且定義

$$\varphi(X,Y) = \sum_{i=1}^{n} x_i y_i \quad \forall\ X = \begin{pmatrix} x_1 \\ \cdots \\ \cdots \\ x_n \end{pmatrix}, Y = \begin{pmatrix} y_1 \\ \cdots \\ \cdots \\ y_n \end{pmatrix} \in V$$

則，由 5.1，5.2 節我們發現，φ is an inner product on V。

由於，內積的值域是一個純量，所以我們亦將其稱為 "scalar product"。另外，假若 φ is an inner product on V 的話，往後我們將簡單以符號，$\langle X, Y \rangle$，替代 $\varphi(X, Y)$，用以表示 X 與 Y 之內積。其實，仔細的看一下，我們也可以將內積的定義重新敘述如下。

An inner product, $\langle\ ,\ \rangle$, on V is precisely a map from $V \times V$ into R that satisfies
1. $\langle a\alpha + b\beta, \gamma \rangle = a \langle \alpha, \gamma \rangle + b \langle \beta, \gamma \rangle$, $\forall \alpha, \beta, \gamma \in V$ and $a, b \in R$.
2. $\langle \alpha, b\beta + c\gamma \rangle = b \langle \alpha, \beta \rangle + c \langle \alpha, \gamma \rangle$, $\forall \alpha, \beta, \gamma \in V$ and $b, c \in R$.
3. $\langle \alpha, \beta \rangle = \langle \beta, \alpha \rangle$, $\forall \alpha, \beta \in V$.
4. $\langle \alpha, \alpha \rangle \geq 0$, $\forall \alpha \in V$ and $\langle \alpha, \alpha \rangle > 0$, if $\alpha \neq 0$.

前述 1.與 2.代表著 $\langle,\rangle$ 是一個 bilinear form，第 3.代表著 $\langle,\rangle$ 是 symmetric，第 4. 則說明 $\langle,\rangle$ 是 positive definite。再看看，5.1 節我們曾經大概提過的一個例子。

Exercise

Let $V = \{ f : [0,1] \to R ; f$ is continuous $\}$.(In general, the collection is always denoted by $C[0,1]$). Defining, for $f, g \in V$, $\langle f, g \rangle = \int_0^1 f(t)g(t)\,dt$, then $\langle,\rangle$ is an inner product on V.

當然啦，以 complex case 而言，上述定義則應該更改為，

A hermitian bilinear form, φ, on a complex vector space, V, is called an "inner product" on V, if the quadratic form, $q(\alpha) = \varphi(\alpha, \alpha)$, $\forall \alpha \in V$, is positive definite.

或者說，將上述之第 2.以及第 3.的敘述更改為

2. $\langle \alpha, b\beta + c\gamma \rangle = \overline{b} \langle \alpha, \beta \rangle + \overline{c} \langle \alpha, \gamma \rangle$, $\forall \alpha, \beta, \gamma \in V$ and $b, c \in C$.

3. $\langle \alpha, \beta \rangle = \overline{\langle \beta, \alpha \rangle}$, $\forall \alpha, \beta \in V$

附註：一個向量空間 V，當它被賦予一個 inner product $\langle\,,\,\rangle$ 時，我們稱 $\{V; \langle,\rangle\}$ 為一內積空間 (inner product space)。同樣的，在前面所提過的例子當中，

若，$V = C^n$，且定義

$$\langle X, Y \rangle = ({}^t X)\overline{Y} = \sum_{i=1}^n x_i \overline{y}_i, \quad \forall\; X = \begin{pmatrix} x_1 \\ \cdots \\ \cdots \\ x_n \end{pmatrix},\; Y = \begin{pmatrix} y_1 \\ \cdots \\ \cdots \\ y_n \end{pmatrix} \in C^n$$

那麼，$\{V; \langle,\rangle\}$ 是一個 inner product space。

前面分別在 R^n 以及 C^n 中所介紹的一個實數、一個複數的例子，一般

而言，我們都將它們稱爲**標準內積** (standard inner product)。另外，假若 $\langle\,,\,\rangle$ 是 semi-positive definite 的話，那麼它所定義出來的純量，我們將其稱爲 semi-inner product，關於這一點，姑且略過不談。下面，我們再看一個 inner product 的例子。

Example 1

Let $V = \{\, f : [0,1] \to C \;;\; f \text{ is continuous} \,\}$. Defining, for $f, g \in V$, $\langle f, g \rangle = \int_0^1 f(t)\overline{g(t)}\,dt$, then $\langle\,,\,\rangle$ is an inner product on V.

Proof

Consider that
$$\langle f, g \rangle = \int_0^1 f(t)\overline{g(t)}\,dt = \overline{\int_0^1 \overline{f(t)}g(t)\,dt} = \overline{\langle g, f \rangle}$$
We see, $\langle\,,\,\rangle$ is Hermitian.
And that
$$\langle f, f \rangle = \int_0^1 f(t)\overline{f(t)}\,dt = \int_0^1 |f(t)|^2\,dt > 0,\ \forall\ \text{non-zero continuous function},\ f.$$
and $\int_0^1 |f(t)|^2\,dt = 0$, if $f \equiv 0$ on $[0,1]$.
so $\langle\,,\,\rangle$ is positive definite.
對於，檢定其爲一雙線型態，留給讀者當作 exercise。∎

Exercise

Let $V = \{\, f : [0, \infty) \to R \;;\; f \text{ is continuous} \,\}$. Show that
$$\langle f, g \rangle = \int_0^\infty f(x)g(x)e^{-x}\,dx,\ \forall\ f, g \in V \text{ is an inner product.}$$

談到這兒，我們順便和讀者共同研讀，分析學裡常使用的一種度量，叫做 **norm**（模），以及它的一些特性。

Definition 5.7

Given an inner product space, $\{V;\langle\,,\,\rangle\}$. The "norm", $\|\alpha\|$, of $\alpha \in V$ is defined to be $\|\alpha\| = \langle \alpha, \alpha \rangle^{1/2}$.

譬如說,

$$\langle X, Y \rangle = \sum_{i=1}^{n} x_i y_i, \quad \forall\ X, Y \in V = R^n$$ is an inner product on R^n. So, the norm of each vector, $X \in R^n$, is $\|X\| = \sqrt{\sum_{i=1}^{n} x_i^2}$.

注意,這正是大家所熟悉的,傳統的歐幾里得空間長度。又,For $V = C^n$, and $\langle X, Y \rangle = ({}^t X)\overline{Y} = \sum_{i=1}^{n} x_i \overline{y_i}$, $\forall\ X, Y \in C^n$, the norm of $X \in V$ is then

$$\|X\| = \sqrt{\sum_{i=1}^{n} x_i \overline{x_i}} = \sqrt{\sum_{i=1}^{n} |x_i|^2}$$

當然,有關連續函數部分,也讓我們來看一看。

For $V = C[0,1]$ and $\langle f, g \rangle = \int_0^1 f(t)\overline{g(t)}\,dt$, the norm of $f \in V$ is then

$$\|f\| = (\int_0^1 |f(t)|^2\, dt)^{1/2}$$

兩件事情補充說明一下。由於,$\langle \alpha, \alpha \rangle \geq 0, \forall \alpha$,所以,$\|\alpha\| \geq 0$。而且,由於,$\langle \alpha, \alpha \rangle = 0$ implies $\alpha = 0$。所以,$\|\alpha\| = 0$ if and only if $\alpha = 0$。此外,有關 norm 還有下列更多的特性,讓我們以定理和大家介紹。

Theorem 5.3

Let $\langle\,,\,\rangle$ be an inner product on a vector space, V. The norm, $\|\alpha\|$, defined by $\|\alpha\| = \langle \alpha, \alpha \rangle^{1/2}$, has the following properties.
1. $\|c\alpha\| = |c|\|\alpha\|$, $\forall\ \alpha \in V$ and a scalar c.

2. $|\langle \alpha, \beta \rangle| \leq \|\alpha\|\|\beta\|$, $\forall \alpha, \beta \in V$. (Cauchy-Bunyakowsky-Schwarz inequality)
3. $\|\alpha + \beta\| \leq \|\alpha\| + \|\beta\|$, $\forall \alpha, \beta \in V$. (Minkowski's inequality)
4. $\|\alpha + \beta\|^2 + \|\alpha - \beta\|^2 = 2\|\alpha\|^2 + 2\|\beta\|^2$, $\forall \alpha, \beta \in V$. (parallelogram law)

Proof

1. Consider that (we prove for the complex case)
$$\|c\alpha\|^2 = \langle c\alpha, c\alpha \rangle = c\bar{c} \langle \alpha, \alpha \rangle = |c|^2 \|\alpha\|^2 \text{, so}$$
$$\|c\alpha\| = |c|\|\alpha\|$$

2. If either $\alpha = 0$ or $\beta = 0$, then the result is clear. And, if $\alpha = c\beta$, for some scalar c, then $\langle \alpha, \beta \rangle = \langle c\beta, \beta \rangle = c\|\beta\|^2$ and hence, $|\langle \alpha, \beta \rangle| = |c|\|\beta\|\|\beta\| = \|\alpha\|\|\beta\|$. Now, for $\alpha \neq 0$, $\beta \neq 0$ and $\alpha \neq c\beta$, we put $d = \dfrac{\langle \alpha, \beta \rangle}{\langle \beta, \beta \rangle}$, then
$$\langle \alpha - d\beta, \beta \rangle = \langle \alpha, \beta \rangle - d \langle \beta, \beta \rangle = 0 \text{, and thus}$$
$$\langle \alpha - d\beta, \alpha - d\beta \rangle = \langle \alpha - d\beta, \alpha \rangle = \langle \alpha, \alpha \rangle - d \langle \beta, \alpha \rangle$$
Since $\langle \alpha - d\beta, \alpha - d\beta \rangle \geq 0$, $\langle \alpha, \alpha \rangle - d \langle \beta, \alpha \rangle \geq 0$. Hence,
$$\langle \alpha, \alpha \rangle \geq d \langle \beta, \alpha \rangle = \dfrac{\langle \alpha, \beta \rangle}{\langle \beta, \beta \rangle} \langle \beta, \alpha \rangle$$
$$\Rightarrow \quad \langle \alpha, \alpha \rangle \langle \beta, \beta \rangle \geq |\langle \alpha, \beta \rangle|^2$$
$$\Rightarrow \quad \|\alpha\|\|\beta\| \geq |\langle \alpha, \beta \rangle|$$
This proved Cauchy-Bunyakowsky-Schwarz inequality.

3. Consider that
$$\|\alpha + \beta\|^2 = \langle \alpha + \beta, \alpha + \beta \rangle$$
$$= \langle \alpha, \alpha \rangle + \langle \alpha, \beta \rangle + \langle \beta, \alpha \rangle + \langle \beta, \beta \rangle$$
$$= \langle \alpha, \alpha \rangle + \langle \alpha, \beta \rangle + \overline{\langle \alpha, \beta \rangle} + \langle \beta, \beta \rangle$$
$$= \langle \alpha, \alpha \rangle + 2\text{Re}\langle \alpha, \beta \rangle + \langle \beta, \beta \rangle$$
$$\leq \|\alpha\|^2 + 2|\langle \alpha, \beta \rangle| + \|\beta\|^2$$
$$\leq \|\alpha\|^2 + 2\|\alpha\|\|\beta\| + \|\beta\|^2 = (\|\alpha\| + \|\beta\|)^2,$$
we see that $\|\alpha + \beta\| \leq \|\alpha\| + \|\beta\|$.

4. 平行四邊形定律的證明較為簡單，由上面的證明中得知

$$\|\alpha + \beta\|^2 + \|\alpha - \beta\|^2$$
$$= \|\alpha\|^2 + 2\operatorname{Re}\langle \alpha, \beta \rangle + \|\beta\|^2 + \|\alpha\|^2 - 2\operatorname{Re}\langle \alpha, \beta \rangle + \|\beta\|^2$$
$$= 2\|\alpha\|^2 + 2\|\beta\|^2$$

談到 inner product，必然會提起向量的正交性，或者俗稱的垂直性。下面先請同學練習兩個簡單的習題。習題當中，所提及的 $\langle \alpha_0, \beta \rangle = 0$ 或是 $\langle \alpha_i, \alpha_j \rangle = 0$, $\forall i \neq j$，即所謂的 "正交"。

Exercises

1. Given an inner product space, $\{V; \langle\, ,\, \rangle\}$, and a vector, $\beta \in V$. Show that if $\langle \beta, \alpha \rangle = 0$, $\forall \alpha \in V$, then $\beta = 0$. (*Hint*: consider, $\alpha = \beta$)

2. Let $\{V; \langle\, ,\, \rangle\}$ be an inner product space with a basis, $\{\alpha_1, \alpha_2, ..., \alpha_n\}$, such that $\langle \alpha_i, \alpha_j \rangle = 0$, $\forall i \neq j$. If $c_1, c_2, ..., c_n$ are any n scalars, show that there exists exactly one $\alpha \in V$ such that $\langle \alpha, \alpha_i \rangle = c_i$, $\forall i = 1, 2, ..., n$. (*Hint*: consider the vector, $\alpha = \sum_{i=1}^{n} c_i \alpha_i$.)

Definition 5.8

Let $\{V; \langle\, ,\, \rangle\}$ be an inner product space. Two vectors, α, β in V are said to be "orthogonal", if $\langle \alpha, \beta \rangle = 0$. And we sometimes write it as $\alpha \perp \beta$.

看到這個定義，馬上想到，$\langle \alpha, 0 \rangle = 0, \forall \alpha \in V$，所以得知，$\alpha \perp 0, \forall \alpha \in V$。此外，我們用傳統垂直的概念來想正交性，將不難體會，兩兩相互垂直的向量，它們之間是線性獨立的。

Theorem 5.4

Let $\{V; \langle\, ,\, \rangle\}$ be an inner product space. If $\{\alpha_1, \alpha_2, ..., \alpha_n\}$ is a set of non-zero vectors in V such that $\langle \alpha_i, \alpha_j \rangle = 0$, $\forall i \neq j$, then $\alpha_1, \alpha_2, ..., \alpha_n$ are linearly independent.

Proof

Let $c_1\alpha_1 + c_2\alpha_2 + \cdots + c_n\alpha_n = 0$, then for each $i = 1,...,n$,

$\langle c_1\alpha_1 + c_2\alpha_2 + \cdots + c_n\alpha_n, \alpha_i \rangle = \langle 0, \alpha_i \rangle = 0$

$\Rightarrow c_i \langle \alpha_i, \alpha_i \rangle = 0$

$\Rightarrow c_i = 0$

Therefore, $\alpha_1, \alpha_2, ..., \alpha_n$ are linearly independent.

Theorem 5.5

Let $\{V; \langle , \rangle\}$ be an inner product space. If $\{\alpha_1, \alpha_2, ..., \alpha_n\}$ is a basis for V and if $\beta \in V$ such that $\beta \perp \alpha_i$, $\forall i = 1, 2, ..., n$, then $\beta = 0$.

請讀者將此定理與上一頁之 Exercise 1.做個比較，看看有何差異。

Proof

$\beta \in V$ implies that $\beta = \sum_{i=1}^{n} x_i \alpha_i$ for some scalars, $x_1, ..., x_n$.

So, we have, $\langle \beta, \beta \rangle = \langle \beta, \sum_{i=1}^{n} x_i \alpha_i \rangle = \sum_{i=1}^{n} \overline{x_i} \langle \beta, \alpha_i \rangle = 0$, by the hypothesis. and therefore, $\beta = 0$ follows.

Definition 5.9

Let $\{\alpha_1, \alpha_2, ..., \alpha_n\}$ be a basis for an inner product space, $\{V; \langle , \rangle\}$.
(i) If $\langle \alpha_i, \alpha_j \rangle = 0$, $\forall i \neq j$, then $\{\alpha_1, \alpha_2, ..., \alpha_n\}$ is called an "orthogonal basis".
(ii) If $\langle \alpha_i, \alpha_j \rangle = 0$, $\forall i \neq j$ and for each $i = 1, 2, ..., n$, $\|\alpha_i\| = 1$, then $\{\alpha_1, \alpha_2, ..., \alpha_n\}$ is called an "orthonormal basis".

當然，或有同學早已發現：

假若，$\{\alpha_1, \alpha_2, ..., \alpha_n\}$ 是 $\{V; \langle\ ,\ \rangle\}$ 的一組 orthogonal basis，那麼毫無疑問的，$\{\frac{\alpha_1}{\|\alpha_1\|}, \frac{\alpha_2}{\|\alpha_2\|}, ..., \frac{\alpha_n}{\|\alpha_n\|}\}$ 是一組 orthonormal basis。

e.g.

For $X = \begin{pmatrix} x_1 \\ x_2 \\ ... \\ x_n \end{pmatrix}, Y = \begin{pmatrix} y_1 \\ y_2 \\ ... \\ y_n \end{pmatrix} \in R^n$, defining $\langle X, Y \rangle = \sum_{i=1}^{n} x_i y_i$, then the standard basis, $\{E_1, E_2, ..., E_n\}$, is an orthonormal basis.

Exercise

Show that if $\alpha \perp \beta$, then the parallelogram law becomes, $\|\alpha + \beta\|^2 = \|\alpha\|^2 + \|\beta\|^2$. (This is called the "Pythagoras law".) (*Hint*: 計算 $\langle \alpha + \beta, \alpha + \beta \rangle$)

假設，$\{\alpha_1, \alpha_2, ..., \alpha_n\}$ 是內積空間 V 的任一基底，那麼若 $\beta \in V$，則 β 可以被寫為 $\beta = \sum_{i=1}^{n} x_i \alpha_i$。一般而言，以這種型態出現的話，要計算 β 的長度，$\|\beta\|$，是不容易的。但是，假若，$\{\alpha_1, \alpha_2, ..., \alpha_n\}$ 是一組 orthonormal basis 的話，$\|\beta\|$ 的計算就變的簡單許多了。看下一個定理。

Theorem 5.6

Let $\{\alpha_1, \alpha_2, ..., \alpha_n\}$ be an orthonormal basis for an inner product space, $\{V; \langle\ ,\ \rangle\}$, and let $\beta \in V$. If $\beta = \sum_{i=1}^{n} x_i \alpha_i$, then

1. $x_i = \langle \beta, \alpha_i \rangle$, $\forall i = 1, 2, ..., n$.
2. $\|\beta\| = (\sum_{i=1}^{n} |x_i|^2)^{1/2}$.

Proof

這個定理證明不會太難。

1. 若 $\beta = \sum_{i=1}^{n} x_i \alpha_i$，則對任意 $i = 1, 2, ..., n$ 而言，

$$\langle \beta, \alpha_i \rangle = \sum_{j=1}^{n} x_j \langle \alpha_j, \alpha_i \rangle = x_i \langle \alpha_i, \alpha_i \rangle = x_i$$

2. $\|\beta\|^2 = \langle \beta, \beta \rangle = \langle \sum_{i=1}^{n} x_i \alpha_i, \sum_{j=1}^{n} x_j \alpha_j \rangle = \sum_{i=1}^{n} x_i \langle \alpha_i, \sum_{j=1}^{n} x_j \alpha_j \rangle$

$$= \sum_{i=1}^{n} x_i \langle \alpha_i, x_i \alpha_i \rangle = \sum_{i=1}^{n} x_i \overline{x_i} \langle \alpha_i, \alpha_i \rangle = \sum_{i=1}^{n} |x_i|^2$$

假若，$\{\alpha_1, \alpha_2, ..., \alpha_n\}$ 只是一個 orthogonal basis 而非 orthonormal 的話，那麼，由前述證明，我們發現，$\|\beta\|^2 = \sum_{i=1}^{n} |x_i|^2 \|\alpha_i\|^2$。在 Fourier series 的課程裡，上述定理中的係數，$x_i = \langle \beta, \alpha_i \rangle$，一般被稱為 Fourier coefficients of β with respect to the orthogonal basis, $\{\alpha_1, \alpha_2, ..., \alpha_n\}$。既然，orthonormal basis 可以帶來那麼大的方便，就讓我們來介紹，如何建立一個內積空間中的 orthonormal basis。

Theorem 5.7

Let $\{V; \langle , \rangle\}$ be an inner product space with $\dim V = n$. If W is a proper subspace of V, and $\{w_1, ..., w_k\}$ is an orthogonal basis for W, then there exist vectors, $w_{k+1}, ..., w_n$ of V such that $\{w_1, ..., w_k, w_{k+1}, ..., w_n\}$ is an orthogonal basis for V.

Proof

第 1.2 節時，我們曾經介紹過，將子空間 W 的基底 $\{w_1, ..., w_k\}$，擴建為整個向量空間 V 的基底。也就是說，我們可以在 V 中找到 $n-k$ 個向量 $v_{k+1}, ..., v_n$，使得 $\{w_1, ..., w_k, v_{k+1}, ..., v_n\}$ 成為 V 的一個基底。(注意：當時的

$w_1,...,w_k,v_{k+1},...,v_n$ 不一定是正交的。)

現在,我們要如何處理正交基底呢?

我們令, $w_{k+1} = v_{k+1} - \dfrac{\langle v_{k+1}, w_1 \rangle}{\langle w_1, w_1 \rangle} w_1 - \dfrac{\langle v_{k+1}, w_2 \rangle}{\langle w_2, w_2 \rangle} w_2 - \cdots - \dfrac{\langle v_{k+1}, w_k \rangle}{\langle w_k, w_k \rangle} w_k$, 則

$$\langle w_{k+1}, w_i \rangle = \langle v_{k+1}, w_i \rangle - \dfrac{\langle v_{k+1}, w_i \rangle}{\langle w_i, w_i \rangle} \langle w_i, w_i \rangle = 0, \ \forall i = 1,...,k$$

而且, $v_{k+1} \in Sp\{w_1,...,w_k, w_{k+1}\}$

因此,我們得知 $\{w_1,...,w_k, w_{k+1}\}$ 形成了由向量組 $\{w_1,...,w_k, v_{k+1}\}$ 所構成的子空間的正交基底。按此要領繼續做下去,我們將能找到 $n-k$ 個向量 $w_{k+1}, w_{k+2},..., w_n$,使得 $\{w_1,...,w_k, w_{k+1},..., w_n\}$ 成為向量空間 $V = Sp\{w_1,...,w_k, v_{k+1},...,v_n\}$ 的一個正交基底。

這個定理的方法被稱為 Gram-Schmidt orthogonalization。從定理的結果得知,每一個 inner product space 有正交基底。當我們把正交基底的向量給 normalize 之後,它們就形成了 orthonormal basis。上述定理是從一個子空間開始,而擴建出整個向量空間的正交基底。當然,我們也可以從沒有開始,而擴建出整個正交基底如下。

Suppose that $\{\alpha_1, \alpha_2,...,\alpha_n\}$ is a basis for an inner product space, V. Taking that $v_1 = \alpha_1$

$$v_2 = \alpha_2 - \dfrac{\langle \alpha_2, v_1 \rangle}{\langle v_1, v_1 \rangle} v_1$$

$$\cdots\cdots\cdots\cdots\cdots\cdots\cdots$$

$$v_n = \alpha_n - \dfrac{\langle \alpha_n, v_1 \rangle}{\langle v_1, v_1 \rangle} v_1 - \cdots - \dfrac{\langle \alpha_n, v_{n-1} \rangle}{\langle v_{n-1}, v_{n-1} \rangle} v_{n-1}$$

then $\{v_1,...,v_n\}$ is an orthogonal basis for V.

Example 2

Let $V = R^3$ with the inner product,

$$\langle X, Y \rangle = \sum_{i=1}^{3} x_i y_i, \quad \forall\, X = \begin{pmatrix} x_1 \\ x_2 \\ x_3 \end{pmatrix}, Y = \begin{pmatrix} y_1 \\ y_2 \\ y_3 \end{pmatrix} \in V$$

Given a basis, $\{\alpha_1 = \begin{pmatrix} 1 \\ 1 \\ 1 \end{pmatrix}, \alpha_2 = \begin{pmatrix} 2 \\ 0 \\ 1 \end{pmatrix}, \alpha_3 = \begin{pmatrix} 0 \\ 1 \\ 2 \end{pmatrix}\}$ of V. Find an orthonormal basis by the Gram-Schmidt orthogonalization.

Solution

Let $v_1 = \alpha_1 = \begin{pmatrix} 1 \\ 1 \\ 1 \end{pmatrix}$. Taking

$$v_2 = \alpha_2 - \frac{\langle \alpha_2, v_1 \rangle}{\langle v_1, v_1 \rangle} v_1 = \begin{pmatrix} 2 \\ 0 \\ 1 \end{pmatrix} - \frac{3}{3} \begin{pmatrix} 1 \\ 1 \\ 1 \end{pmatrix} = \begin{pmatrix} 1 \\ -1 \\ 0 \end{pmatrix}$$

$$v_3 = \alpha_3 - \frac{\langle \alpha_3, v_1 \rangle}{\langle v_1, v_1 \rangle} v_1 - \frac{\langle \alpha_3, v_2 \rangle}{\langle v_2, v_2 \rangle} v_2 = \begin{pmatrix} 0 \\ 1 \\ 2 \end{pmatrix} - \frac{3}{3} \begin{pmatrix} 1 \\ 1 \\ 1 \end{pmatrix} - \frac{-1}{2} \begin{pmatrix} 1 \\ -1 \\ 0 \end{pmatrix} = \begin{pmatrix} -1/2 \\ -1/2 \\ 1 \end{pmatrix}$$

Then

$\{v_1, v_2, v_3\}$ is an orthogonal basis for V

However,

$\|v_1\| = \sqrt{3}$, $\|v_2\| = \sqrt{2}$, $\|v_3\| = \sqrt{3/2}$

So, $\{\frac{v_1}{\sqrt{3}}, \frac{v_2}{\sqrt{2}}, \frac{v_3}{\sqrt{3/2}}\}$ is an orthonormal basis. ∎

下列例子是富氏級數的一個應用，僅提供讀者參考。若有感覺進度不足之時，本例題可以略過而不致影響本節的完整性。

Example 3

Let $V = C[-\pi, \pi] = \{f: [-\pi, \pi] \to C; f$ is continuous.$\}$.
Defining, for $f, g \in V, \langle f, g \rangle = \int_{-\pi}^{\pi} f(t)\overline{g(t)}\, dt$, then
$\{e^{int} = \cos nt + i\sin nt : n \in Z\}$ is a set of orthogonal elements in V.

Proof

$$\langle e^{int}, e^{imt} \rangle = \int_{-\pi}^{\pi} e^{int} e^{-imt}\, dt = \int_{-\pi}^{\pi} e^{i(n-m)t}\, dt = \begin{cases} 0, & \text{if } n \neq m \\ 2\pi, & \text{if } n = m \end{cases}$$ ∎

Remarks

1. Actually, $\{e^{int} = \cos nt + i \sin nt : n \in Z\}$ is an orthogonal basis for V, and this can be seen in more advanced mathematics.

2. If $f \in V$, then $\dfrac{\langle f, e^{int} \rangle}{\langle e^{int}, e^{int} \rangle} = \dfrac{1}{2\pi} \int_{-\pi}^{\pi} f(t) e^{-int}\, dt$ is called the "Fourier coefficients" of f.

3. Writing $\tilde{f}(n) = \dfrac{\langle f, e^{int} \rangle}{\langle e^{int}, e^{int} \rangle}$, then $\sum_{n=-\infty}^{\infty} \tilde{f}(n) e^{int}$ is called the "Fourier series" of f.

Exercises

1. Given linearly independent vectors, $\alpha_1 = \begin{pmatrix} 1 \\ 3 \\ 2 \end{pmatrix}, \alpha_2 = \begin{pmatrix} 2 \\ 1 \\ 3 \end{pmatrix}, \alpha_3 = \begin{pmatrix} 0 \\ 1 \\ 1 \end{pmatrix}$, in R^3. Find an orthonormal basis by the Gram-Schmidt orthogonalization with respect to the standard inner product.

2. Let $V = \{f : R \to R \; ; \; f(x) = a + bx + cx^2 + cx^3\}$ and define for $f, g \in V$,
$$\langle f, g \rangle = \int_{-1}^{1} f(x) g(x)\, dx.$$
Use the basis, $\{1, x, x^2, x^3\}$, to find an orthogonal basis for V.
(These orthogonal polynomials are called "Legendre polynomials".)

3. Suppose that $\langle \; , \; \rangle_1$ and $\langle \; , \; \rangle_2$ are two inner products on a vector space, V. Show that $\langle \; , \; \rangle = \langle \; , \; \rangle_1 + \langle \; , \; \rangle_2$ is also an inner product on V. (*Hint*: check step by step.)

4. Let S be a non-empty subset of an inner product space, $\{V; \langle \; , \; \rangle\}$. The "orthogonal complement", $S^\perp$, of S is defined by $S^\perp = \{v \in V : \langle v, u \rangle = 0, \; \forall u \in S\}$.

a. Show that $S^\perp$ is a subspace of V.
b. Show that $\{0\}^\perp = V$ and $V^\perp = \{0\}$.
c. Show that if $V = R^3$ and $S = \{E_1\}$ (the standard unit vector with 1 at the first coordinate and zeros at the others.), then $S^\perp$ equals the yz-plane with respect to the standard inner product.

5. Let W be a finite dimensional subspace of an inner product space, $\{V; \langle\,,\,\rangle\}$, and let $v \in V$. Show that there exist unique vectors, $u \in W^\perp$ and $w \in W$ such that $v = u + w$. (*Hint*: Let $\{\alpha_1, \alpha_2, \ldots, \alpha_l\}$ be an orthonormal basis for W. Consider the vectors, $w = \sum_{i=1}^{l} \langle v, \alpha_i \rangle \alpha_i$ and $u = v - w$.)

6. Let $\{\alpha_1, \alpha_2, \ldots, \alpha_n\}$ be an orthonormal basis for an inner product space, $\{V; \langle\,,\,\rangle\}$. Show that if $u, v \in V$, then $\langle u, v \rangle = \sum_{i=1}^{n} \langle u, \alpha_i \rangle \overline{\langle v, \alpha_i \rangle}$. (*Hint*: 利用 157 頁的定理 5.6，for $u \in V$, $u = \sum_{i=1}^{n} \langle u, \alpha_i \rangle \alpha_i$。) (這個等式叫做，Parseval's identity。)

7. Let A be an $n \times n$ complex matrix. Show that $AA^* = I_n$ if and only if the rows of A form an orthonormal basis for C^n.

§5.3* Appendixes for orthogonal complements

在 5.3 節最後的 exercises 中，第 4. 及第 5. 題稍微提到了 orthogonal complement 這個主題。現在，我們想借用點篇幅，把它給清楚的再介紹一番。首先，回憶一下前述 exercise 5.。由於，$W \cap W^\perp = \{0\}$，所以，$V = W \oplus W^\perp$。也因此得知，$\dim V = \dim W + \dim W^\perp$。此外，下面還有兩個簡單的基本特性，給同學當作 exercises。

Exercises

1. Given an inner product space, $\{V ; \langle\ ,\ \rangle\}$. Let U and W be two subspaces of V. Show that $(U+W)^\perp = U^\perp \cap W^\perp$.

2. Show that if U is a subspace of an inner product space, $\{V ; \langle\ ,\ \rangle\}$, then $(U^\perp)^\perp = U$. (*Hint*: 利用 5.3 節 Exercise 5.)

觀察前述 exercise 2.，我們發現 subspace U，完全被決定於它的 orthogonal complement。也就是說，假若，U 是 $\{V ; \langle\ ,\ \rangle\}$ 的一個子空間的話，那麼，$U = \{v \in V : \langle v, w \rangle = 0 , \forall\ w \in U^\perp\}$。或者更進一步的說，假設 $\{\alpha_1, \alpha_2, ..., \alpha_l\}$ 是 $U^\perp$ 的一組基底的話，那麼

$$U = \{v \in V : \langle v, \alpha_i \rangle = 0 , \forall\ i = 1, 2, ..., l\}$$

最後這個式子被稱為，U 的 normal form。而方程式，$\langle v, \alpha_i \rangle = 0$ 則被稱為，U 的 normal equation。想想看在基本的代數學裡面，我們曾經以 normal equations 來表示，一個平面方程式如下。

假設 U 是一個在 R^3 中，通過原點的平面。那麼，它的 orthogonal complement，$U^\perp$ is a one-dimensional subspace of R^3。令向量 $\left\{\begin{pmatrix}a\\b\\c\end{pmatrix}\right\}$，為 $U^\perp$ 之一基底，則按照 normal form 而言，$U = \left\{X = \begin{pmatrix}x\\y\\z\end{pmatrix} \in R^3 : ax + by + cz = 0\right\}$。

其中，$\begin{pmatrix} a \\ b \\ c \end{pmatrix}$ 被稱為 U 的 normal vector。

Exercise

Find the normal form of a one-dimensional subspace of R^3 and interpret these geometrically. (*Hint*: Find a basis for $U^\perp$ and write down the normal form of U.)

我們再清楚的回憶 5.3 節的 orthogonal complement，假若，W is a subspace of an inner product space, $\{V \,;\, \langle\,,\,\rangle\}$，then for every $v \in V$, there exist unique $u \in W^\perp$ and $w \in W$ such that $v = u + w$。其中，假設 $\{\alpha_1, \alpha_2, ..., \alpha_l\}$ 是 W 的 orthonormal basis 的話，那麼，$w = \sum_{i=1}^{l} \langle v, \alpha_i \rangle \alpha_i$。而，$w = \sum_{i=1}^{l} \langle v, \alpha_i \rangle \alpha_i$ 一般被稱為，向量 v 在 W 上的**投射** (projection)。

Proposition 5.8

> Let W be a subspace of an inner product space, $\{V \,;\, \langle\,,\,\rangle\}$. If $\{\alpha_1, \alpha_2, ..., \alpha_l\}$ is an orthonormal basis for W and $\{\beta_1, \beta_2, ..., \beta_{n-l}\}$ is an orthonormal basis for $W^\perp$, then $\{\alpha_1, \alpha_2, ..., \alpha_l, \beta_1, \beta_2, ..., \beta_{n-l}\}$ is an orthonormal basis for V.

這個定理的結果很清楚。現在我們看看，若 $v \in V$，則 v 可以被表示為

$$v = \sum_{i=1}^{l} \langle v, \alpha_i \rangle \alpha_i + \sum_{j=1}^{n-l} \langle v, \beta_j \rangle \beta_j$$

今令，$x_i = \langle v, \alpha_i \rangle, i = 1, 2, ..., l$ 且 $y_j = \langle v, \beta_j \rangle, j = 1, 2, ..., n-l$，由 5.3 節第 157 頁之定理 5.6，得知，$\|v\|^2 = \sum_{i=1}^{l} |x_i|^2 + \sum_{j=1}^{n-l} |y_j|^2$。我們將這個結果重新敘述如下。

Proposition 5.9

Let W be a subspace of an inner product space, $\{V; \langle , \rangle\}$ and $v \in V$. If w and u are projections of v on W and $W^\perp$, respectively, then $\|v\|^2 = \|w\|^2 + \|u\|^2$.

通常，我們喜歡以符號，v_w 與 $v_{w^\perp}$ 分別表示定理中，向量，v，在 W 和 $W^\perp$ 上的投射，w 與 u。也因此，定理中的等式常被寫為，$\|v\|^2 = \|v_w\|^2 + \|v_{w^\perp}\|^2$。而這個等式的最大用處是，它提供了**最小誤差問題** (least error problem) 的解。

Theorem 5.10

Let W be a subspace of an inner product space, $\{V; \langle , \rangle\}$ and $v \in V$. Then $\|v - v_w\| \leq \|v - u\|$, $\forall u \in W$. (i.e. v_w minimize the distance from v to W.)

Proof

Consider, for $u \in W$, $v - u = v_w + v_{w^\perp} - u = (v_w - u) + v_{w^\perp}$. Since $v_w - u \in W$ and $v_{w^\perp} \in W^\perp$, we have, $\|v - u\|^2 = \|v_w - u\|^2 + \|v_{w^\perp}\|^2$. But then $\|v - u\|^2 \geq \|v_{w^\perp}\|^2 = \|v - v_w\|^2$

由於，$v_w \in W$，所以，$\|v - u\|$ 的最小值是，$\|v_{w^\perp}\|$。

§5.4 Gram-Schmidt and orthogonal matrices

我們想要利用 5.3 節 Gram-Schmidt 的方法，應用在矩陣的行向量上，然後和大家介紹矩陣的正交性質。首先，拿一個矩陣以實數系的標準內積來論，開宗明義和大家先認識所謂的**正交三角矩陣分解** (ortho-triangular factorization)。

假設矩陣 B 的三個 columns，B^1, B^2, B^3，是矩陣 $A = \begin{pmatrix} 1 & 0 & -1 \\ 0 & 1 & -1 \\ 1 & 2 & -1 \end{pmatrix}$ 的三個 columns 所演繹出來的 orthogonal columns 的話，那麼

$$B^1 = A^1 = \begin{pmatrix} 1 \\ 0 \\ 1 \end{pmatrix}, \text{ 而且,}$$

$$B^2 = A^2 - \frac{\langle A^2, B^1 \rangle}{\langle B^1, B^1 \rangle} B^1, \quad B^3 = A^3 - \frac{\langle A^3, B^1 \rangle}{\langle B^1, B^1 \rangle} B^1 - \frac{\langle A^3, B^2 \rangle}{\langle B^2, B^2 \rangle} B^2$$

或者說

$$A^2 = B^2 + \frac{\langle A^2, B^1 \rangle}{\langle B^1, B^1 \rangle} B^1, \quad A^3 = B^3 + \frac{\langle A^3, B^1 \rangle}{\langle B^1, B^1 \rangle} B^1 + \frac{\langle A^3, B^2 \rangle}{\langle B^2, B^2 \rangle} B^2$$

今令另一矩陣，$C = \begin{pmatrix} 1 & c_{12} & c_{13} \\ 0 & 1 & c_{23} \\ 0 & 0 & 1 \end{pmatrix}$，其中

$$c_{12} = \frac{\langle A^2, B^1 \rangle}{\langle B^1, B^1 \rangle}, \quad c_{13} = \frac{\langle A^3, B^1 \rangle}{\langle B^1, B^1 \rangle}, \quad c_{23} = \frac{\langle A^3, B^2 \rangle}{\langle B^2, B^2 \rangle}$$

且令，$D = BC$，則

$$D^1 = B^1 = A^1$$

$$D^2 = \begin{pmatrix} b_{11}c_{12} + b_{12} \\ b_{21}c_{12} + b_{22} \\ b_{31}c_{12} + b_{32} \end{pmatrix} = c_{12}B^1 + B^2 = A^2$$

$$D^3 = \begin{pmatrix} b_{11}c_{13} + b_{12}c_{23} + b_{13} \\ b_{21}c_{13} + b_{22}c_{23} + b_{23} \\ b_{31}c_{13} + b_{32}c_{23} + b_{33} \end{pmatrix} = c_{13}B^1 + c_{23}B^2 + B^3 = A^3$$

因此我們發現，

$$A = BC$$

也就是說，矩陣 A 可以被分解成，一個有正交行向量的矩陣 B 和一個 upper unit-triangular matrix C 的乘積。這種因式分解，我們將其稱為 "ortho-triangular factorization" (正交三角化)。有關這個因式分解，我們利用下面定理，重複敘述一遍。不過，在此之前讓我們先完成上述矩陣 A 的分解。

首先，$B^1 = A^1 = \begin{pmatrix} 1 \\ 0 \\ 1 \end{pmatrix}$，所以

$$c_{12} = \frac{\langle A^2, B^1 \rangle}{\langle B^1, B^1 \rangle} = \frac{2}{2} = 1 \quad \text{而且} \quad c_{13} = \frac{\langle A^3, B^1 \rangle}{\langle B^1, B^1 \rangle} = \frac{-2}{2} = -1$$

接著，$B^2 = A^2 - \dfrac{\langle A^2, B^1 \rangle}{\langle B^1, B^1 \rangle} B^1 = \begin{pmatrix} 0 \\ 1 \\ 2 \end{pmatrix} - \begin{pmatrix} 1 \\ 0 \\ 1 \end{pmatrix} = \begin{pmatrix} -1 \\ 1 \\ 1 \end{pmatrix}$，也因此

$$c_{23} = \frac{\langle A^3, B^2 \rangle}{\langle B^2, B^2 \rangle} = \frac{-1}{3}$$

最後，$B^3 = A^3 - \dfrac{\langle A^3, B^1 \rangle}{\langle B^1, B^1 \rangle} B^1 - \dfrac{\langle A^3, B^2 \rangle}{\langle B^2, B^2 \rangle} B^2 = \begin{pmatrix} -1 \\ -1 \\ -1 \end{pmatrix} - \begin{pmatrix} -1 \\ 0 \\ -1 \end{pmatrix} - \begin{pmatrix} 1/3 \\ -1/3 \\ -1/3 \end{pmatrix} = \begin{pmatrix} -1/3 \\ -2/3 \\ 1/3 \end{pmatrix}$

到此，我們得出 the ortho-triangular factorization of A is

$$A = \begin{pmatrix} 1 & 0 & -1 \\ 0 & 1 & -1 \\ 1 & 2 & -1 \end{pmatrix} = BC = \begin{pmatrix} 1 & -1 & -1/3 \\ 0 & 1 & -2/3 \\ 1 & 1 & 1/3 \end{pmatrix} \begin{pmatrix} 1 & 1 & -1 \\ 0 & 1 & -1/3 \\ 0 & 0 & 1 \end{pmatrix}.$$

Theorem 5.11 (Ortho-triangular factorization)

If A is an $n \times n$ matrix with it's n columns are linearly independent, then there exist an $n \times n$ matrix, B, with it's n columns are orthogonal and an upper unit-triangular matrix, C, such that $A = BC$.

我們再看下面一個例子。

Example 1

Given, $A = \begin{pmatrix} 1 & 2 & 1 \\ 1 & 4 & 3 \\ 1 & -1 & -2 \end{pmatrix}$. Find B and C such that $A = BC$.

Solution

先是矩陣 B 的第一行，$B^1 = \begin{pmatrix} 1 \\ 1 \\ 1 \end{pmatrix}$，然後矩陣 C 的第一列，

$$c_{12} = \frac{\langle A^2, B^1 \rangle}{\langle B^1, B^1 \rangle} = \frac{5}{3} \quad \text{以及} \quad c_{13} = \frac{\langle A^3, B^1 \rangle}{\langle B^1, B^1 \rangle} = \frac{2}{3}$$

接著矩陣 B 的第二行，$B^2 = A^2 - \dfrac{\langle A^2, B^1 \rangle}{\langle B^1, B^1 \rangle} B^1 = \begin{pmatrix} 2 \\ 4 \\ -1 \end{pmatrix} - \dfrac{5}{3}\begin{pmatrix} 1 \\ 1 \\ 1 \end{pmatrix} = \begin{pmatrix} 1/3 \\ 7/3 \\ -8/3 \end{pmatrix}$

矩陣 C 的第二列，$c_{23} = \dfrac{\langle A^3, B^2 \rangle}{\langle B^2, B^2 \rangle} = \dfrac{38/3}{38/3} = 1$

最後矩陣 B 的第三行，

$$B^3 = A^3 - \frac{\langle A^3, B^1 \rangle}{\langle B^1, B^1 \rangle} B^1 - \frac{\langle A^3, B^2 \rangle}{\langle B^2, B^2 \rangle} B^2 = \begin{pmatrix} 1 \\ 3 \\ -2 \end{pmatrix} - \begin{pmatrix} 2/3 \\ 2/3 \\ 2/3 \end{pmatrix} - \begin{pmatrix} 1/3 \\ 7/3 \\ -8/3 \end{pmatrix} = \begin{pmatrix} 0 \\ 0 \\ 0 \end{pmatrix}$$

所以答案是

$$\begin{pmatrix} 1 & 2 & 1 \\ 1 & 4 & 3 \\ 1 & -1 & -2 \end{pmatrix} = \begin{pmatrix} 1 & 1/3 & 0 \\ 1 & 7/3 & 0 \\ 1 & -8/3 & 0 \end{pmatrix} \begin{pmatrix} 1 & 5/3 & 2/3 \\ 0 & 1 & 1 \\ 0 & 0 & 1 \end{pmatrix}$$

上述結果要特別注意了，所得出來的矩陣

$$B = \begin{pmatrix} 1 & 1/3 & 0 \\ 1 & 7/3 & 0 \\ 1 & -8/3 & 0 \end{pmatrix}$$

它的第三行是一個 zero column，為什麼會造成這樣的結果呢？只因為原矩陣 A 的三個 columns are not linearly independent。所以，為了避免產生這不必要的困擾，我們在上一個定理中，特別要求矩陣 A 的 columns 為 linearly independent。

Exercises

練習 1~2，求出矩陣 A 的 ortho-triangular factorization。

1. $A = \begin{pmatrix} 2 & 0 \\ -1 & 1 \end{pmatrix}$

2. $A = \begin{pmatrix} 1 & 2 & 1 \\ -1 & -2 & 3 \\ 0 & 1 & -2 \end{pmatrix}$

3. Let B be an $n \times k$ matrix with orthogonal columns. Show that if B is a real matrix, then $D = ({}^t B)B$ is a diagonal matrix and for each
$i = 1, 2, 3, ..., k$, $d_{ii} = \langle B^i, B^i \rangle$.

4. if B is a complex matrix, then $D = ({}^t B)\overline{B}$ is a diagonal matrix and for each
$i = 1, 2, 3, ..., k$, $d_{ii} = \langle B^i, B^i \rangle$.

5. Let B be a non-singular $n \times n$ matrix with orthogonal columns and let $D = ({}^t B)\overline{B}$. Show that $B^{-1} = (\overline{D})^{-1}(B^*)$. (*Hint*: Consider, $(\overline{D})^{-1} = (B^{-1})(B^*)^{-1}$.)

無論如何，上述習題第 5. 題告訴我們，如何輕鬆的求出一個 non-singular

matrix with orthogonal columns 的逆矩陣。我們馬上練習一個例子看看。

Example 2

Find the inverse of the orthogonal non-singular matrix

$$B = \begin{pmatrix} 1 & -1 & 1 & -1 \\ 1 & 1 & 1 & 1 \\ -1 & 1 & 1 & -1 \\ -1 & -1 & 1 & 1 \end{pmatrix}$$

Solution

$$D = (^tB)B = \begin{pmatrix} 1 & 1 & -1 & -1 \\ -1 & 1 & 1 & -1 \\ 1 & 1 & 1 & 1 \\ -1 & 1 & -1 & 1 \end{pmatrix} \begin{pmatrix} 1 & -1 & 1 & -1 \\ 1 & 1 & 1 & 1 \\ -1 & 1 & 1 & -1 \\ -1 & -1 & 1 & 1 \end{pmatrix}$$

$$= \begin{pmatrix} 4 & 0 & 0 & 0 \\ 0 & 4 & 0 & 0 \\ 0 & 0 & 4 & 0 \\ 0 & 0 & 0 & 4 \end{pmatrix}$$

$$B^{-1} = (D^{-1})(^tB) = \begin{pmatrix} 1/4 & 0 & 0 & 0 \\ 0 & 1/4 & 0 & 0 \\ 0 & 0 & 1/4 & 0 \\ 0 & 0 & 0 & 1/4 \end{pmatrix} \begin{pmatrix} 1 & 1 & -1 & -1 \\ -1 & 1 & 1 & -1 \\ 1 & 1 & 1 & 1 \\ -1 & 1 & -1 & 1 \end{pmatrix}$$

$$= \frac{1}{4} \begin{pmatrix} 1 & 1 & -1 & -1 \\ -1 & 1 & 1 & -1 \\ 1 & 1 & 1 & 1 \\ -1 & 1 & -1 & 1 \end{pmatrix}$$

由例題中我們發現，假設矩陣 B 的每一個 column 的長度，都一樣而為 d 的話，那麼我們得出，$B^{-1} = \frac{1}{d}(B^*)$。也就是說，$B(\frac{1}{d}B^*) = (\frac{1}{d}B^*)B = I$，或者說，$BB^* = B^*B = dI$。

Theorem 5.12

Let B be an $n \times n$ non-singular matrix such that the columns are orthogonal and have the same length. Then the rows of B are orthogonal and the length of the rows equals the length of the columns.

Proof

這個定理的證明由式子，$BB^* = B^*B = dI$，可輕鬆得出。因為，$BB^* = dI$ implies that

$$\langle B_i, B_j \rangle = \begin{cases} d, & \text{if } i = j \\ 0, & \text{if } i \neq j \end{cases}$$

也就是說，矩陣 B 的 rows are orthogonal，而且它們的長度皆為 d。

附註：

一般而言，這個定理證明是這樣子的。首先假設，$(A^*)(A) = dI$。由於，矩陣 A 是 non-singular，所以 A 是 surjective from C^n onto C^n。也因此，對每一 $Y \in C^n$ 而言，存在一 $X \in C^n$，使得

$$AX = Y$$
$$\Rightarrow \quad dAX = dY$$
$$\Rightarrow \quad AdX = dY$$
$$\Rightarrow \quad A(A^*A)X = dY$$
$$\Rightarrow \quad (A)(A^*)AX = dY$$
$$\Rightarrow \quad (A)(A^*)Y = dY$$

注意前面，最後式子是對任一 $Y \in C^n$ 都成立的。所以，$(A)(A^*) = dI$。這就是我們所要的結果。

我們稱一個 square real matrix，A，為 orthonormal matrix，假設，矩陣 A 的 columns form an orthonormal set (with respect to the standard inner product)。在 complex case 而言，我們亦稱矩陣 A 為 unitary，假設 A has orthonormal columns。下面，讓我們稍微整理一下，弄清楚一點。

若 A 是一個 real $n \times n$ matrix 的話，則下列敘述都是等價的。

1. A is an orthonormal matrix.
2. A has orthonormal columns.
3. A has orthonormal rows.
4. $A^{-1} = {}^t A$.

又若 A 是一個 complex $n \times n$ matrix 的話，那麼下列敘述都是等價的

1. A is a unitary matrix.
2. A has orthonormal columns.
3. A has orthonormal rows.
4. $A^{-1} = A^*$.

Theorem 5.13

> Let A be a real (complex) $n \times n$ matrix and let $\langle\ ,\ \rangle$ be the standard real (complex) inner product. Then A is orthonormal (unitary) if and only if for each pair $\alpha, \beta \in R^n(C^n)$, $\langle A\alpha, A\beta \rangle = \langle \alpha, \beta \rangle$.

Proof

我們拿 complex case 來說明。We first note that
$$\langle A\alpha, A\beta \rangle = {}^t(A\alpha)\overline{(A\beta)} = ({}^t\alpha)({}^tA)(\overline{A})(\overline{\beta}).$$
Since A is unitary, $({}^tA)\overline{A} = I$. So, we see that $\langle A\alpha, A\beta \rangle = \langle \alpha, \beta \rangle$.

To the contrary, if $\langle A\alpha, A\beta \rangle = \langle \alpha, \beta \rangle$, then $({}^t\alpha)({}^tA)(\overline{A})(\overline{\beta}) = ({}^t\alpha)(\overline{\beta})$. This is true for all $\alpha, \beta \in C^n$. By section 5.1, we obtain that $({}^tA)\overline{A} = I$. That says, A is unitary.

定理中，一個特別的情況是，當 $\alpha = \beta$ 時。也就是說，若 A is unitary 的話，則 $\langle A\alpha, A\alpha \rangle = \langle \alpha, \alpha \rangle$。或者說，$\|A\alpha\| = \|\alpha\|$, $\forall\ \alpha \in C^n$。這樣一個特性，我們將其稱為"保存長度的"，A preserves the length of a vector。其實，這個敘述反過來說也是成立的。看看下面 Exercise。

Exercise

Prove that if for every vector, $\alpha \in C^n$, $\|A\alpha\| = \|\alpha\|$, then A is unitary.

(*Hint*: Use the following two equations;

    a. $\langle A(\alpha - \beta), A(\alpha - \beta)\rangle = \langle \alpha - \beta, \alpha - \beta \rangle$

    b. $\langle A(\alpha - i\beta), A(\alpha - i\beta)\rangle = \langle \alpha - i\beta, \alpha - i\beta \rangle$

Deduce $\operatorname{Re}\langle A\alpha, A\beta\rangle = \operatorname{Re}\langle \alpha, \beta\rangle$ and $\operatorname{Im}\langle A\alpha, A\beta\rangle = \operatorname{Im}\langle \alpha, \beta\rangle$)

注意，這裡的 $\operatorname{Re}\langle \alpha, \beta\rangle$ 代表 $\langle \alpha, \beta\rangle$ 的實數部分；$\operatorname{Im}\langle \alpha, \beta\rangle$ 則代表 $\langle \alpha, \beta\rangle$ 的虛數部分。

Chapter Six

Determinants

6.1 Determinants of 2×2 matrices
6.2 N-linear functions and the determinant
6.3 Permutations and uniqueness of the determinant
6.4 Computing determinants
6.5 Cramer's rule and the inverse of a matrix

　　二十一世紀以來，行列式的角色似乎逐漸失去了它早期的主導地位。尤其，在數值時代開始之後，無論是求"linear systems"的解方面，或者是求"矩陣的逆矩陣"方面也是如此，電腦工程師們為了使用計算機，他們早已，以"numerical method"取代了傳統的純代數式的算法。不過儘管如此，"行列式"在計算或是在建立"特徵值"的特性以及一些相關領域上，仍然有它少不了的機動地位。此外，在處理矩陣的 column vectors 之線性獨立以及 rank 方面，"行列式"也仍然扮演著關鍵性的角色。所以，到目前為止，還沒有任何一位線性代數或是矩陣理論的作者，在著作完稿出版之前，未將"行列式"給予有系統的處理的。

　　雖然，determinant 不具有 linearity 的特性，可是它確擁有某方面的 linearity 的個性。這個 chapter 我們就將以該所謂的 n-linearity 的個性著手，從 2×2 矩陣行列式的原始定義開始，大家一起來研讀一般所謂的"determinant"。

§6.1　Determinants of 2×2 matrices

介紹一般的行列式之前，我們要先從 2×2 矩陣開始談起。主要去檢驗一下，我們所定義的 2×2 矩陣之行列式，它所擁有的一些特性。然後，根據這些基本的特性發展出，一般 $n \times n$ 矩陣行列式的作法。

Definition　6.1

Given a 2×2 real (or complex) matrix, $A = \begin{pmatrix} a & b \\ c & d \end{pmatrix}$. The determinant, $\det(A)$, of A is defined by $\det(A) = ad - bc$.

譬如說

若 $A = \begin{pmatrix} 2 & -1 \\ 3 & 1 \end{pmatrix}$，則 $\det(A) = 2 - (-3) = 5$。

根據這個定義，下列有三件簡單的事情，我們說明一下。
1. $\det(I_2) = 1$
2. If the two columns of A are equal, then $\det(A) = 0$.
3. $\det \begin{pmatrix} a & tb + b' \\ c & td + d' \end{pmatrix} = t \det \begin{pmatrix} a & b \\ c & d \end{pmatrix} + \det \begin{pmatrix} a & b' \\ c & d' \end{pmatrix}$ and
$\det \begin{pmatrix} ta + a' & b \\ tc + c' & d \end{pmatrix} = t \det \begin{pmatrix} a & b \\ c & d \end{pmatrix} + \det \begin{pmatrix} a' & b \\ c' & d \end{pmatrix}$

前述特性第 3. 說明，若以矩陣 A 的 column vectors 而言，行列式的運算是雙線性的。讓我們先把這個概念記住，一般行列式的定義就是依此概念出發的。除此之外，2×2 矩陣的行列式還有下列一些簡單的特性，將其一一敘述如下：

Properties 1

1. $\det(A^1 + tA^2, A^2) = \det(A^1, A^2) + t\det(A^2, A^2) = \det(A^1, A^2)$

    這說明，if the first column is replaced by the sum of a scalar multiple of the second column and itself, then the determinant keeps unchanged.

2. $\det(A^1, A^2) = ad - bc = -(bc - ad) = -\det(A^2, A^1)$

    This says that the determinant changes its sign, when two columns are switched.

3. $\det(A) = ad - bc = \det(^tA)$

除了這些特性之外，好發問的讀者或許老早想問，"對任意兩個 2×2 矩陣， A 與 B ， $\det(A+B)$ 與 $\det(A) + \det(B)$ 是否相等？" 很不幸的，有關這個問題的答案是否定的。我們隨便舉個例子，若

$A = \begin{pmatrix} 1 & 2 \\ 3 & 4 \end{pmatrix}$ ， $B = \begin{pmatrix} 2 & 4 \\ 6 & 8 \end{pmatrix}$ ，則， $\det(A) = 4 - 6 = -2$ ， $\det(B) = 16 - 24 = -8$ 且

$\det(A+B) = \det\begin{pmatrix} 3 & 6 \\ 9 & 12 \end{pmatrix} = 36 - 54 = -18$ ，然而， $\det(A) + \det(B) = -2 - 8 = -10$ 。

所以，一般來說， $\det(A+B) \neq \det(A) + \det(B)$ 。

矩陣的行列式，其是否為 0，關係著該矩陣是否為可逆的。這是一個關鍵性的判讀，我們就先以 2×2 矩陣為例，搶先欣賞這個重要的理論。

Theorem 6.1

Let A be a 2×2 matrix. Then $\det(A) \neq 0$ if and only if A is invertible.

Proof

($\Rightarrow$)

Write, $A = \begin{pmatrix} a & b \\ c & d \end{pmatrix}$ and suppose that $\det(A) \neq 0$. Consider the matrix, $B = \dfrac{1}{\det(A)} \begin{pmatrix} d & -b \\ -c & a \end{pmatrix}$, we find out that

$$AB = \begin{pmatrix} a & b \\ c & d \end{pmatrix} (\frac{1}{\det(A)}) \begin{pmatrix} d & -b \\ -c & a \end{pmatrix} = \frac{1}{ad-bc} \begin{pmatrix} ad-bc & 0 \\ 0 & ad-bc \end{pmatrix} = \begin{pmatrix} 1 & 0 \\ 0 & 1 \end{pmatrix}$$

and that $BA = \begin{pmatrix} 1 & 0 \\ 0 & 1 \end{pmatrix}$. So, A is invertible and $A^{-1} = \frac{1}{\det(A)} \begin{pmatrix} d & -b \\ -c & a \end{pmatrix}$.

($\Leftarrow$)

Suppose that A is invertible, then $rank(A) = 2$. Use section 3.1, the two columns of A have to be linearly independent. Hence the first column, A^1, must not be a zero column. That is either $a \neq 0$ or $c \neq 0$. Without loss of generality, we assume $a \neq 0$, and use basic row operations to deduce A into $\begin{pmatrix} 1 & b/a \\ 0 & (ad-bc)/a \end{pmatrix}$. As we knew in chapter two and three, basic row operations are rank-preserving, so the rank of $\begin{pmatrix} 1 & b/a \\ 0 & (ad-bc)/a \end{pmatrix}$ is 2. This says, $\frac{ad-bc}{a} \neq 0$. And hence, $\det(A) = ad - bc \neq 0$.

本節剩下的篇幅，我們和讀者來探討一下，2×2 矩陣的行列式所能給我們的幾何印象，而下面的定理是理解這個幾何印象的起始道路。它說明，行列式是定義在矩陣上的，某種形式的線性函數。這些特性將被用以做為下一節，我們定義一般行列式的依據。

Theorem 6.2

Let $M_{2 \times 2}$ denote the collection of all 2×2 matrices. If F is a scalar-valued function defined on $M_{2 \times 2}$ having the following three properties,

(i) F is linear to either column of a 2×2 matrix, when the other column is held fixed.
(ii) If the two columns of a 2×2 matrix, A, are identical, then $F(A) = 0$.
(iii) $F(I_2) = 1$

then $F(A) = \det(A)$, $\forall A \in M_{2 \times 2}$.

Proof

Let $A = \begin{pmatrix} a & b \\ c & d \end{pmatrix}$. We shall show that $F(A) = ad - bc$.

Writing, $A^1 = \begin{pmatrix} a \\ c \end{pmatrix}$ and $A^2 = \begin{pmatrix} b \\ d \end{pmatrix}$, then $F(A) = F(a(\frac{1}{a})A^1, A^2) = aF\begin{pmatrix} 1 & b \\ c/a & d \end{pmatrix}$.

Setting, $B^1 = \begin{pmatrix} 1 \\ c/a \end{pmatrix}$ and $B^2 = \begin{pmatrix} b \\ d \end{pmatrix}$, then

$F(A) = aF(B^1, B^2)$
$= aF(B^1, B^2 + (-b)B^1)$, by properties (i) and (ii).
$= aF\begin{pmatrix} 1 & 0 \\ \frac{c}{a} & -\frac{bc}{a} + d \end{pmatrix} = aF(B^1, \frac{ad-bc}{a}\begin{pmatrix} 0 \\ 1 \end{pmatrix}) = a(\frac{ad-bc}{a})F\begin{pmatrix} 1 & 0 \\ c/a & 1 \end{pmatrix}$

Setting, $C^1 = \begin{pmatrix} 1 \\ c/a \end{pmatrix}$ and $C^2 = \begin{pmatrix} 0 \\ 1 \end{pmatrix}$, then the last equation becomes,

$F(A) = (ad-bc)F(C^1 + (-c/a)C^2, C^2) = (ad-bc)F\begin{pmatrix} 1 & 0 \\ 0 & 1 \end{pmatrix} = ad - bc$

We completed the proof.

現在，將一個 2×2 實數矩陣，A，之兩個 columns，A^1, A^2，看成 xy-平面上從原點往外射出的兩個向量。若 A^1, A^2 不共線的話，它們形成了以 A^1, A^2 為鄰邊的平行四邊形 (如右圖)。這平行四邊形的面積與 determinant 的關係如何呢？請看下面的推論。

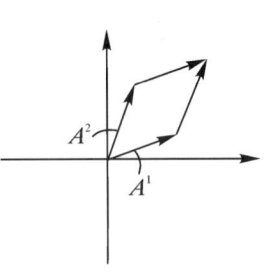

Corollary

As mentioned above, the area, $area(A^1, A^2)$, of the parallelogram is equal to $|\det(A)|$.

Proof

這個純粹代數式的證明有點冗長，請讀者集中精神、忍耐的欣賞一下。

We first define the orientation of the two columns of a 2×2 real matrix, A, by $O(A^1, A^2) = \dfrac{\det(A)}{|\det(A)|}$, if A^1 and A^2 are linearly independent. And, also define, for convenience, $O(A^1, A^2) = 1$, if A^1 and A^2 are linearly dependent. Then, we see that $O(A^1, A^2) = \pm 1$.

We now need to show, $area(A^1, A^2) = O(A^1, A^2) \cdot \det(A)$. To do this, we define,

$$F(A^1, A^2) = O(A^1, A^2) \cdot area(A^1, A^2) \ , \ \forall \ A \in M_{2 \times 2}$$

And the proof will be done, if F satisfies the 3 properties of the previous theorem.

Firstly, consider for a real number, c. If $c > 0$, then $area(cA^1, A^2) = c \cdot area(A^1, A^2)$
and
$$O(cA^1, A^2) = \frac{\det(cA^1, A^2)}{|\det(cA^1, A^2)|} = \frac{c\det(A^1, A^2)}{|c||\det(A^1, A^2)|}$$
$$= \frac{\det(A^1, A^2)}{|\det(A^1, A^2)|} = O(A^1, A^2)$$

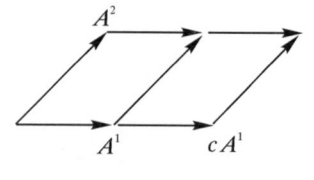

Hence
$$F(cA^1, A^2) = c \cdot O(A^1, A^2) \cdot area(A^1, A^2) = cF(A^1, A^2)$$

If $c < 0$, then $area(cA^1, A^2) = (-c) \cdot area(A^1, A^2)$
and $O(cA^1, A^2) = \dfrac{\det(cA^1, A^2)}{|\det(cA^1, A^2)|} = \dfrac{c\det(A^1, A^2)}{|c||\det(A^1, A^2)|} = -O(A^1, A^2)$

Hence
$$F(cA^1, A^2) = -O(A^1, A^2) \cdot (-c)area(A^1, A^2) = c \cdot O(A^1, A^2) \cdot area(A^1, A^2)$$
$$= cF(A^1, A^2)$$

If $c = 0$, then $area(cA^1, A^2) = area(0, A^2) = 0$

and $O(cA^1, A^2) = O(0, A^2) = 1$

Hence
$$F(0A^1, A^2) = 1 \cdot 0 = 0 \cdot F(A^1, A^2)$$
So, in general, $F(cA^1, A^2) = cF(A^1, A^2)$ is true for every number, c.
Similarly, $F(A^1, cA^2) = cF(A^1, A^2)$ is also true for every number, c.

Again, consider the graph to the right.
Since $area(A^1 + A^2, A^2) = area(A^1, A^2)$ and
$det(A^1 + A^2, A^2) = det(A^1, A^2)$,
$O(A^1 + A^2, A^2) = O(A^1, A^2)$ and
$F(A^1 + A^2, A^2) = F(A^1, A^2)$.

Now, if $a, b \in R$ and $b \neq 0$, then, by the previous result,

$$F(aA^1 + bA^2, A^2) = bF(\frac{a}{b}A^1 + A^2, A^2) = bF(\frac{a}{b}A^1, A^2) = aF(A^1, A^2).$$

(Of course, we will have the same result, if $b = 0$.)
We are now ready to show that for any column vectors, $A, B, C \in R^2$,
$$F(A + B, C) = F(A, C) + F(B, C).$$
The result is clear, if $C = 0$. So, we assume, $C \neq 0$. Taking a vector, $D \in R^2$, such that C and D are linearly independent. Then there are scalars, a_1, a_2 and b_1, b_2, such that $A = a_1C + a_2D$, $B = b_1C + b_2D$ and hence that

$$F(A + B, C) = F((a_1 + b_1)C + (a_2 + b_2)D, C) = (a_2 + b_2)F(D, C)$$
$$= F(a_2D, C) + F(b_2D, C)$$
$$= F(a_1C + a_2D, C) + F(b_1C + b_2D, C)$$
$$= F(A, C) + F(B, C)$$

Similarly, $F(A, B + C) = F(A, B) + F(A, C)$, $\forall A, B, C \in R^2$.

These proved, F is linear to both the first and the second coordinates. (Which is the first property of the last theorem.)

To the second property, we consider the area of the parallelogram formed by the same vector, A. Since $area(A, A) = 0$, $F(A, A) = 0$.

To the 3rd property, if $A^1 = \begin{pmatrix} 1 \\ 0 \end{pmatrix}$, $A^2 = \begin{pmatrix} 0 \\ 1 \end{pmatrix}$, then $area(A^1, A^2) = 1 = O(A^1, A^2)$

and hence, $F(I_2) = 1$. Combining all these 3 properties, we see, by the last theorem, $F(A) = \det(A)$, $\forall A \in M_{2\times 2}$. Therefore,
$$\det(A) = O(A^1, A^2) \cdot area(A^1, A^2) , \quad \forall A \in M_{2\times 2}$$
$$area(A^1, A^2) = |\det(A)|, \quad \forall A \in M_{2\times 2}$$

眼尖的讀者或許早已發現，以向量分析的角度來證明前述定理就簡單多了。以向量的 cross product 而言，若 U 與 V 為 xyz-space 上的兩個向量的話，那麼 $\|U \times V\|$ 代表的就是，U 與 V 所形成的平行四邊形的面積。此外，我們也知道，若 $U = a_1 i + a_2 j + a_3 k$，$V = b_1 i + b_2 j + b_3 k$ 的話，$U \times V = \det \begin{pmatrix} i & j & k \\ a_1 & a_2 & a_3 \\ b_1 & b_2 & b_3 \end{pmatrix}$（有關 3×3 矩陣的行列式，本章 6.4 節即將介紹。）現在，我們令前述定理中的 2×2 矩陣為 $A = \begin{pmatrix} a & b \\ c & d \end{pmatrix}$，那麼向量 $ai + cj$ 與 $bi + dj$ 在 xy-平面上所形成的平行四邊形和向量 $W = ai + cj + 0k$ 與 $S = bi + dj + 0k$ 在 xyz-空間上所形成的平行四邊形是一樣的。所以，該定理中之平行四邊形的面積，依據前面的說法應為 $\|W \times S\|$。也就是說，

$$area = \left|\det \begin{pmatrix} i & j & k \\ a & c & 0 \\ b & d & 0 \end{pmatrix}\right| = \left|\det \begin{pmatrix} a & c \\ b & d \end{pmatrix} k\right| = \left|\det \begin{pmatrix} a & c \\ b & d \end{pmatrix}\right| = |\det(A)|$$。

最後，我們隨便看兩個例子，以為本節的結尾。由向量，$A^1 = \begin{pmatrix} 3 \\ 1 \end{pmatrix}$ 與 $A^2 = \begin{pmatrix} -1 \\ 2 \end{pmatrix}$，所形成的平行四邊形，它的面積為，$|\det(A^1, A^2)| = 6 - (-1) = 7$。而向量 $A^1 = \begin{pmatrix} -1 \\ 5 \end{pmatrix}$ 與向量 $A^2 = \begin{pmatrix} 4 \\ -2 \end{pmatrix}$，所形成的平行四邊形，它的面積則為，$|\det(A^1, A^2)| = |2 - 20| = 18$。

Exercises

1. Compute the area of the parallelogram determined by each of the following pairs of vectors, A and B, in R^2.

  a. $A = \begin{pmatrix} 3 \\ -2 \end{pmatrix}$ and $B = \begin{pmatrix} 1 \\ 3 \end{pmatrix}$

  b. $A = \begin{pmatrix} 2 \\ 4 \end{pmatrix}$ and $B = \begin{pmatrix} -1 \\ 2 \end{pmatrix}$

  c. $A = \begin{pmatrix} 4 \\ -1 \end{pmatrix}$ and $B = \begin{pmatrix} -6 \\ -2 \end{pmatrix}$

  d. $A = \begin{pmatrix} -2 \\ 3 \end{pmatrix}$ and $B = \begin{pmatrix} 1 \\ 4 \end{pmatrix}$

2. Prove that if $A, B \in M_{2\times 2}$, then $\det(AB) = \det(A) \cdot \det(B)$.

(Hint：令 $C = AB$，以 $C^1 = b_{11}A^1 + b_{21}A^2$, $C^2 = b_{12}A^1 + b_{22}A^2$ 代入，利用 linear 的特性計算 $\det(C)$，可得。)

§6.2　*N*-linear functions and the determinant

這一節，我們將以 2×2 矩陣的行列式為基礎，介紹 n-linear functions，從而介紹一般 $n\times n$ 矩陣的行列式之存在性。

Definition　6.2

> A function, $F : R^n \times R^n \times \cdots \times R^n \to R$, is said to be n-linear (or multilinear), if F is linear with respect to each coordinate, while the other $(n-1)$ coordinates hold fixed. That is, if $B, C \in R^n$ and $i = 1, \ldots, n$, then
> $$F(A^1, \ldots, A^{i-1}, B+tC, A^{i+1}, \ldots, A^n)$$
> $$= F(A^1, \ldots, A^{i-1}, B, A^{i+1}, \ldots, A^n) + tF(A^1, \ldots, A^{i-1}, C, A^{i+1}, \ldots, A^n) \quad , \quad \forall\ t \in R$$

至於 complex case 而言，我們只要將定義中的實數體，R，改為複數體，C，即可。按照這樣一個定義，我們發現，6.1 節有關 2×2 矩陣的行列式是一個 2-linear function。下面，我們看一個簡單的例題。

Example　1

Defining for every $n\times n$ matrix, A, $F(A) = \prod_{i=1}^{n} a_{ii}$, then F is n-linear.

Proof

我們只證明 real case，有關 complex case 同理可證。首先，我們把矩陣 A 的 n 個 columns 看成 $R^n \times R^n \times \cdots \times R^n$ 中的一個 elements。

For each $i = 1, 2, \ldots, n$，$B = \begin{pmatrix} b_1 \\ b_2 \\ \cdots \\ b_n \end{pmatrix}$, $C = \begin{pmatrix} c_1 \\ c_2 \\ \cdots \\ c_n \end{pmatrix}$ in R^n and a real number, t,

$$F(A^1, \ldots, A^{i-1}, B+tC, A^{i+1}, \ldots, A^n)$$
$$= (\prod_{j \neq i} a_{jj}) \cdot (b_i + tc_i)$$

$$= (\prod_{j \neq i} a_{jj}) \cdot b_i + t(\prod_{j \neq i} a_{jj}) c_i$$
$$= F(A^1,...,A^{i-1},B,A^{i+1},...,A^n) + tF(A^1,...,A^{i-1},C,A^{i+1},...,A^n)$$

確實沒錯，F is n-linear。

Lemma 6.3

A linear combination of n-linear functions is also n-linear.

Proof

It is sufficient to show that a linear combination of two n-linear functions, F_1, F_2, is n-linear. To do this, we let a, b be two scalars, we shall show that $F = aF_1 + bF_2$ is n-linear.

For $B, C, A^i \in R^n (or\ C^n)$, $i = 1, 2, ..., n$ and a scalar, t,
$$F(A^1,...,A^{i-1},B+tC,A^{i+1},...,A^n)$$
$$= aF_1(A^1,...,A^{i-1},B+tC,A^{i+1},...,A^n) + bF_2(A^1,...,A^{i-1},B+tC,A^{i+1},...,A^n)$$
$$= aF_1(A^1,...,A^{i-1},B,A^{i+1},...,A^n) + bF_2(A^1,...,A^{i-1},B,A^{i+1},...,A^n) +$$
$$t(aF_1(A^1,...,A^{i-1},C,A^{i+1},...,A^n) + bF_2(A^1,...,A^{i-1},C,A^{i+1},...,A^n))$$
$$= F(A^1,...,A^{i-1},B,A^{i+1},...,A^n) + tF(A^1,...,A^{i-1},C,A^{i+1},...,A^n)$$

上述結果是推導行列式定義的過程中，一個必要的預備定理。除了這個預備定理之外，還有一個相關的 terminology，我們得先弄清楚，那就是，

Definition 6.3

An n-linear function, F, is said to be "alternating", if $F(A^1,...,A^n) = 0$, whenever $A^j = A^{j+1}$, for some j.

這意思是說，若有任意兩個相鄰的 columns 相等的話，則 $F(A^1,...,A^n) = 0$ 時，我們稱 F 為 alternating。譬如說，6.1 節的 2×2 矩陣之行列式，若將其看成 2-linear function 的話，那麼它就是 alternating。可以了，現在已經到了定義，所謂**行列式** (the determinant) 的時候了。

Definition 6.4

> A function, F, mapping from $n \times n$ matrices (according to their column vectors in R^n or C^n) into R (or C) is called a "determinant function", if F is n-linear, alternating and $F(I_n) = 1$.

很顯然的，我們 6.1 節所看的 2×2 矩陣之行列式，滿足了定義中的三要件。所以，它確實是一個 determinant function (讀者可以自行檢定一下)。在如此定義的情況之下，我們實在急著想要知道，到底一個 $n \times n$ 矩陣的行列式是什麼東西呢？它又要如何計算呢？慢慢來，我們還有一個東西要看。那就是所謂的**餘因子** (the cofactor)。

Definition 6.5

> Let A be an $n \times n$ matrix, and let M_{ij} be the $(n-1) \times (n-1)$ matrix obtained from A by deleting the ith row and the jth column. If D is a determinant function defined on $(n-1) \times (n-1)$ matrices, then the number, $A_{ij} = (-1)^{i+j} D(M_{ij})$, is called the "cofactor" of a_{ij}.

定義中，矩陣 M_{ij} 被稱為原 $n \times n$ 矩陣 A 的 ij^{th} 降階矩陣 (M_{ij} is the ij^{th} minor of A)。

Example 2

Given that $A = \begin{pmatrix} 1 & 0 & 2 \\ -1 & -3 & -1 \\ 5 & 7 & 2 \end{pmatrix}$. Then the minors,

$$M_{12} = \begin{pmatrix} -1 & -1 \\ 5 & 2 \end{pmatrix}, \quad M_{22} = \begin{pmatrix} 1 & 2 \\ 5 & 2 \end{pmatrix}, \quad M_{32} = \begin{pmatrix} 1 & 2 \\ -1 & -1 \end{pmatrix}$$

Hence,

$$\det(M_{12}) = 3, \ \det(M_{22}) = -8, \ \det(M_{32}) = 1$$

and hence the cofactors, A_{12}, A_{22} and A_{32}, of a_{12}, a_{22} and a_{32} are -3, 8 and -1, respectively. ∎

Theorem 6.4

Let $n > 1$ and let D be a determinant function on $(n-1) \times (n-1)$ matrices. Defining functions, F_i, $i = 1, 2, ..., n$, on $n \times n$ matrices by

$$F_i(A) = \sum_{j=1}^{n} (-1)^{i+j} a_{ij} D(M_{ij}), \text{ for every } n \times n \text{ matrix, } A \in M_{n \times n}$$

then each F_i is a determinant function on $n \times n$ matrices.

Proof

We need to show that each F_i acting on $n \times n$ matrices is n-linear, alternating and $F_i(I_n) = 1$. The n-linearity of F_i follows from the previous Lemma 6.3.

To show that F_i is alternating, we assume that $A^k = A^{k+1}$.

If $j \neq k$ and also $j \neq k+1$, then the minor, M_{ij}, has two equal adjacent columns and thus $D(M_{ij}) = 0$. Hence, we have that

$$F_i(A) = \sum_{j=1}^{n} (-1)^{i+j} a_{ij} D(M_{ij})$$

$$= (-1)^{i+k} a_{ik} D(M_{ik}) + (-1)^{i+k+1} a_{i,k+1} D(M_{i,k+1})$$

Now, $A^k = A^{k+1}$ implies that $a_{ik} = a_{i,k+1}$ and $M_{ik} = M_{i,k+1}$

Thus

$$F_i(A) = (-1)^{i+k} a_{ik} D(A_{ik}) + (-1)^{i+k+1} a_{i,k+1} D(A_{i,k+1}) = 0$$

This proved that each F_i is alternating.

For the identity matrix, I_n, $F(I_n) = (-1)^{i+i} D(I_{n-1}) = 1$ is clear. Therefore, we conclude that each F_i is a determinant function.

Remarks

1. 定理中的 determinant function，F_i，我們稱其為 "the expansion according to the ith row"。
2. 前面我們已經確立了 2×2 矩陣行列式的存在，所以從上述定理，我們可以歸納出 $n \times n$ 矩陣行列式的**存在性** (existence)。
3. 可無論如何，定理中所定義出來的 determinant function，是唯一的。有關這個**唯一性** (uniqueness)，下一個 section 我們再詳細解說。
4. 另外，在下一個 section 中，我們也會提到，以 column 來當作 determinant function 的**展開式** (expansion according to columns)。譬如說，以 j^{th} column 而言，$F(A) = \sum_{i=1}^{n} (-1)^{i+j} a_{ij} D(M_{ij})$。

Example 3

Given that $A = \begin{pmatrix} 1 & 2 & 1 \\ -1 & 3 & 1 \\ 0 & 1 & 5 \end{pmatrix}$. Find the determinant of A expanded according to the first, the second and the third row, respectively.

Solution

$$F_1(A) = 1\begin{vmatrix} 3 & 1 \\ 1 & 5 \end{vmatrix} - 2\begin{vmatrix} -1 & 1 \\ 0 & 5 \end{vmatrix} + 1\begin{vmatrix} -1 & 3 \\ 0 & 1 \end{vmatrix} = 14 + 10 - 1 = 23$$

$$F_2(A) = +1\begin{vmatrix} 2 & 1 \\ 1 & 5 \end{vmatrix} + 3\begin{vmatrix} 1 & 1 \\ 0 & 5 \end{vmatrix} - 1\begin{vmatrix} 1 & 2 \\ 0 & 1 \end{vmatrix} = 9 + 15 - 1 = 23$$

$$F_3(A) = 0 - 1\begin{vmatrix} 1 & 1 \\ -1 & 1 \end{vmatrix} + 5\begin{vmatrix} 1 & 2 \\ -1 & 3 \end{vmatrix} = -2 + 25 = 23$$

∎

不錯，不管以那一個 row 來展開，所得出來的行列式都一樣。這個結果給了我們有關 "行列式完整定義" 的信心。例題中，我們有時使用符號 $\begin{vmatrix} a & b \\ c & d \end{vmatrix}$ 代表矩陣 $\begin{pmatrix} a & b \\ c & d \end{pmatrix}$ 的行列式。其實，這也是常被使用的符號。下面是一個簡單而重要的運算公式，讀者可自行練習。

Exercise

Given an $n \times n$ matrix, A, and a scalar c. Show that if D is a determinant function on $n \times n$ matrices, then $D(cA) = c^n D(A)$. (*Hint*：利用行列式 n -linear 的特性。)

一個 determinant function，除了定義中的三個基本特性之外，我們也不難推出下列實用度較高的一些特性。

Theorem 6.5

Let F be a determinant function on $n \times n$ matrices, and A an $n \times n$ matrix.

1. If A' is obtained from A by interchanging any two adjacent columns, then $F(A') = -F(A)$.
2. If $A^i = A^j$ for some $i \neq j$, then $F(A) = 0$.
3. If A' is obtained from A by interchanging any two columns, then $F(A') = -F(A)$.
4. If A' is obtained from A by adding a scalar multiple of one column to any other column, then $F(A') = F(A)$.

Proof

1. Suppose that we interchange columns, A^j and A^{j+1}, and consider that
$$0 = F(..., A^{j-1}, A^j + A^{j+1}, A^j + A^{j+1}, A^{j+2},...)$$
$$= F(..., A^{j-1}, A^j, A^j, A^{j+2},...) + F(..., A^{j-1}, A^j, A^{j+1}, A^{j+2},...) +$$
$$F(..., A^{j-1}, A^{j+1}, A^j, A^{j+2},...) + F(..., A^{j-1}, A^{j+1}, A^{j+1}, A^{j+2},...)$$
$$= F(..., A^{j-1}, A^j, A^{j+1}, A^{j+2},...) + F(..., A^{j-1}, A^{j+1}, A^j, A^{j+2},...)$$
$$= F(A) + F(A')$$
We see that $F(A') = -F(A)$.

2. 首先，我們對調矩陣，A，的每兩相鄰行，直到 A^i 與 A^j 為緊相鄰的兩行為止。而且，假設矩陣，A'，即為最後的結果矩陣。那麼，由行列式的特性 1.，我們發現，$F(A') = \pm F(A)$。而且，由於 A' 有相同的相鄰兩行，所以，$F(A') = 0$。也因此，$F(A) = 0$。

3. 假設，我們所對調的兩行為，A^i 與 A^j。現在，我們以 $A^i + A^j$ 分別取代矩陣，A，的第 i 及第 j 行。如此由 2.得出，

$0 = F(..., A^i + A^j, ..., A^i + A^j, ...)$
$= F(..., A^i, ..., A^i, ...) + F(..., A^i, ..., A^j, ...) + F(..., A^j, ..., A^i, ...) +$
$\quad F(..., A^j, ..., A^j, ...)$
$= F(..., A^i, ..., A^j, ...) + F(..., A^j, ..., A^i, ...)$
$= F(A) + F(A')$

所以，$F(A') = -F(A)$。

4. 假設，我們將第 j 行的 t 倍加到第 i 行。也就是說，
$A' = (..., A^i + tA^j, ..., A^j, ...)$，那麼
$F(A') = F(..., A^i, ..., A^j, ...) + tF(..., A^j, ..., A^j, ...)$
$\quad\quad = F(..., A^i, ..., A^j, ...) = F(A)$

再看一下，前述第 3. 和第 4. 個特性

3. Interchanging any two columns of a matrix.

4. Adding a scalar multiple of one column to any other column.

如同第二章的列運算一樣，此時我們將這兩個運算稱為**行運算** (column operations)。而且，我們稱矩陣 A 與矩陣 B 為 column equivalent，if B is obtained from A by applying a finite number of column operations to A。

Example 4

Given, $A = \begin{pmatrix} 3 & 0 & 1 \\ 1 & 2 & 5 \\ -1 & 4 & 2 \end{pmatrix}$. Find a determinant of A.

(*Note*: We shall reduce A into a simpler matrix by applying column operations.)

Solution

Letting $A' = (A^1 - 3A^3, A^2, A^3) = \begin{pmatrix} 0 & 0 & 1 \\ -14 & 2 & 5 \\ -7 & 4 & 2 \end{pmatrix}$ and using the expansion according to the first row, we obtain that

$$F(A') = \begin{vmatrix} -14 & 2 \\ -7 & 4 \end{vmatrix} = -56 + 14 = -42$$

and hence that

$$F(A) = F(A') = -42.$$

Example 5

Given that $A = \begin{pmatrix} 1 & 3 & -1 & 1 \\ 2 & 0 & 3 & 4 \\ 0 & 1 & 2 & 3 \\ -1 & -2 & -2 & 0 \end{pmatrix}$. Find a determinant of A.

Solution

Letting $A' = (A^1, A^2 - 3A^1, A^3 + A^1, A^4 - A^1) = \begin{pmatrix} 1 & 0 & 0 & 0 \\ 2 & -6 & 5 & 2 \\ 0 & 1 & 2 & 3 \\ -1 & 1 & -3 & 1 \end{pmatrix}$

then

$$F(A') = \det \begin{pmatrix} -6 & 5 & 2 \\ 1 & 2 & 3 \\ 1 & -3 & 1 \end{pmatrix}$$

Writing

$$B = \begin{pmatrix} -6 & 5 & 2 \\ 1 & 2 & 3 \\ 1 & -3 & 1 \end{pmatrix}$$

and letting

$$B' = (B^1, B^2 + 3B^1, B^3 - B^1) = \begin{pmatrix} -6 & -13 & 8 \\ 1 & 5 & 2 \\ 1 & 0 & 0 \end{pmatrix}$$

then

$$F(B') = \begin{vmatrix} -13 & 8 \\ 5 & 2 \end{vmatrix} = -26 - 40 = -66$$

Therefore,
$$F(A) = F(A') = F(B') = -66.$$

∎

Exercises

1. Find a determinant of each of the following matrices.

   a. $\begin{pmatrix} 2 & 4 & 3 \\ -1 & 2 & 1 \\ 0 & -3 & 4 \end{pmatrix}$

   b. $\begin{pmatrix} 1 & 2 & -1 & 4 \\ 0 & 1 & -2 & 3 \\ 2 & -1 & 1 & 0 \\ 3 & 1 & 0 & 5 \end{pmatrix}$

2. a. Find a determinant of the following diagonal matrix.

   $\begin{pmatrix} a_{11} & 0 & \cdots & \cdots & 0 \\ 0 & a_{22} & 0 & \cdots & 0 \\ \cdots & 0 & \cdots & \cdots & \cdots \\ \cdots & \cdots & \cdots & \cdots & 0 \\ 0 & \cdots & \cdots & 0 & a_{nn} \end{pmatrix}$

   b. Find a determinant of an upper triangular matrix.

   $\begin{pmatrix} a_{11} & a_{12} & \cdots & \cdots & a_{1n} \\ 0 & a_{22} & \cdots & \cdots & a_{2n} \\ \cdots & 0 & \cdots & \cdots & \cdots \\ \cdots & \cdots & \cdots & \cdots & \cdots \\ 0 & \cdots & \cdots & 0 & a_{nn} \end{pmatrix}$

3. Show that a determinant of $\begin{pmatrix} 1 & x_1 & x_1^2 \\ 1 & x_2 & x_2^2 \\ 1 & x_3 & x_3^2 \end{pmatrix}$ is equal to $(x_2 - x_1)(x_3 - x_1)(x_3 - x_2)$, for every real numbers, x_1, x_2, x_3.

4. Let $f(t)$ and $g(t)$ be two functions having derivatives of all orders. Define

$$\phi(t) = \det \begin{pmatrix} f(t) & g(t) \\ f'(t) & g'(t) \end{pmatrix}$$

Show that $\quad \phi'(t) = \det \begin{pmatrix} f(t) & g(t) \\ f''(t) & g''(t) \end{pmatrix}$

§6.3 Permutations and uniqueness of the determinant

為了要介紹矩陣行列式的另一種表示方式，以及確定行列式的唯一性，本節將先給大家介紹**正整數的排列** (permutations of the set of positive integers)，$\{1, 2,...,n\}$。同時，我們將以符號，$J_n = \{1, 2,...,n\}$，來表示從 1 到 n 的正整數集合。而且，$n \geq 2$ 的情形是我們的主要訴求對象。

Definition 6.6

> A "permutation", σ, of J_n is a bijection from J_n onto J_n.

譬如說，
1. $\sigma : J_n \to J_n$ defined by
   $\sigma(k) = k+1, \ \forall\ k = 1,2,...,n-1$ and $\sigma(n) = 1$
   is a permutation on J_n.
2. $\sigma : J_3 \to J_3$ defined by
   $\sigma(1) = 1, \sigma(2) = 3, \sigma(3) = 2$
   is a permutation on J_3.

在如此定義之下，一些有關排列的基本特性，我們先把它給整理出來，和大家一起先睹為快。
1. $\sigma(i) \neq \sigma(j) \Leftrightarrow i \neq j, \forall\ i, j \in J_n$
2. The inverse permutation, σ^{-1}, of σ is defined by
   $\sigma^{-1}(k) = i \Leftrightarrow \sigma(i) = k,\ \forall\ i, k \in J_n$
   Note, σ^{-1} is indeed a permutation.
3. If σ and τ are permutations of J_n, then the composites, $\sigma \circ \tau$ and $\tau \circ \sigma$, are also permutations of J_n. (We shall simply write $\sigma\tau$ instead of $\sigma \circ \tau$)
4. $\sigma\sigma^{-1} = \sigma^{-1}\sigma = id$, the identity map of J_n.
5. If $\sigma_1,...,\sigma_k$ are permutations of J_n, then $(\sigma_1 \cdots \sigma_k)^{-1} = \sigma_k^{-1} \cdots \sigma_1^{-1}$

Definition 6.7

> A permutation, σ, of J_n is called a "transposition" (對換) of J_n, if $\sigma(i) = j$ and $\sigma(j) = i$, for some $i \neq j \in J_n$, and $\sigma(k) = k$ for every other $k \neq i, j$.

假設 σ 是一個 transposition，那麼我們不難得知，$\sigma\sigma = \sigma^2 = id$。也就是說，$\sigma^{-1} = \sigma$。也因此，一個 transposition，其之逆應射，σ^{-1}，也當然是一個 transposition。

譬如說，

1→1, 2→3, 3→2 就是，$J_3 = \{1, 2, 3\}$ 上的一個 transposition。認識對換有它一個非常大的好處，就是 "任何一個 permutation，都可以被表示為幾個 transpositions 的合成。" 請看下面定理。

Theorem 6.6

> Every permutation of J_n can be expressed as a composite of transpositions.

Proof

我們將使用數學歸納法來證明。

For $n = 2$, a permutation of J_2 is either a transposition or an identity map, and it can be realized that either one of them is a composite of transposition(s).

We now assume that the theorem is true for $n = k$. And, consider for $n = k+1$. If $\sigma \in J_{k+1}$ and $\sigma(k+1) = k+1$, then σ is nothing more than a permutation in J_k, and the result is clear by the assumption. So, we suppose that $\sigma(k+1) = j$, where $j \neq k+1$. Taking τ as the transposition of J_{k+1} such that $\tau(k+1) = j$ and $\tau(j) = k+1$, then we see that $\tau\sigma(k+1) = \tau(j) = k+1$, and hence, $\tau\sigma$ is again a permutation in J_k, thus there are transpositions, $\tau_1, \ldots, \tau_s$, such that $\tau\sigma = \tau_1 \cdots \tau_s$, and therefore, $\sigma = \tau^{-1} \tau_1 \cdots \tau_s$. And the proof is completed.

為了方便起見，往後我們將以下列符號，來表示 J_n 上的一個 permutation σ。

$$\begin{pmatrix} 1 & 2 & \cdots & \cdots & n \\ \sigma(1) & \sigma(2) & \cdots & \cdots & \sigma(n) \end{pmatrix}$$

Example 1

Express the permutation, $\sigma = \begin{pmatrix} 1 & 2 & 3 \\ 3 & 1 & 2 \end{pmatrix}$, as a product of transpositions.

Solution

Let τ be the transposition which interchanges 1 and 3, then $\tau\sigma = \begin{pmatrix} 1 & 2 & 3 \\ 1 & 3 & 2 \end{pmatrix}$. And let τ_1 be the one that interchanges 2 and 3, then $\tau\sigma = \tau_1$ and $\sigma = \tau^{-1}\tau_1 = \tau\tau_1 = \begin{pmatrix} 1 & 2 & 3 \\ 3 & 2 & 1 \end{pmatrix}\begin{pmatrix} 1 & 2 & 3 \\ 1 & 3 & 2 \end{pmatrix}$ ∎

Example 2

Express the permutation, $\sigma = \begin{pmatrix} 1 & 2 & 3 & 4 \\ 2 & 3 & 4 & 1 \end{pmatrix}$, as a product of transpositions.

Solution

Let τ be the transposition which interchanges 1 and 2, then $\tau\sigma = \begin{pmatrix} 1 & 2 & 3 & 4 \\ 1 & 3 & 4 & 2 \end{pmatrix}$. Let τ_1 be the transposition which interchanges 2 and 3, then $\tau_1\tau\sigma = \begin{pmatrix} 1 & 2 & 3 & 4 \\ 1 & 2 & 4 & 3 \end{pmatrix}$. Finally, let τ_2 be the one that interchanges 3 and 4, then $\tau_1\tau\sigma = \tau_2$ and

$$\sigma = \tau^{-1}\tau_1^{-1}\tau_2 = \tau\tau_1\tau_2 = \begin{pmatrix} 1 & 2 & 3 & 4 \\ 2 & 1 & 3 & 4 \end{pmatrix}\begin{pmatrix} 1 & 2 & 3 & 4 \\ 1 & 3 & 2 & 4 \end{pmatrix}\begin{pmatrix} 1 & 2 & 3 & 4 \\ 1 & 2 & 4 & 3 \end{pmatrix}$$ ∎

下面我們介紹更爲精簡的符號。例如說，假設

$$\sigma=\begin{pmatrix}1&2&3\\3&1&2\end{pmatrix}$$

則此後，我們打算將其簡單的表示爲，$\sigma=(1\quad 3\quad 2)$，即，$1\to3,\ 3\to2$，$2\to1$的意思。

又若

$$\sigma=\begin{pmatrix}1&2&3&4\\3&2&4&1\end{pmatrix}$$

則簡單寫爲，$\sigma=(1\quad 3\quad 4)$，即，$1\to3,\ 3\to4$，$4\to1$，其中，數字2保持不變之意。此外，對於 transposition，我們也將以下列簡單的符號表示。例如，

$$\tau=\begin{pmatrix}1&2&3\\3&2&1\end{pmatrix}$$

則我們將其記爲，$\tau=(1\quad 3)$或$(3\quad 1)$。也就是，1 與 3 對調，而數字2保持不變的意思。

有了這些符號之後，任何一個在J_n上的 permutation，它的 transpositions 的合成，就可以輕鬆簡單而又明瞭了。例如前面的例題，

$$\sigma=(1\quad 3\quad 2)=(1\quad 3)(3\quad 2)$$

$$\sigma=(1\quad 2\quad 3\quad 4)=(1\quad 2)(2\quad 3)(3\quad 4)$$

注意，有關 transpositions 的合成，我們要特別小心它的先後順序。譬如說，這個符號$\sigma=(1\quad 3\quad 2)=(1\quad 3)(3\quad 2)$，按照慣例，我們先從右邊的(3 2)看起，然後，再看左邊的(1 3)。也就是說，

$1\to1$，之後，再$1\to3$；此兩合成之後，即爲$1\to3$之意。

$2\to3$，之後，再$3\to1$；此兩合成之後，即爲$2\to1$之意。

$3\to2$，之後，再$2\to2$；此兩合成之後，即爲$3\to2$之意。

通常 transpositions 的合成並非唯一的。譬如說，以前面的 permutation，$\sigma=(1\quad 3\quad 2)$來說，它也可以被寫爲$\sigma=(3\quad 2\quad 1)=(3\quad 2)(2\quad 1)$。

Definition 6.8

Given an $n \times n$ matrix, $A = (A^1, ..., A^n)$, with $\det(A) \neq 0$. If σ is a permutation of J_n, then the "sign" of σ, denoted by $\varepsilon(\sigma)$, is defined to be
$$\varepsilon(\sigma) = \det(A^{\sigma(1)}, ..., A^{\sigma(n)}) / \det(A^1, ..., A^n)$$

至於，$\det(A) = 0$ 的情況，我們則定義其任意排列之符號，$\varepsilon(\sigma)$，為 1。按照這樣一個符號定義，我們可以馬上得知，對任一 transposition，τ，而言，假如 $\det(A) \neq 0$，則 $\varepsilon(\tau) = -1$。而且，對任意排列 σ，其之符號，$\varepsilon(\sigma) = \pm 1$。

Theorem 6.7

If σ and σ' are permutations of J_n, then $\varepsilon(\sigma\sigma') = \varepsilon(\sigma)\varepsilon(\sigma')$.

Proof

這個證明不會太難，首先我們看看
$$\det(A^{\sigma\sigma'(1)}, ..., A^{\sigma\sigma'(n)}) = \varepsilon(\sigma)\det(A^{\sigma'(1)}, ..., A^{\sigma'(n)})$$
$$= \varepsilon(\sigma)\varepsilon(\sigma')\det(A^1, ..., A^n)$$

另外，$\sigma\sigma'$ 也是一個 permutation，所以，我們也有下面結果
$$\det(A^{\sigma\sigma'(1)}, ..., A^{\sigma\sigma'(n)}) = \varepsilon(\sigma\sigma')\det(A^1, ..., A^n)$$

上述兩式相比，得出
$$\varepsilon(\sigma\sigma') = \varepsilon(\sigma)\varepsilon(\sigma')$$

Corollary 1

If σ is a permutation of J_n with $\sigma = \tau_1 \cdots \tau_s$, where each τ_i is a transposition of J_n, then $\varepsilon(\sigma) = (-1)^s$.

Proof

這是前述定理的一個簡單推論。因為 $\sigma = \tau_1...\tau_s$ 而且 $\varepsilon(\tau_i) = -1$，$\forall\, i = 1,...,s$。所以，

$$\varepsilon(\sigma) = \varepsilon(\tau_1) \cdots \varepsilon(\tau_s) = (-1)^s$$

Corollary 2

If $\sigma = \tau_1 \cdots \tau_s$ and also $\sigma = \tau'_1 \cdots \tau'_r$, then either both r and s are even or odd.

Proof

這是一個顯然的結果。因為由前述推論得知，

$$\varepsilon(\sigma) = (-1)^s = (-1)^r$$

一個排列 σ，我們將其稱為，"even"，假若，$\varepsilon(\sigma) = 1$，又為，"odd"，假若，$\varepsilon(\sigma) = -1$。另外，由於，

$$id = \sigma\sigma^{-1} \;\Rightarrow\; \varepsilon(id) = \varepsilon(\sigma)\varepsilon(\sigma^{-1}) \;\Rightarrow\; 1 = \varepsilon(\sigma)\varepsilon(\sigma^{-1})$$

所以，$\varepsilon(\sigma) = \varepsilon(\sigma^{-1})$。也就是說，$\sigma$ 與 σ^{-1} 同時為 even，或同時為 odd。

Example 3

Find $\varepsilon(\sigma)$, if $\sigma = \begin{pmatrix} 1 & 2 & 3 & 4 \\ 2 & 3 & 4 & 1 \end{pmatrix}$.

Solution

$$\sigma = \begin{pmatrix} 1 & 2 & 3 & 4 \\ 2 & 3 & 4 & 1 \end{pmatrix} = \begin{pmatrix} 1 & 2 & 3 & 4 \end{pmatrix} = \begin{pmatrix} 1 & 2 \end{pmatrix}\begin{pmatrix} 2 & 3 \end{pmatrix}\begin{pmatrix} 3 & 4 \end{pmatrix}$$

$$\therefore \quad \varepsilon(\sigma) = (-1)^3 = -1$$

Example 4

Find $\varepsilon(\sigma)$, if $\sigma = \begin{pmatrix} 1 & 2 & 3 & 4 \\ 3 & 2 & 4 & 1 \end{pmatrix}$

Solution

$$\sigma = \begin{pmatrix} 1 & 2 & 3 & 4 \\ 3 & 2 & 4 & 1 \end{pmatrix} = (1 \quad 3 \quad 4) = (1 \quad 3)(3 \quad 4)$$

$\Rightarrow \quad \varepsilon(\sigma) = (-1)^2 = 1$

下面,請同學們練習幾個題目,以便快速進入狀況。

Exercises

Express each of the following permutations as a product of transpositions and determine the sign of each of them.

1. $\begin{pmatrix} 1 & 2 & 3 \\ 3 & 1 & 2 \end{pmatrix}$

2. $\begin{pmatrix} 1 & 2 & 3 \\ 3 & 2 & 1 \end{pmatrix}$

3. $\begin{pmatrix} 1 & 2 & 3 & 4 \\ 3 & 1 & 4 & 2 \end{pmatrix}$

4. $\begin{pmatrix} 1 & 2 & 3 & 4 \\ 4 & 1 & 3 & 2 \end{pmatrix}$

了解了 permutations 之後,別忘了我們的主要目標,是在研究矩陣的行列式。首先,回憶一下行列式的一些特性,然後用以導出行列式的另一種表示法。從而,我們將證明,一個 square matrix 的行列式是唯一的。

我們從一個 3×3 矩陣談起。已知有一 3×3 矩陣 A,以及 R^3 上的一組基底向量,X^1, X^2, X^3。我們知道,矩陣 A 的三個 column vectors,A^1, A^2 以及 A^3 可以被表示為如下之線性組合,

$$A^1 = b_{11}X^1 + b_{21}X^2 + b_{31}X^3 = \sum_{i=1}^{3} b_{i1}X^i$$

$$A^2 = b_{12}X^1 + b_{22}X^2 + b_{32}X^3 = \sum_{j=1}^{3} b_{j2}X^j$$

$$A^3 = b_{13}X^1 + b_{23}X^2 + b_{33}X^3 = \sum_{l=1}^{3} b_{l3}X^l$$

若 D 為 3×3 矩陣的行列式，則我們發現
$$D(A) = D(A^1, A^2, A^3) = D(\sum_{i=1}^{3} b_{i1}X^i, \sum_{j=1}^{3} b_{j2}X^j, \sum_{l=1}^{3} b_{l3}X^l)$$

由行列式的特性得知

$$D(A) = \sum_{i=1}^{3} b_{i1} D(X^i, \sum_{j=1}^{3} b_{j2}X^j, \sum_{l=1}^{3} b_{l3}X^l)$$

$$= \sum_{l=1}^{3} \sum_{j=1}^{3} \sum_{i=1}^{3} b_{i1} b_{j2} b_{l3} D(X^i, X^j, X^l)$$

$$= \sum_{l=1}^{3} \sum_{j=1}^{3} (b_{11} b_{j2} b_{l3} D(X^1, X^j, X^l) + b_{21} b_{j2} b_{l3} D(X^2, X^j, X^l) + b_{31} b_{j2} b_{l3} D(X^3, X^j, X^l))$$

$$= \sum_{l=1}^{3} (b_{11} b_{22} b_{l3} D(X^1, X^2, X^l) + b_{11} b_{32} b_{l3} D(X^1, X^3, X^l) + b_{21} b_{12} b_{l3} D(X^2, X^1, X^l) + b_{21} b_{32} b_{l3} D(X^2, X^3, X^l) + b_{31} b_{12} b_{l3} D(X^3, X^1, X^l) + b_{31} b_{22} b_{l3} D(X^3, X^2, X^l))$$

(注意，由於 $D(X^1, X^1, X^l) = D(X^2, X^2, X^l) = D(X^3, X^3, X^l) = 0$，所以，前式只剩六項。)一般來說，只要 X^i, X^j 與 X^l 中有任意兩個相同，則 $D(X^i, X^j, X^l) = 0$。因此，前式更進一步的等於

$$b_{11} b_{22} b_{33} D(X^1, X^2, X^3) + b_{11} b_{32} b_{23} D(X^1, X^3, X^2) +$$
$$b_{21} b_{12} b_{33} D(X^2, X^1, X^3) + b_{21} b_{32} b_{13} D(X^2, X^3, X^1) +$$
$$b_{31} b_{12} b_{23} D(X^3, X^1, X^2) + b_{31} b_{22} b_{13} D(X^3, X^2, X^1)$$
$$= \sum_{\sigma} b_{\sigma(1)1} b_{\sigma(2)2} b_{\sigma(3)3} D(X^{\sigma(1)}, X^{\sigma(2)}, X^{\sigma(3)})$$
$$= \sum_{\sigma} \varepsilon(\sigma) b_{\sigma(1)1} b_{\sigma(2)2} b_{\sigma(3)3} D(X^1, X^2, X^3)$$

前述式子中，要是我們所選取的基底是標準基底的話，那麼 $A^1 = a_{11}E^1 + a_{21}E^2 + a_{31}E^3$，$A^2 = a_{12}E^1 + a_{22}E^2 + a_{32}E^3$，$A^3 = a_{13}E^1 + a_{23}E^2 + a_{33}E^3$，而且，$D(X^1, X^2, X^3) = D(E^1, E^2, E^3) = D(I_3) = 1$。

因此，$D(A) = \sum_\sigma \varepsilon(\sigma) a_{\sigma(1)1} a_{\sigma(2)2} a_{\sigma(3)3}$。利用這個結果，我們先來檢查一下，$2 \times 2$ 矩陣的行列式。

Example 5

Let $A = \begin{pmatrix} a & b \\ c & d \end{pmatrix}$, then $D(A) = ad - bc$.

Proof

Let $E^1 = \begin{pmatrix} 1 \\ 0 \end{pmatrix}, E^2 = \begin{pmatrix} 0 \\ 1 \end{pmatrix}$ be the standard basis of R^2. Then

$$A^1 = aE^1 + cE^2 \text{ and } A^2 = bE^1 + dE^2$$

因此，
$$\begin{aligned} D(A) &= D(aE^1 + cE^2, bE^1 + dE^2) \\ &= abD(E^1, E^1) + adD(E^1, E^2) + bcD(E^2, E^1) + cdD(E^2, E^2) \\ &= ad - bc. \end{aligned}$$
∎

太好了，這個結果正如同 6.1 節一開始所定義的，2×2 矩陣的行列式一樣。可無論如何，我們目前主要的工作還是要導出 $n \times n$ 矩陣的行列式，那就是下面這個定理。

Theorem 6.8

Let D be a determinant function on $n \times n$ matrices. Suppose that A is an $n \times n$ matrix with columns, $A^1, ..., A^n$, and if $X^1, ..., X^n$ are column vectors in R^n such that $A^i = b_{1i} X^1 + b_{2i} X^2 + \cdots + b_{ni} X^n$, $\forall\ i = 1, 2, ..., n$, then
$$D(A) = \sum_\sigma \varepsilon(\sigma) b_{\sigma(1)1} b_{\sigma(2)2} \cdots b_{\sigma(n)n} D(X^1, X^2, ..., X^n)$$

Proof

本定理證明與前述 3×3 矩陣的解說過程相類似，我們不打算再重複此一動作。現在，我們所要強調的是行列式的唯一性質。

Corollary (the uniqueness of the determinant)

Let $A = (a_{ij})$ be a real $n \times n$ matrix. The determinant of A is uniquely determined and $D(A) = \sum_{\sigma} \varepsilon(\sigma) a_{\sigma(1)1} a_{\sigma(2)2} \cdots a_{\sigma(n)n}$

Proof

From the last theorem, if $\{E^1, E^2, ..., E^n\}$ is the standard basis for R^n, then we have, for each $i = 1, 2, 3, ..., n$, $A^i = \sum_{j=1}^{n} a_{ji} E^j$, and hence,

$$D(A) = \sum_{\sigma} \varepsilon(\sigma) a_{\sigma(1)1} a_{\sigma(2)2} \cdots a_{\sigma(n)n} D(I_n)$$
$$= \sum_{\sigma} \varepsilon(\sigma) a_{\sigma(1)1} a_{\sigma(2)2} \cdots a_{\sigma(n)n}$$

上個定理與推論告訴我們，只要 D 是一個 determinant function，那麼結論一定是，對任意 $n \times n$ 矩陣 A 而言，$D(A) = \sum_{\sigma} \varepsilon(\sigma) a_{\sigma(1)1} a_{\sigma(2)2} \cdots a_{\sigma(n)n}$。這個結果說明了，行列式的唯一性質。有關行列式的符號，除了 $D(A)$ 之外，$\det(A)$ 或 $|A|$ 也常被使用。下面是行列式的幾個重要特性。

Theorem 6.9

Let A and B be two $n \times n$ matrices. Then $D(AB) = D(A)D(B)$.

Proof

Let $C = AB$. Then by the definition of matrix multiplication,
$$C^k = b_{1k} A^1 + b_{2k} A^2 + \cdots + b_{nk} A^n, \quad \forall \, k = 1, 2, ..., n$$
and hence by the previous theorem,
$$D(C) = \sum_{\sigma} \varepsilon(\sigma) b_{\sigma(1)1} b_{\sigma(2)2} \cdots b_{\sigma(n)n} D(A^1, ..., A^n)$$
$$= \sum_{\sigma} \varepsilon(\sigma) b_{\sigma(1)1} b_{\sigma(2)2} \cdots b_{\sigma(n)n} D(A)$$
$$= D(B) D(A)$$

Corollary

> Let A be a non-singular matrix. Then $D(A^{-1}) = \dfrac{1}{D(A)}$.

Proof

$$1 = D(I_n) = D(A^{-1}A) = D(A^{-1})D(A)$$

$$\Rightarrow \quad D(A^{-1}) = \frac{1}{D(A)}$$

前述推論不僅簡單明瞭，而且我們也可以從中得知，假若矩陣 A 是可逆的話，則其行列式是不為 0 的。也就是說，$D(A) \neq 0$。下面的定理是我們在 6.1 節時曾經提及的，矩陣在轉置之後，其行列式值不變。

Theorem 6.10

> Let A be a square matrix. Then $D({}^tA) = D(A)$.

Proof

這個證明需要費點心思，我們得沉著應戰。We first write S as the set of permutations of J_n, and consider the map, M, mapping from S into S defined by $M(\sigma) = \sigma^{-1}$, $\forall\ \sigma \in S$. (It can be easily seen that the map, M, is a bijection.)

Now, if $\sigma \in S$ with $\sigma(j) = k$, then $\sigma^{-1}(k) = j$, and then $a_{\sigma(j)j} = a_{k\sigma^{-1}(k)}$. Since the integer k occurs exactly once among $\sigma(1), \sigma(2), \ldots, \sigma(n)$, we thus have,

$$a_{\sigma(1)1} a_{\sigma(2)2} \cdots a_{\sigma(n)n} = a_{1\sigma^{-1}(1)} a_{2\sigma^{-1}(2)} \cdots a_{n\sigma^{-1}(n)}$$

Hence,

$$D(A) = \sum_\sigma \varepsilon(\sigma) a_{\sigma(1)1} a_{\sigma(2)2} \cdots a_{\sigma(n)n}$$

$$= \sum_\sigma \varepsilon(\sigma) a_{1\sigma^{-1}(1)} a_{2\sigma^{-1}(2)} \cdots a_{n\sigma^{-1}(n)}$$

$$= \sum_{\sigma^{-1}} \varepsilon(\sigma^{-1}) a_{1\sigma^{-1}(1)} a_{2\sigma^{-1}(2)} \cdots a_{n\sigma^{-1}(n)}, \quad \text{since } \varepsilon(\sigma^{-1}) = \varepsilon(\sigma).$$

Consequently, since the map, $M(\sigma) = \sigma^{-1}$, $\forall\, \sigma \in S$, is a bijection, we must have,

$$\sum_{\sigma^{-1}} \varepsilon(\sigma^{-1}) a_{1\sigma^{-1}(1)} a_{2\sigma^{-1}(2)} \cdots a_{n\sigma^{-1}(n)} = \sum_{\sigma} \varepsilon(\sigma) a_{1\sigma(1)} a_{2\sigma(2)} \cdots a_{n\sigma(n)}$$

Therefore,

$$D(A) = \sum_{\sigma^{-1}} \varepsilon(\sigma^{-1}) a_{1\sigma^{-1}(1)} a_{2\sigma^{-1}(2)} \cdots a_{n\sigma^{-1}(n)}$$

$$= \sum_{\sigma} \varepsilon(\sigma) a_{1\sigma(1)} a_{2\sigma(2)} \cdots a_{n\sigma(n)}$$

$$= D({}^t A)$$

Exercises

1. Let A be a non-singular $n \times n$ matrix, show that if B is any $n \times n$ matrix, then $\det(ABA^{-1}) = \det(B)$.

2. Let A be an $n \times n$ matrix with linearly dependent columns. Show that $\det(A) = 0$.

   (*Hint*: Some column of A is a linear combination of all the other columns.)

3. A square matrix, C, of the form, $C = \begin{pmatrix} A & 0 \\ 0 & B \end{pmatrix}$, where A and B are square matrices and each 0 denotes a zero-matrix, is called a block-diagonal matrix with diagonal blocks A and B. Show that $\det(C) = \det(A)\det(B)$.

   (*Hint*: $\begin{pmatrix} A & 0 \\ 0 & B \end{pmatrix} = \begin{pmatrix} A & 0 \\ 0 & I_m \end{pmatrix}\begin{pmatrix} I_n & 0 \\ 0 & B \end{pmatrix}$)

4. Prove that if A is an orthonormal matrix, then the determinant of A is either 1 or -1.
   (*Hint*: ${}^t AA = I$)(In complex case, A is unitary, if $(A^*)A = I$. And the determinant of A is also either 1 or -1.)

§6.4 Computing determinants

這節我們想利用前面所學過的公式，給大家介紹行列式的幾個計算方法。譬如說，3.2 節的 *LU*-factorization、5.4 節的 ortho-triangular factorization、以及 block triangular factorization。首先看看，*LU*-factorization。

Let A be an $n \times n$ matrix which has an *LU*-factorization $A = LU$. 譬如說，A is a 3×3 matrix, then

$$A = \begin{pmatrix} l_{11} & 0 & 0 \\ l_{21} & l_{22} & 0 \\ l_{31} & l_{32} & l_{33} \end{pmatrix} \begin{pmatrix} 1 & u_{12} & u_{13} \\ 0 & 1 & u_{23} \\ 0 & 0 & 1 \end{pmatrix}$$

Hence, we have that

$$\det(A) = \det(L)\det(U) = \prod_{i=1}^{3} l_{ii}$$

又譬如說

$$A = \begin{pmatrix} 3 & 4 & 5 \\ 2 & 1 & -1 \\ -2 & 2 & 8 \end{pmatrix} = \begin{pmatrix} 3 & 0 & 0 \\ 2 & -5/3 & 0 \\ -2 & 14/3 & -4/5 \end{pmatrix} \begin{pmatrix} 1 & 4/3 & 5/3 \\ 0 & 1 & 13/5 \\ 0 & 0 & 1 \end{pmatrix}$$

We find out that

$$\det(A) = 3 \cdot \frac{-5}{3} \cdot \frac{-4}{5} = 4$$

這樣一個方便的方法，主要的還是要靠 *LU*-factorization 的動作。下面我們不妨從 *LU*-factorization 開始，來求一個 $n \times n$ 矩陣的行列式。

Example 1

Use *LU*-factorization to find the determinant of

$$A = \begin{pmatrix} 2 & 1 & -2 \\ 1 & 2 & 3 \\ 5 & -1/2 & 0 \end{pmatrix}$$

Solution

同 3.2 節一樣，我們令

$$A = \begin{pmatrix} 2 & 1 & -2 \\ 1 & 2 & 3 \\ 5 & -1/2 & 0 \end{pmatrix} = \begin{pmatrix} l_{11} & 0 & 0 \\ l_{21} & l_{22} & 0 \\ l_{31} & l_{32} & l_{33} \end{pmatrix} \begin{pmatrix} 1 & u_{12} & u_{13} \\ 0 & 1 & u_{23} \\ 0 & 0 & 1 \end{pmatrix}$$

從而得知，

$$L^1 = \begin{pmatrix} l_{11} \\ l_{21} \\ l_{31} \end{pmatrix} = \begin{pmatrix} 2 \\ 1 \\ 5 \end{pmatrix}$$

以及 $(1 \quad u_{12} \quad u_{13}) = (1 \quad 1/2 \quad -1)$

接著，$\frac{1}{2} + l_{22} = 2$，以及，$\frac{5}{2} + l_{32} = \frac{-1}{2}$。所以，$l_{22} = \frac{3}{2}$，$l_{32} = -3$

又由，$-1 + \frac{3}{2}u_{23} = 3$，得出，$u_{23} = \frac{8}{3}$

最後再由，$-5 - 3 \cdot \frac{8}{3} + l_{33} = 0$，算出，$l_{33} = 13$。結果得到

$$A = \begin{pmatrix} 2 & 1 & -2 \\ 1 & 2 & 3 \\ 5 & -1/2 & 0 \end{pmatrix} = \begin{pmatrix} 2 & 0 & 0 \\ 1 & 3/2 & 0 \\ 5 & -3 & 13 \end{pmatrix} \begin{pmatrix} 1 & 1/2 & -1 \\ 0 & 1 & 8/3 \\ 0 & 0 & 1 \end{pmatrix}$$

所以矩陣 A 的行列式為

$$\det(A) = 2 \cdot \frac{3}{2} \cdot 13 = 39$$

顯然的，這不是一個很快的方法，但畢竟也是一個方法。請同學不要怕麻煩，以下先練習一個題目。

Exercise

Use *LU*-factorization to find the determinant of

$$A = \begin{pmatrix} 3 & 4 & 5 \\ 2 & 1 & -1 \\ -2 & 2 & 8 \end{pmatrix}$$

下面要看的是，5.4 節的 ortho-triangular factorization。

Suppose that $A = BC$ is an ortho-triangular factorization of A, where B has orthogonal columns and C is an upper unit-triangular matrix. Say,

$$A = B \begin{pmatrix} 1 & \cdots & \cdots & \cdots \\ 0 & 1 & \cdots & \cdots \\ 0 & 0 & \cdots & \cdots \\ 0 & 0 & 0 & 1 \end{pmatrix}$$

From this, we see that $\det(A) = \det(B)$.

Since B is an orthogonal matrix, ${}^t BB = D$, where D is a diagonal matrix. So, we have that $\det({}^t B)\det(B) = \prod_{i=1}^{n} d_{ii}$, or that $(\det(B))^2 = \prod_{i=1}^{n} d_{ii}$. Hence, we see that $(\det(A))^2 = \prod_{i=1}^{n} d_{ii}$ or that $\det(A) = \pm\sqrt{\prod_{i=1}^{n} d_{ii}}$.

在這兒很可惜的是，目前為止我們似乎沒有辦法從中決定 $\det(A)$ 的符號。但是，無論如何也有它值得一看的地方。我們拿 5.4 節一開始的例子來看看

$$A = \begin{pmatrix} 1 & 0 & -1 \\ 0 & 1 & -1 \\ 1 & 2 & -1 \end{pmatrix} = BC = \begin{pmatrix} 1 & -1 & -1/3 \\ 0 & 1 & -2/3 \\ 1 & 1 & 1/3 \end{pmatrix} \begin{pmatrix} 1 & 1 & -1 \\ 0 & 1 & -1/3 \\ 0 & 0 & 1 \end{pmatrix}$$

$${}^t BB = \begin{pmatrix} 1 & 0 & 1 \\ -1 & 1 & 1 \\ -1/3 & -2/3 & 1/3 \end{pmatrix} \begin{pmatrix} 1 & -1 & -1/3 \\ 0 & 1 & -2/3 \\ 1 & 1 & 1/3 \end{pmatrix} = \begin{pmatrix} 2 & 0 & 0 \\ 0 & 3 & 0 \\ 0 & 0 & 2/3 \end{pmatrix}$$

$$\det(A) = \pm\sqrt{2 \cdot 3 \cdot \frac{2}{3}} = \pm 2$$

最後我們所要介紹的是，method of block matrices。也就是，把一個矩陣 A 給分解成

$$A = \begin{pmatrix} I & 0 \\ X & I \end{pmatrix} \begin{pmatrix} G & 0 \\ 0 & F \end{pmatrix} \begin{pmatrix} I & Y \\ 0 & I \end{pmatrix}$$

其中，$\begin{pmatrix} I & 0 \\ X & I \end{pmatrix}$, $\begin{pmatrix} G & 0 \\ 0 & F \end{pmatrix}$, $\begin{pmatrix} I & Y \\ 0 & I \end{pmatrix}$ 都為 block matrices，$\begin{pmatrix} I & 0 \\ X & I \end{pmatrix}$, $\begin{pmatrix} I & Y \\ 0 & I \end{pmatrix}$ 叫做 block triangular matrices，而 $\begin{pmatrix} G & 0 \\ 0 & F \end{pmatrix}$ 則稱為 block diagonal matrix。那麼，在這種情形之下，$\det(A) = \det(G)\det(F)$。

註：參看 6.3 節，

$$\det\begin{pmatrix} I & 0 \\ X & I \end{pmatrix} = 1, \quad \det\begin{pmatrix} I & Y \\ 0 & I \end{pmatrix} = 1$$

而且，$\det\begin{pmatrix} G & 0 \\ 0 & F \end{pmatrix} = \det(G)\det(F)$

所以，現在的問題是，如何把矩陣 A 化成

$$A = \begin{pmatrix} I & 0 \\ X & I \end{pmatrix} \begin{pmatrix} G & 0 \\ 0 & F \end{pmatrix} \begin{pmatrix} I & Y \\ 0 & I \end{pmatrix}$$

的型式。這個型式我們姑且稱之為 *LDU*-factorization。其中，L 是 block lower triangular matrix，D 是 block diagonal matrix，U 是 block upper triangular matrix。

Theorem 6.11

Given a block matrix, $A = \begin{pmatrix} M & N \\ U & P \end{pmatrix}$. If M is a non-singular matrix, then
$$A = \begin{pmatrix} I & 0 \\ UM^{-1} & I \end{pmatrix} \begin{pmatrix} M & 0 \\ 0 & P - UM^{-1}N \end{pmatrix} \begin{pmatrix} I & M^{-1}N \\ 0 & I \end{pmatrix}$$

Proof

這個證明沒什麼，只要做一下矩陣的乘法即可以得證。

在這種情形之下，$\det(A) = \det(M)\det(P - UM^{-1}N)$。其實，前述公式不僅在計算 $\det(A)$ 有它的方便性，就連求矩陣 A 的乘法反元素方面，也幫了不少忙。(因為，矩陣 $\begin{pmatrix} I & 0 \\ X & I \end{pmatrix}$ 與矩陣 $\begin{pmatrix} I & Y \\ 0 & I \end{pmatrix}$ 的乘法反元素分別為 $\begin{pmatrix} I & 0 \\ -X & I \end{pmatrix}$ 與 $\begin{pmatrix} I & -Y \\ 0 & I \end{pmatrix}$。而矩陣 $\begin{pmatrix} G & 0 \\ 0 & F \end{pmatrix}$ 的乘法反元素則為 $\begin{pmatrix} G^{-1} & 0 \\ 0 & F^{-1} \end{pmatrix}$。)

有一件事情要特別提醒注意的是，定理中的 M 及 P 務必是 square matrices，至於 N 和 U 則可不一定是 square matrices。下面，我們練習一個簡單的例子。

Example 2

Find $\det(A)$ and A^{-1}, given that
$$A = \begin{pmatrix} 1 & 2 & 1 & 1 \\ 0 & 1 & 0 & 1 \\ 1 & 1 & 1 & 1 \\ 0 & 0 & -1 & 0 \end{pmatrix}$$

Solution

按照虛線所隔離出來的 block matrices，得知
$$M = \begin{pmatrix} 1 & 2 \\ 0 & 1 \end{pmatrix}, N = \begin{pmatrix} 1 & 1 \\ 0 & 1 \end{pmatrix}, U = \begin{pmatrix} 1 & 1 \\ 0 & 0 \end{pmatrix}, P = \begin{pmatrix} 1 & 1 \\ -1 & 0 \end{pmatrix}$$

現在我們分別計算，
$$M^{-1} = \begin{pmatrix} 1 & -2 \\ 0 & 1 \end{pmatrix}, UM^{-1} = \begin{pmatrix} 1 & -1 \\ 0 & 0 \end{pmatrix}, M^{-1}N = \begin{pmatrix} 1 & -1 \\ 0 & 1 \end{pmatrix},\text{ 以及}$$
$$UM^{-1}N = \begin{pmatrix} 1 & 0 \\ 0 & 0 \end{pmatrix}, P - UM^{-1}N = \begin{pmatrix} 0 & 1 \\ -1 & 0 \end{pmatrix}, (P - UM^{-1}N)^{-1} = \begin{pmatrix} 0 & -1 \\ 1 & 0 \end{pmatrix}$$

所以，
$$\det(A) = \det(M)\det(P - UM^{-1}N) = 1$$

又，
$$A = \begin{pmatrix} I & 0 \\ UM^{-1} & I \end{pmatrix} \begin{pmatrix} M & 0 \\ 0 & P-UM^{-1}N \end{pmatrix} \begin{pmatrix} I & M^{-1}N \\ 0 & I \end{pmatrix}$$

所以
$$A^{-1} = (\begin{pmatrix} I & 0 \\ UM^{-1} & I \end{pmatrix} \begin{pmatrix} M & 0 \\ 0 & P-UM^{-1}N \end{pmatrix} \begin{pmatrix} I & M^{-1}N \\ 0 & I \end{pmatrix})^{-1}$$

$$= \begin{pmatrix} I & M^{-1}N \\ 0 & I \end{pmatrix}^{-1} \begin{pmatrix} M & 0 \\ 0 & P-UM^{-1}N \end{pmatrix}^{-1} \begin{pmatrix} I & 0 \\ UM^{-1} & I \end{pmatrix}^{-1}$$

$$= \begin{pmatrix} I & -M^{-1}N \\ 0 & I \end{pmatrix} \begin{pmatrix} M^{-1} & 0 \\ 0 & (P-UM^{-1}N)^{-1} \end{pmatrix} \begin{pmatrix} I & 0 \\ -UM^{-1} & I \end{pmatrix}$$

最後得出，
$$A^{-1} = \begin{pmatrix} 1 & 0 & -1 & 1 \\ 0 & 1 & 0 & -1 \\ 0 & 0 & 1 & 0 \\ 0 & 0 & 0 & 1 \end{pmatrix} \begin{pmatrix} 1 & -2 & 0 & 0 \\ 0 & 1 & 0 & 0 \\ 0 & 0 & 0 & -1 \\ 0 & 0 & 1 & 0 \end{pmatrix} \begin{pmatrix} 1 & 0 & 0 & 0 \\ 0 & 1 & 0 & 0 \\ -1 & 1 & 1 & 0 \\ 0 & 0 & 0 & 1 \end{pmatrix}$$

$$= \begin{pmatrix} 0 & -1 & 1 & 1 \\ 1 & 0 & -1 & 0 \\ 0 & 0 & 0 & -1 \\ -1 & 1 & 1 & 0 \end{pmatrix}$$

Exercise

Find $\det(A)$ and A^{-1}, given that
$$A = \begin{pmatrix} 2 & 0 & 0 & 1 \\ 1 & 1 & 0 & -2 \\ \hdashline 1 & -1 & 1 & 0 \\ 0 & 0 & 2 & -1 \end{pmatrix}$$

這個方法有一個比較不方便的地方，在於矩陣 M 的選取，所選出來的 M 必須是可逆的。因此之故，我們通常選擇比較小的 M，例如，2×2 矩陣，甚至於1×1矩陣。也就是說，只要 $M = (a_{11})$ 而 $a_{11} \neq 0$，那麼 M 就是可逆的。根據這個說法，我們再看一個例題。

Example 3

Find $\det(A)$ and A^{-1}, given that
$$A = \begin{pmatrix} 2 & 4 & 7 \\ 1 & 1 & -1 \\ 3 & 2 & 4 \end{pmatrix}$$

Solution

$$M = (2),\ N = (4\ \ 7),\ U = \begin{pmatrix} 1 \\ 3 \end{pmatrix},\ P = \begin{pmatrix} 1 & -1 \\ 2 & 4 \end{pmatrix}$$

$$M^{-1} = \begin{pmatrix} \frac{1}{2} \end{pmatrix},\ UM^{-1} = \begin{pmatrix} 1/2 \\ 3/2 \end{pmatrix},\ M^{-1}N = (2\ \ 7/2),\ UM^{-1}N = \begin{pmatrix} 2 & 7/2 \\ 6 & 21/2 \end{pmatrix}$$

$$P - UM^{-1}N = \begin{pmatrix} -1 & -9/2 \\ -4 & -13/2 \end{pmatrix}$$

$$A = \begin{pmatrix} I & 0 \\ UM^{-1} & I \end{pmatrix} \begin{pmatrix} M & 0 \\ 0 & P - UM^{-1}N \end{pmatrix} \begin{pmatrix} I & M^{-1}N \\ 0 & I \end{pmatrix}$$

$$= \begin{pmatrix} 1 & 0 & 0 \\ 1/2 & 1 & 0 \\ 3/2 & 0 & 1 \end{pmatrix} \begin{pmatrix} 2 & 0 & 0 \\ 0 & -1 & -9/2 \\ 0 & -4 & -13/2 \end{pmatrix} \begin{pmatrix} 1 & 2 & 7/2 \\ 0 & 1 & 0 \\ 0 & 0 & 1 \end{pmatrix}$$

$$\det(A) = \det(M)\det(P - UM^{-1}N) = 2 \cdot \frac{-23}{2} = -23$$

$$A^{-1} = \begin{pmatrix} 1 & -2 & -7/2 \\ 0 & 1 & 0 \\ 0 & 0 & 1 \end{pmatrix} \begin{pmatrix} 1/2 & 0 & 0 \\ 0 & 13/23 & -9/23 \\ 0 & -8/23 & 2/23 \end{pmatrix} \begin{pmatrix} 1 & 0 & 0 \\ -1/2 & 1 & 0 \\ -3/2 & 0 & 1 \end{pmatrix}$$

$$= \begin{pmatrix} -6/23 & 2/23 & 11/23 \\ 7/23 & 13/23 & -9/23 \\ 1/23 & -8/23 & 2/23 \end{pmatrix}$$

由上述例題當中，我們發現，使用這個選擇 $M = (a_{11})$，而求出 $\det(A)$ 的方式，似乎比以列展開式來的簡單。

Theorem 6.12

Let A be an $n \times n$ matrix with $a_{11} \neq 0$ and if A is in the partitioned form

$$A = \left(\begin{array}{c|c} a_{11} & N \\ \hline U & P \end{array} \right)$$

then $\det(A) = (\dfrac{1}{a_{11}})^{n-2} \det(a_{11}P - UN)$.

Proof

由式子

$$A = \begin{pmatrix} I & 0 \\ UM^{-1} & I \end{pmatrix} \begin{pmatrix} M & 0 \\ 0 & P - UM^{-1}N \end{pmatrix} \begin{pmatrix} I & M^{-1}N \\ 0 & I \end{pmatrix}$$

得知

$$\det(A) = \det(M)\det(P - UM^{-1}N) = a_{11}\det(P - U\dfrac{1}{a_{11}}N)$$

$$= a_{11}(\dfrac{1}{a_{11}})^{n-1}\det(a_{11}P - UN) = (\dfrac{1}{a_{11}})^{n-2}\det(a_{11}P - UN)$$

這個公式叫做**軸元壓縮法** (pivotal condensation)，也就是將 $n \times n$ 矩陣的行列式展開而為 $(n-1) \times (n-1)$ 矩陣的行列式表示法。

Example 4

Find the determinant of $A = \begin{pmatrix} 3 & 0 & 1 \\ -1 & 4 & 1 \\ 2 & 0 & -5 \end{pmatrix}$ by using pivotal condensation.

Solution

$$M = (3),\ N = (0\ \ 1),\ U = \begin{pmatrix} -1 \\ 2 \end{pmatrix},\ P = \begin{pmatrix} 4 & 1 \\ 0 & -5 \end{pmatrix}$$

$$\det(A) = (\frac{1}{3})\det(3\begin{pmatrix} 4 & 1 \\ 0 & -5 \end{pmatrix} - \begin{pmatrix} 0 & -1 \\ 0 & 2 \end{pmatrix}) = \frac{1}{3}\det\begin{pmatrix} 12 & 4 \\ 0 & -17 \end{pmatrix} = -68 \qquad \blacksquare$$

我們鼓起勇氣，試試看一個 4×4 矩陣的例子，如何？

Example 5

Find the determinant of $A = \begin{pmatrix} 3 & -1 & 0 & 1 \\ 0 & 2 & -1 & 3 \\ 1 & 1 & 4 & 4 \\ 2 & 0 & -2 & 0 \end{pmatrix}$ by using pivotal condensation.

Solution

$$M = (3),\ N = (-1\ \ 0\ \ 1),\ U = \begin{pmatrix} 0 \\ 1 \\ 2 \end{pmatrix},\ P = \begin{pmatrix} 2 & -1 & 3 \\ 1 & 4 & 4 \\ 0 & -2 & 0 \end{pmatrix}$$

則 $\quad UN = \begin{pmatrix} 0 & 0 & 0 \\ -1 & 0 & 1 \\ -2 & 0 & 2 \end{pmatrix},\quad 3P = \begin{pmatrix} 6 & -3 & 9 \\ 3 & 12 & 12 \\ 0 & -6 & 0 \end{pmatrix}$

$$\det(A) = (\frac{1}{3})^2 \det(3P - UN) = \frac{1}{9}\det\begin{pmatrix} 6 & -3 & 9 \\ 4 & 12 & 11 \\ 2 & -6 & -2 \end{pmatrix}$$

接著再來一次，計算最後面那個 3×3 矩陣的行列式。同樣的符號

$$M = (6),\ N = (-3\ \ 9),\ U = \begin{pmatrix} 4 \\ 2 \end{pmatrix},\ P = \begin{pmatrix} 12 & 11 \\ -6 & -2 \end{pmatrix}$$

$$UN = \begin{pmatrix} -12 & 36 \\ -6 & 18 \end{pmatrix},\quad 6P = \begin{pmatrix} 72 & 66 \\ -36 & -12 \end{pmatrix}$$

最後得出，$\det(A) = \dfrac{1}{9} \cdot \dfrac{1}{6} \det(6P - UN) = \dfrac{1}{54} \det\begin{pmatrix} 84 & 30 \\ -30 & -30 \end{pmatrix} = -30$ ■

Exercise

利用軸元壓縮法求出矩陣 A 的行列式

1. $A = \begin{pmatrix} 1 & 1 & 3 \\ 2 & 0 & 3 \\ -1 & 3 & 1 \end{pmatrix}$

2. $A = \begin{pmatrix} 2 & 2 & 0 & -1 \\ 4 & 1 & -2 & 0 \\ -3 & 2 & 0 & 1 \\ 0 & -3 & 1 & 2 \end{pmatrix}$

§6.5 Cramer's rule and the inverse of a matrix

從一開始以來,我們一直談過很多有關逆矩陣的事情,不過當時我們都以數值的計算為主要方法。現在,我們則打算透過 Cramer's rule 解線性系統的方式,和大家介紹一個求逆矩陣的代數方法。下面,我們就先來看看所謂的 Cramer's rule。

Theorem 6.13

Given a linear system,
$$\begin{cases} a_{11}x_1 + a_{12}x_2 + \ldots + a_{1n}x_n = b_1 \\ a_{21}x_1 + a_{22}x_2 + \ldots + a_{2n}x_n = b_2 \\ \vdots \\ a_{m1}x_1 + a_{m2}x_2 + \ldots + a_{mn}x_n = b_m \end{cases}$$

If $A = \begin{pmatrix} a_{11} & a_{12} & \ldots & a_{1n} \\ a_{21} & a_{22} & \ldots & a_{2n} \\ \ldots & \ldots & \ldots & \ldots \\ a_{m1} & a_{m2} & \ldots & a_{mn} \end{pmatrix}$, $X = \begin{pmatrix} x_1 \\ x_2 \\ \ldots \\ x_n \end{pmatrix}$, $B = \begin{pmatrix} b_1 \\ b_2 \\ \ldots \\ b_m \end{pmatrix}$, and if $\det(A) \neq 0$,

then for each $i = 1, 2, 3, \ldots, n$,
$$x_i = \frac{\det(A^1, A^2, \ldots, B, \ldots, A^n)}{\det(A)}$$

Where the numerator is the determinant of the matrix obtained from A by replacing the ith column, A^i, by B.

Proof

只要利用行列式的特性,我們就可以很簡單的完成這個證明。按照 2.2 節式子,$x_1 A^1 + x_2 A^2 + \ldots + x_n A^n = B$ 的表示,把它代入前面最後式子的分子,我們得到

$\det(A^1, A^2, \ldots, B, \ldots, A^n)$
$= \det(A^1, A^2, \ldots, x_1 A^1 + x_2 A^2 + \cdots + x_n A^n, \ldots, A^n)$

$$= x_1 \det(A^1, A^2,...,A^1,...,A^n) + x_2 \det(A^1, A^2,...,A^2,...,A^n) + \cdots +$$
$$x_i \det(A^1, A^2,...,A^i,...,A^n) + \cdots + x_n \det(A^1, A^2,...,A^n,...,A^n)$$
$$= x_i \det(A^1, A^2,...,A^i,...,A^n)$$
$$= x_i \det(A)$$

這正是我們所要的結果。

Example 1

Solve the linear system, $\begin{cases} 3x + z = 1 \\ -y + z = 0 \\ x + 2y = 1 \end{cases}$, by using the Cramer's rule.

Solution

We first compute the determinant of the coefficient matrix, $A = \begin{pmatrix} 3 & 0 & 1 \\ 0 & -1 & 1 \\ 1 & 2 & 0 \end{pmatrix}$,

$\det(A) = 3(-6) + 1(1) = -5 \neq 0$.

And the solutions are

$$x = \frac{\det\begin{pmatrix} 1 & 0 & 1 \\ 0 & -1 & 1 \\ 1 & 2 & 0 \end{pmatrix}}{-5} = \frac{-2+1}{-5} = \frac{1}{5},\quad y = \frac{\det\begin{pmatrix} 3 & 1 & 1 \\ 0 & 0 & 1 \\ 1 & 1 & 0 \end{pmatrix}}{-5} = \frac{3(-1)+1}{-5} = \frac{2}{5}$$

$$z = \frac{\det\begin{pmatrix} 3 & 0 & 1 \\ 0 & -1 & 0 \\ 1 & 2 & 1 \end{pmatrix}}{-5} = \frac{3(-1)+1}{-5} = \frac{2}{5}$$ ∎

Cramer's rule 中有一個不可或缺的要求 $\det(A) \neq 0$。這個要求在矩陣的可逆性當中，扮演著極為關鍵性的角色，那就是接下來的定理。

Theorem 6.14

An $n \times n$ matrix, A, is non-singular if and only if $\det(A) \neq 0$.

Proof

($\Rightarrow$)

這個方向簡單。假若 A 是一個 non-singular matrix，那麼存在一個 $n \times n$ 矩陣 B，使得 $AB = I$。接著由行列式的特性得知，$\det(A)\det(B) = 1$。所以，肯定 $\det(A) \neq 0$。

($\Leftarrow$)

這個方向，我們用反證法。也就是假設，A is singular。由第 3.1 節得知，columns, $A^1, A^2, ..., A^n$ are linearly dependent. 也就是存在不全為 0 的係數，$x_1, x_2, ..., x_n$，使得

$$x_1 A^1 + x_2 A^2 + \cdots + x_n A^n = 0$$

假若，$x_j \neq 0$，那麼

$$A^j = \frac{x_1}{x_j} A^1 + \cdots + \frac{x_{j-1}}{x_j} A^{j-1} + \frac{x_{j+1}}{x_j} A^{j+1} + \cdots + \frac{x_n}{x_j} A^n$$

因此，若 $y_i = \frac{x_i}{x_j}, i \neq j$，則

$$\begin{aligned}\det(A) &= \det(A^1, A^2, ..., y_1 A^1 + \cdots + y_{j-1} A^{j-1} + y_{j+1} A^{j+1} + \cdots + y_n A^n, ..., A^n) \\ &= y_1 \det(A^1, A^2, ..., A^1, ..., A^n) + y_2 \det(A^1, A^2, ..., A^2, ..., A^n) + \cdots \\ &\quad + y_n \det(A^1, A^2, ..., A^n, ..., A^n) \\ &= 0\end{aligned}$$

這個結果與原題意相矛盾。所以，假設錯誤。也就是說，A is non-singular。

現在，我們就假設 $A = (a_{ij})_{n \times n}$ 是一個 $n \times n$ 矩陣，且 $\det(A) \neq 0$。那麼，前述定理告訴我們，存在一 $n \times n$ 矩陣 $B = (b_{ij})_{n \times n}$，使得 $AB = I_n$。若 $B^1, B^2, ..., B^n$ 代表的是矩陣 B 的 n 個 columns，且 $E^1, E^2, ..., E^n$ 代表單位矩陣 I_n 的 n 個 unit column vectors。則，$AB = I_n$ implies,

$$AB^1 = E^1, \quad AB^2 = E^2, \quad \ldots, \quad AB^n = E^n$$

此時，這個式子已經很清楚的告訴了我們下列求逆矩陣的定理。

Theorem 6.15

Let A be an $n \times n$ matrix with $\det(A) \neq 0$. If $B = (b_{ij})_{n \times n}$ is the inverse matrix of A, then for each $j = 1, 2, 3, \ldots, n$,

$$b_{ij} = \frac{\det(A^1, A^2, \ldots, E^j, \ldots, A^n)}{\det(A)}$$

where E^j occurs in the ith place of $(A^1, A^2, \ldots, E^j, \ldots, A^n)$.

Proof

For $j = 1$, $AB^1 = E^1$ and this implies, by the Cramer's rule,

$$b_{i1} = \frac{\det(A^1, A^2, \ldots, E^1, \ldots, A^n)}{\det(A)}$$

For $j = 2$, $AB^2 = E^2$ and this implies, by the Cramer's rule,

$$b_{i2} = \frac{\det(A^1, A^2, \ldots, E^2, \ldots, A^n)}{\det(A)}$$

....

....

For $j = n$, $AB^n = E^n$ and this implies, by the Cramer's rule,

$$b_{in} = \frac{\det(A^1, A^2, \ldots, E^n, \ldots, A^n)}{\det(A)}$$

我們靠近一點，詳細的再看一下式子，

$$b_{ij} = \frac{\det(A^1, A^2, \ldots, E^j, \ldots, A^n)}{\det(A)}$$

注意，由 6.2 節行列式的特性得知，

$$\det(A^1, A^2, \ldots, E^j, \ldots, A^n) = (-1)^{i+j} \det(M_{ji})$$

還記得吧！它正是 6.2 節，我們曾經提過的，a_{ji} 的 cofactor。所以，上式也可以被改寫為

$$b_{ij} = \frac{\det(A^1, A^2, \ldots, E^j, \ldots, A^n)}{\det(A)} = \frac{(-1)^{i+j} \det(M_{ji})}{\det(A)}$$

或者說，如 6.2 節所言，$A_{ij} = (-1)^{i+j} \det(M_{ij})$ 的話，那麼矩陣 A 的逆矩陣就是

$$A^{-1} = B = \frac{1}{\det(A)} (^t(A_{ij}))$$

Example 2

Given, $A = \begin{pmatrix} 2 & 1 \\ 0 & -1 \end{pmatrix}$. Find the inverse matrix of A.

Solution

首先，$\det(A) = -2$。以及所有 cofactors，A_{ij}。
$A_{11} = (-1)^{1+1}(-1) = -1$，$A_{12} = (-1)^{1+2}(0) = 0$，$A_{21} = (-1)^{2+1}(1) = -1$，
$A_{22} = (-1)^{2+2}(2) = 2$。
所以，$A^{-1} = \frac{1}{\det(A)} \begin{pmatrix} A_{11} & A_{21} \\ A_{12} & A_{22} \end{pmatrix} = \frac{1}{-2} \begin{pmatrix} -1 & -1 \\ 0 & 2 \end{pmatrix}$。

Example 3

Given, $A = \begin{pmatrix} 1 & 1 & 3 \\ 2 & 0 & 3 \\ -1 & 3 & 1 \end{pmatrix}$. Find the inverse matrix of A.

Solution

$$\det(A) = \begin{vmatrix} 0 & 3 \\ 3 & 1 \end{vmatrix} - \begin{vmatrix} 2 & 3 \\ -1 & 1 \end{vmatrix} + 3 \begin{vmatrix} 2 & 0 \\ -1 & 3 \end{vmatrix} = -9 - 5 + 18 = 4 \text{ and all the cofactors,}$$

$$A_{11} = \det\begin{pmatrix} 0 & 3 \\ 3 & 1 \end{pmatrix} = -9 \ , \ A_{12} = -\det\begin{pmatrix} 2 & 3 \\ -1 & 1 \end{pmatrix} = -5 \ , \ A_{13} = \det\begin{pmatrix} 2 & 0 \\ -1 & 3 \end{pmatrix} = 6$$

$$A_{21} = -\det\begin{pmatrix} 1 & 3 \\ 3 & 1 \end{pmatrix} = 8 \ , \ A_{22} = \det\begin{pmatrix} 1 & 3 \\ -1 & 1 \end{pmatrix} = 4 \ , \ A_{23} = -\det\begin{pmatrix} 1 & 1 \\ -1 & 3 \end{pmatrix} = -4$$

$$A_{31} = \det\begin{pmatrix} 1 & 3 \\ 0 & 3 \end{pmatrix} = 3 \ , \ A_{32} = -\det\begin{pmatrix} 1 & 3 \\ 2 & 3 \end{pmatrix} = 3 \ , \ A_{33} = \det\begin{pmatrix} 1 & 1 \\ 2 & 0 \end{pmatrix} = -2$$

Therefore,

$$A^{-1} = \frac{1}{\det(A)} \begin{pmatrix} A_{11} & A_{21} & A_{31} \\ A_{12} & A_{22} & A_{32} \\ A_{13} & A_{23} & A_{33} \end{pmatrix} = \frac{1}{4}\begin{pmatrix} -9 & 8 & 3 \\ -5 & 4 & 3 \\ 6 & -4 & -2 \end{pmatrix}$$

Example 4

Solve the following linear system by finding the inverse of its coefficient matrix.

$$\begin{cases} x + y + 3z = -2 \\ 2x + 3z = 1 \\ -x + 3y + z = 4 \end{cases}$$

Solution

The coefficient matrix is $A = \begin{pmatrix} 1 & 1 & 3 \\ 2 & 0 & 3 \\ -1 & 3 & 1 \end{pmatrix}$ and which is the matrix of the last example. So, $A^{-1} = \frac{1}{4}\begin{pmatrix} -9 & 8 & 3 \\ -5 & 4 & 3 \\ 6 & -4 & -2 \end{pmatrix}$.

Writing $X = \begin{pmatrix} x \\ y \\ z \end{pmatrix}$, $B = \begin{pmatrix} -2 \\ 1 \\ 4 \end{pmatrix}$, then $AX = B$ implies that

$$X = \begin{pmatrix} x \\ y \\ z \end{pmatrix} = A^{-1}B = \frac{1}{4}\begin{pmatrix} -9 & 8 & 3 \\ -5 & 4 & 3 \\ 6 & -4 & -2 \end{pmatrix}\begin{pmatrix} -2 \\ 1 \\ 4 \end{pmatrix} = \frac{1}{4}\begin{pmatrix} 38 \\ 26 \\ -24 \end{pmatrix} = \begin{pmatrix} 19/2 \\ 13/2 \\ -6 \end{pmatrix}$$ ∎

Exercises

1. Use Cramer's rule to solve each of the following linear systems

   a. $\begin{cases} 2x - y + z = 1 \\ x + 3y - 2z = 0 \\ 4x - 3y + z = 0 \end{cases}$

   b. $\begin{cases} 2x - y + z = 0 \\ x + 3y - 2z = 1 \\ 4x - 3y + z = 0 \end{cases}$

   c. $\begin{cases} 2x - y + z = 0 \\ x + 3y - 2z = 0 \\ 4x - 3y + z = 1 \end{cases}$

2. If $A = \begin{pmatrix} 2 & -1 & 1 \\ 1 & 3 & -2 \\ 4 & -3 & 1 \end{pmatrix}$, use the results of exercise 1 to obtain A^{-1}.

3. Use Exercise 2 to solve each of the following linear systems

   a. $\begin{cases} 2x - y + z = 1 \\ x + 3y - 2z = 2 \\ 4x - 3y + z = 3 \end{cases}$

   b. $\begin{cases} 2x - y + z = 3 \\ x + 3y - 2z = 2 \\ 4x - 3y + z = 1 \end{cases}$

   c. $\begin{cases} 2x - y + z = 0 \\ x + 3y - 2z = 1 \\ 4x - 3y + z = 2 \end{cases}$

4. Given, $A = \begin{pmatrix} 3 & 1 & 5 \\ 1 & 2 & 2 \\ 1 & 1 & 1 \end{pmatrix}$. Find A^{-1} by finding all its cofactors.

Chapter Seven

Eigenvalues、Diagonalizations、Idempotent Matrices and Householder Matrices

7.1 Definitions and elementary facts
7.2 Diagonalizations and hermitian matrices
7.3 Idempotent matrices
7.4 Householder matrices

　　第七及第八章，我們將從了解特徵值與特徵多項式開始，跟大家介紹一般 $n \times n$ 矩陣的對角化和上三角化的問題，以及**赫密特矩陣** (Hermitian matrices) 與**正規矩陣** (normal matrices) 的對角化過程。然後，介紹一個可對角化矩陣的解構問題。不管是對角化的主題也好，上三角化的問題也罷，甚或是矩陣的解構問題，都有一定的深度，而且牽涉層面廣泛。研讀起來，其精彩的論述頗令人有一種回甘的感覺。所以說，這兩章的內容是矩陣論或線性代數中，比較具有代表性、可看度較高、實用度較廣的主題，當不為過。

§7.1　Definitions and elementary facts

　　這一節我們從介紹特徵值、特徵向量開始，然後認識特徵多項式的義涵及其一些特性，和特徵值的求法。相異的特徵值其之特徵向量之間的線性獨立個性，與相似矩陣之特徵多項式相關要點，當然也是本節必須品嘗的佳餚。

Definition 7.1

> Let T be a linear transformation from a vector space, V, into itself. A scalar, λ, is called an "eigenvalue" of T, if $Ker(T - \lambda I) \neq \{0\}$.
> (Where I is the identity map on V)

從這個角度切入所謂的 eigenvalue，在感覺上，或許會有些許的不適應。但，實質上這個定義和其他教本所慣用的定義是完全一樣的。我們來看看，所謂的 $Ker(T - \lambda I) \neq \{0\}$，意思也就是說，$V$ 中存在一非零向量 α，使得 $(T - \lambda I)\alpha = 0$。換句話說，$T\alpha = \lambda \alpha$ 的意思。那麼，一般而言，這個非零向量 α 就被稱為 "T 的一個 eigenvector"。而 λ 則被稱為是，屬於 α 的一個 eigenvalue。

一個線性映射，T，的**特徵值集合** (the set of eigenvalues) 在分析學上，常被稱為 "T 的 point spectrum"，通常被以符號 $\sigma_p(T)$ 表之 (往後我們或將使用這個符號)。此外，若 $\lambda \in \sigma_p(T)$，則 $V_\lambda = Ker(T - \lambda I)$ 被稱為，"T 的一個 eigenspace"。我們先別廢話那麼多，看看幾個例題吧！

Example 1

Consider a 2×2 matrix, $A = \begin{pmatrix} 1 & 0 \\ 1 & 2 \end{pmatrix}$, as a linear map from R^2 into R^2.

If $\lambda = 1$, then

$$A - \lambda I = \begin{pmatrix} 0 & 0 \\ 1 & 1 \end{pmatrix}, \text{ and } \begin{pmatrix} 0 & 0 \\ 1 & 1 \end{pmatrix}\begin{pmatrix} x \\ -x \end{pmatrix} = \begin{pmatrix} 0 \\ 0 \end{pmatrix}, \quad \forall \begin{pmatrix} x \\ -x \end{pmatrix} \in R^2.$$

So $\lambda = 1$ is an eigenvalue of A, and $\begin{pmatrix} x \\ -x \end{pmatrix}$, $x \in R$ are eigenvectors of A having $\lambda = 1$ as an eigenvalue.

Also if $\lambda = 2$, then

$$A - \lambda I = \begin{pmatrix} -1 & 0 \\ 1 & 0 \end{pmatrix}, \text{ and } \begin{pmatrix} -1 & 0 \\ 1 & 0 \end{pmatrix}\begin{pmatrix} 0 \\ x \end{pmatrix} = \begin{pmatrix} 0 \\ 0 \end{pmatrix}, \quad \forall \begin{pmatrix} 0 \\ x \end{pmatrix} \in R^2.$$

So $\lambda = 2$ is an eigenvalue of A, and $\begin{pmatrix} 0 \\ x \end{pmatrix}$, $\forall x \in R$ are eigenvectors of A having $\lambda = 2$ as an eigenvalue. ∎

Example 2

Let $V = \{ f : R \to R ; f$ is infinitely differentiable$\}$. Defining, for $f \in V$, $D(f(t)) = f'(t)$, then D is no doubt a linear operator on V as you can check.

Consider that $f(t) = e^{\lambda t}$, we have that $D(f(t)) = \lambda e^{\lambda t} = \lambda f(t)$. So, we see that λ is an eigenvalue of D, and $f(t) = e^{\lambda t}$ is an eigenvector of D having λ as eigenvalue. ∎

有時候，characteristic value 以及 characteristic vector，也常被用來取代 eigenvalue 以及 eigenvector 的稱呼。除此之外，有一個特殊現象就是，若 T 本身不是一個1–1的線性映射的話，那麼 $Ker(T) \neq \{0\}$。這也就是說，存在一非零向量 α，使得 $T\alpha = 0 = 0\alpha$。所以，此時 $\lambda = 0$ 是 T 的一個 eigenvalue。我們要說明的是 eigenvalue 有時可能是 0。

前面我們曾經提過，eigenspace，$V_\lambda = Ker(T - \lambda I)$。由於，$Ker(T - \lambda I)$ 是 V 的一個 subspace，所以我們得知，若 α 是 T 的一個 eigenvector 的話，那麼 α 的任意常數倍數，$a\alpha$，也仍舊是 T 的 eigenvector。換句話說，只要，$\beta \in Sp\{\alpha\}$(向量 α 的生成集)，則 $T\beta = \lambda\beta$。也就是說，$T(Sp\{\alpha\}) \subseteq Sp\{\alpha\}$。這種情形之下，我們常稱，$Sp\{\alpha\}$ 為線型映射 T 的一個**固定軸線** (fixed line)。依此觀察，我們或許已經發現，並非每一個線型映射都有 eigenvector 的。譬如說，大家所熟悉的 "xy-平面座標軸旋轉"，

$$A = \begin{pmatrix} \cos\theta & \sin\theta \\ -\sin\theta & \cos\theta \end{pmatrix}$$

當 $\theta \neq 2n\pi$ 的時候，A 沒有固定軸線。所以，A, considered as a linear map from R^2 into R^2, has no eigenvector。有關這一點，我們看了下面定理之後，可以更清楚。不過，為了方便說明起見。從現在開始，我們將把任意一個佈在體 F 上的 $m \times n$ 矩陣看成是一個從 F^n 映射到 F^m 的線型映射，而不再

多加以解釋了。

Theorem 7.1

Let A be an $n \times n$ matrix. A number, λ, is an eigenvalue of A if and only if $\det(A - \lambda I) = 0$.

Proof

這個定理的證明簡單。因為

$Ker(A - \lambda I) \neq \{0\}$ if and only if $A - \lambda I$ is singular and

$A - \lambda I$ is singular if and only if $\det(A - \lambda I) = 0$.

(參閱 204 頁的 Corollary 與 205 頁 Exercise 2)。

有了這個定理，我們可以輕鬆的求出一個矩陣的 eigenvalues 了。可無論如何，我們先回過頭去，看看座標軸的旋轉

$$A = \begin{pmatrix} \cos\theta & \sin\theta \\ -\sin\theta & \cos\theta \end{pmatrix}$$

若 λ 是 A 的 eigenvalue 的話，那麼

$\det(A - \lambda I) = 0$

$\Rightarrow \det\begin{pmatrix} \cos\theta - \lambda & \sin\theta \\ -\sin\theta & \cos\theta - \lambda \end{pmatrix} = 0$

$\Rightarrow \lambda^2 - 2\lambda\cos\theta + 1 = 0$

當 $\theta \neq 2n\pi$ 的時候，最後式子沒有實數解。所以，以 real space 來講，A has no eigenvalue。

假若，A 是一個 $n \times n$ 矩陣，那麼 $\det(A - \lambda I)$ 是一個以 λ 為參數的 n 次多項式。為了方便起見，我們通常以 $\det(A - tI)$，或者以 $\det(tI - A)$ 替代 $\det(A - \lambda I)$，並且稱呼

$$P_A(t) = \det(tI - A) = t^n + \text{terms of lower degree}$$

為**特徵多項式** (the characteristic polynomial)。如此情形之下，我們可以改口

說

λ is an eigenvalue of A if and only if λ is a root of $P_A(t) = 0$

注意，在複數體系來說，for any $n \times n$ complex matrix, A, since the characteristic polynomial $P_A(t)$ has n complex zeros, A has n eigenvalues。而且假設，$\lambda_1,...,\lambda_n$ 是 $P_A(t) = 0$ 的 n 個根的話，那麼我們可以將 $P_A(t)$ 分解成

$$P_A(t) = (t - \lambda_1)(t - \lambda_2) \cdots (t - \lambda_n)$$

當然，在有重根的情形之下，前面式子將變成

$$P_A(t) = (t - \lambda_{i_1})^{k_1}(t - \lambda_{i_2})^{k_2} \cdots (t - \lambda_{i_l})^{k_l}, \text{其中} \sum_{j=1}^{l} k_j = n$$

這種情形在實數體系來說，可就沒那麼順遂了。因為，方程式 $P_A(t) = 0$，在實數體系內不見得有 n 個實數根。所以，$P_A(t)$ 或許無法如前述一樣的被分解。但是無論如何，利用代數基本定理，我們知道，$P_A(t)$ 最起碼可以被分解為，多個線性以及二項式的乘積。話先說到這兒，下面我們來練習兩個簡單的例子。

Example 3

Given $A = \begin{pmatrix} 3 & 4 \\ 2 & 1 \end{pmatrix}$. Find the eigenvalues and their related eigenvectors of A.

Solution

$$P_A(t) = \det(tI - A) = (t-3)(t-1) - 8 = t^2 - 4t - 5 = (t-5)(t+1)$$

So, the eigenvalues are 5 and -1.

To find the eigenvectors, we solve the following simultaneous equations

$$\begin{cases} 3x + 4y = \lambda x \\ 2x + y = \lambda y \end{cases}$$

For $\lambda = 5$，we have

$$\begin{cases} -2x + 4y = 0 \\ 2x - 4y = 0 \end{cases}$$

注意,它們是相依方程。不要懷疑,它們務必是相依方程。因為,前面我們曾經說過,若 α 是 A 的一個 eigenvector 的話,那麼 α 的任意常數倍數,$a\alpha$,也仍舊是 A 的 eigenvector。

現在令 $y = 1$,則得 $x = 2$。所以,$\alpha = \begin{pmatrix} 2 \\ 1 \end{pmatrix}$ 是 $\lambda = 5$ 時的一個 eigenvector。

至於,$\lambda = -1$,we also have,

$$\begin{cases} 4x + 4y = 0 \\ 2x + 2y = 0 \end{cases}$$

令 $x = 1$,得 $y = -1$。所以,$\alpha = \begin{pmatrix} 1 \\ -1 \end{pmatrix}$ 是 $\lambda = -1$ 時的一個 eigenvector。∎

Example 4

Given $A = \begin{pmatrix} 3 & 2 & 1 \\ 0 & 1 & 2 \\ 0 & 1 & -1 \end{pmatrix}$. Find the eigenvalues and their related eigenvectors of A.

Solution

$$P_A(t) = \det(tI - A) = t^3 - 3t^2 - 3t + 9 = (t^2 - 3)(t - 3)$$

So, the eigenvalues are $\lambda = 3, \lambda = \sqrt{3}, \lambda = -\sqrt{3}$

For $\lambda = 3$, we have that

$$\begin{cases} 3x + 2y + z = 3x \\ y + 2z = 3y \\ y - z = 3z \end{cases}$$

And thus, $\begin{cases} 2y + z = 0 \\ 2y - 2z = 0 \\ y - 4z = 0 \end{cases}$

This implies that $y = z = 0$ and x can be any real number.

Taking $x=1$, then $\alpha = \begin{pmatrix} 1 \\ 0 \\ 0 \end{pmatrix}$ is an eigenvector of A with respect to $\lambda = 3$.

For $\lambda = \sqrt{3}$, we then have
$$\begin{cases} 3x + 2y + z = \sqrt{3}x \\ y + 2z = \sqrt{3}y \\ y - z = \sqrt{3}z \end{cases}$$

And hence
$$\begin{cases} 2(3-\sqrt{3})x + 4y + 2z = 0 \\ (1-\sqrt{3})y + 2z = 0 \end{cases}$$

This implies that
$$2(3-\sqrt{3})x + (3+\sqrt{3})y = 0$$
Letting $x=1$, then $y = -4 + 2\sqrt{3}$ and $z = 5 - 3\sqrt{3}$, we see then
$$\alpha = \begin{pmatrix} 1 \\ -4 + 2\sqrt{3} \\ 5 - 3\sqrt{3} \end{pmatrix}$$ is an eigenvector of A with respect to $\lambda = \sqrt{3}$

For $\lambda = -\sqrt{3}$
$$\begin{cases} 3x + 2y + z = -\sqrt{3}x \\ y + 2z = -\sqrt{3}y \\ y - z = -\sqrt{3}z \end{cases}$$

$\Rightarrow \quad \begin{cases} 2(3+\sqrt{3})x + 4y + 2z = 0 \\ (1+\sqrt{3})y + 2z = 0 \end{cases}$

$\Rightarrow \quad 2(3+\sqrt{3})x + (3-\sqrt{3})y = 0$

Letting $x=1$, then $y = -4 - 2\sqrt{3}$ and $z = 5 + 3\sqrt{3}$, we see then
$$\alpha = \begin{pmatrix} 1 \\ -4 - 2\sqrt{3} \\ 5 + 3\sqrt{3} \end{pmatrix}$$ is an eigenvector of A with respect to $\lambda = -\sqrt{3}$ ∎

下面是幾個有關於 eigenvalues、characteristic polynomial 及其相互之間的重要特性。了解這個之後，有一些事情我們還得要和大家補充說明。

Theorem 7.2

Given an $n \times n$ matrix, $A = (a_{ij})_{n \times n}$. Let $P_A(t) = t^n + b_{n-1}t^{n-1} + \cdots + b_1 t + b_0$ be the characteristic polynomial of A, then $b_{n-1} = -(a_{11} + a_{22} + \cdots + a_{nn})$. Moreover, if $\lambda_1, \lambda_2, \ldots, \lambda_n$ are n eigenvalues of A, then $b_0 = \det(-A) = (-1)^n \prod_{i=1}^{n} \lambda_i$.

Proof

Recall that the determinant, $\det(tI - A)$, is a sum of $n!$ terms. And the term corresponding to the identity permutation is $(t - a_{11})(t - a_{22}) \cdots (t - a_{nn})$. All the other $n! - 1$ terms involve at most $n - 2$ diagonal terms. 所以，$P_A(t) = \det(tI - A)$ 中，t^n 以及 t^{n-1} 的係數完全被決定於，$(t - a_{11})(t - a_{22}) \cdots (t - a_{nn})$。這種情形之下，我們清楚的發現，$b_{n-1} = -(a_{11} + a_{22} + \cdots + a_{nn})$。

至於常數項 b_0，則為，

$$b_0 = P_A(0) = \det(0I - A) = \det(-A)$$

然而，由於 $\lambda_1, \lambda_2, \ldots, \lambda_n$ 是 $P_A(t)$ 的 n 個根，所以它可以被寫為

$$P_A(t) = (t - \lambda_1)(t - \lambda_2) \cdots (t - \lambda_n)$$

也因此，

$$\det(-A) = b_0 = P_A(0) = (0 - \lambda_1)(0 - \lambda_2) \cdots (0 - \lambda_n) = (-1)^n \prod_{i=1}^{n} \lambda_i$$

利用這個定理，我們來考慮下列情況。假設，矩陣 A 有 n 個 eigenvalues，$\lambda_1, \lambda_2, \ldots, \lambda_n$，的話，那麼 A 的 characteristic polynomial 可以被寫為

$$P_A(t) = (t - \lambda_1)(t - \lambda_2) \cdots (t - \lambda_n)$$

由這個式子，我們得出 t^{n-1} 項的係數是 $-(\lambda_1 + \lambda_2 + \cdots + \lambda_n)$。將這個結果與上述定理比對之後，我們發現

$$-(a_{11} + a_{22} + \cdots + a_{nn}) = -(\lambda_1 + \lambda_2 + \cdots + \lambda_n)$$

也就是說，
$$trace(A) = \sum_{i=1}^{n} \lambda_i$$

請讀者把這個重要結果記下。

Example 5

Given that two eigenvalues of $A = \begin{pmatrix} -2 & -1 & 3 \\ 6 & 5 & -6 \\ 2 & 2 & -1 \end{pmatrix}$ are $\lambda_1 = 1$ and $\lambda_2 = 2$. Find the third eigenvalue of A.

Solution

首先注意，由於已知 A 有兩實數特徵值，所以 A 必有第三個實數特徵值。接著，利用前面的式子，我們發現

$$trace(A) = -2 + 5 - 1 = 2 = 1 + 2 + \lambda_3$$

所以，第三個 eigenvalue 是 $\lambda_3 = -1$。∎

Theorem 7.3

> If $\lambda_1, \ldots, \lambda_k$ are distinct eigenvalues of A and if for each $i = 1, \ldots, k$, α_i is an eigenvector of A having λ_i as an eigenvalue, then $\alpha_1, \ldots, \alpha_k$ are linearly independent.

Proof

We shall use the method of induction on k.

For $k = 1$,

$\{\alpha_1\}$ is a single non-zero vector set, and is obviously linearly independent.

Assume that the theorem is true for $k = t$, i.e. $\alpha_1, \alpha_2, \ldots, \alpha_t$ are linearly independent.

For $k = t + 1$, we let

$$c_1\alpha_1 + \cdots + c_{t+1}\alpha_{t+1} = 0 \tag{1}$$

multiplying both sides by λ_{t+1}, we have that

$$c_1\lambda_{t+1}\alpha_1 + \cdots + c_{t+1}\lambda_{t+1}\alpha_{t+1} = 0, \tag{2}$$

also applying to both sides of (1) by the map, A, then

$$c_1\lambda_1\alpha_1 + \cdots + c_{t+1}\lambda_{t+1}\alpha_{t+1} = 0. \tag{3}$$

Subtracting (3) from (2), we obtain that

$$c_1(\lambda_{t+1} - \lambda_1)\alpha_1 + \cdots + c_t(\lambda_{t+1} - \lambda_t)\alpha_t = 0$$

From the assumption, we thus have that

$$c_i(\lambda_{t+1} - \lambda_i) = 0, \quad \forall i = 1, \ldots, t$$

So $\quad c_i = 0, \ \forall \ i = 1, \ldots, t$, since $\lambda_1, \lambda_2, \ldots, \lambda_{t+1}$ are distinct.

Bring this to (1), we obtain that $c_{t+1} = 0$

$\therefore \quad \alpha_1, \alpha_2, \ldots, \alpha_{t+1}$ are linearly independent.

Therefore, the result is also true for $k = t+1$, and hence, the proof is completed.

Exercise

Suppose that α_1, α_2 are eigenvectors of A having λ_1, λ_2 as eigenvalues, respectively. Show that if $\lambda_1 \neq \lambda_2$, then $\alpha_1 + \alpha_2$ is not an eigenvector of A. (*Hint:* α_1 and α_2 are linearly independent)

Theorem 7.4

Given two $n \times n$ matrices, A and P. If P is invertible, then the characteristic polynomial of A is equal to the characteristic polynomial of $P^{-1}AP$.

Proof

Consider that
$$\det(P^{-1}(tI - A)P) = \det(P^{-1})\det(tI - A)\det(P) = \det(tI - A)$$

so

$$P_A(t) = \det(tI - A) = \det(P^{-1}(tI - A)P) = \det(tP^{-1}P - P^{-1}AP) = \det(tI - P^{-1}AP)$$

這個定理說明的是，兩個相似矩陣有相同的 characteristic polynomial。同學們還記得吧！在第四章的時候，所謂"兩個矩陣是相似的"，其意思是說，"這兩個矩陣代表著相同的線型映射"。我們現在把那些概念，簡單扼要的重新敘述一遍好了。一個從向量空間 V 映射到向量空間 W 的線型轉換 T，由於 V 和 W 的多個不同的基底而對應出不同的 T 的矩陣表示。(參看下圖)

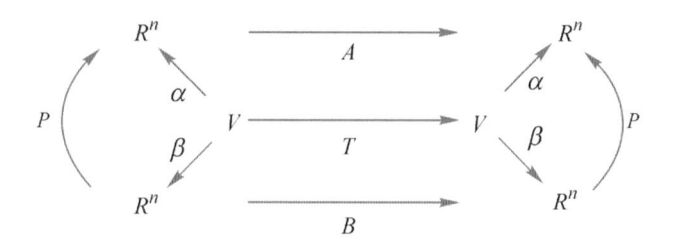

假設 α 和 β 是向量空間 V 的兩組不同的基底，α' 和 β' 是向量空間 W 的兩組不同的基底，且 $P_1 = mat(I, \beta, \alpha)$，$P_2 = mat(I, \beta', \alpha')$，$A = mat(T, \alpha, \alpha')$，$B = mat(T, \beta, \beta')$，則，$B = P_2^{-1}AP_1$。這也就是說，矩陣 A 和 B 是相似的。

當然，在特殊的情形之下，若 $V = W$ 的話，則上圖將變成

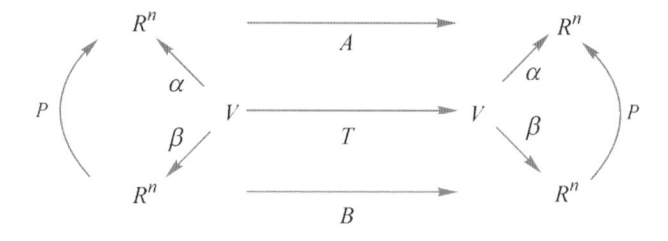

此時，$B = P^{-1}AP$。由前述這樣的論證結果，我們可以確定，一個 linear transformation 的特徵多項式是唯一被決定的。因此，我們似乎可以做如下一個完整的定義。

Definition 7.2

Let T be a linear transformation from V into V. If A is a matrix representation of T corresponding to any basis of V, then the characteristic polynomial of T, $P_T(t)$, is defined to be $P_T(t) = P_A(t)$.

下面我們再檢視兩個例題。

Example 6

Given $A = \begin{pmatrix} 1 & 1 & 2 \\ 0 & 5 & -1 \\ 0 & 0 & 7 \end{pmatrix}$. Show that A is similar to $B = \begin{pmatrix} 1 & 0 & 0 \\ 0 & 5 & 0 \\ 0 & 0 & 7 \end{pmatrix}$.

Solution

我們要找一個 non-singular matrix，P，使得 $P^{-1}AP = B$。首先，看看
$$P_A(t) = \det(tI - A) = (t-1)(t-5)(t-7)。$$
從中，我們發現 $\lambda_1=1, \lambda_2=5, \lambda_3=7$ 是 A 的 eigenvalues。而，$\alpha_1={}^t(1\ 0\ 0)$，$\alpha_2={}^t(1/4\ 1\ 0)$，$\alpha_3={}^t(1/4\ -1/2\ 1)$ 分別為它們的 eigenvectors。然後，由前述定理得知，they are linearly independent in R^3（因而形成了 R^3 的一組基底）。現在，令 P 為一 3×3 矩陣，使其每一個 column 分別為，$P^1=\alpha_1$，$P^2=\alpha_2$，$P^3=\alpha_3$。那麼，由前面的說明，此矩陣 $P = \begin{pmatrix} 1 & 1/4 & 1/4 \\ 0 & 1 & -1/2 \\ 0 & 0 & 1 \end{pmatrix}$ 即為所求。也就是說，$P^{-1}AP = B$。

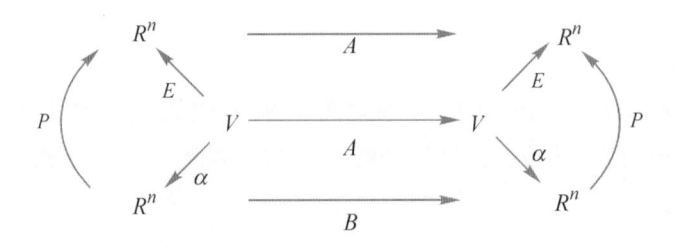

Example 7

Show that $A = \begin{pmatrix} 1 & 0 \\ 0 & 1 \end{pmatrix}$ and $B = \begin{pmatrix} 1 & 1 \\ 0 & 1 \end{pmatrix}$ are not similar, but they have the same characteristic polynomial.

Proof

Consider that for any 2×2 nonsingular matrix, P,

$$P^{-1}AP = P^{-1}P = I = A \text{ and } A \neq B$$

So A and B can not be similar.

However, $P_A(t) = \det(tI - A) = (t-1)^2 = \det(tI - B) = P_B(t)$ ∎

雖然，相似矩陣有相同的特徵多項式。但是，前述例題卻告訴我們，有相同特徵多項式的矩陣不一定是相似的，這一點要特別注意。

看到上面的矩陣，$B = \begin{pmatrix} 1 & 1 \\ 0 & 1 \end{pmatrix}$，我們來看看 $\lambda = 1$ 的時候它的 eigenspace

$$V_{\lambda=1} = Ker(A - I) = Ker\begin{pmatrix} 0 & 1 \\ 0 & 0 \end{pmatrix} = \{\begin{pmatrix} x \\ 0 \end{pmatrix} : x \in R\}$$

注意，$\lambda = 1$ 是特徵多項式 $P_B(t) = 0$ 的重根，但是 $\dim(V_{\lambda=1}) = 1$。這邊要說明的是，特徵多項式重根的次數，並不一定就是 eigenspace 的維度。

下面我們看一個複數體系的例子，以為本節之結尾。

Example 8

Find the eigenvalues of $A = \begin{pmatrix} 1 & i \\ 1 & i \end{pmatrix}$ and a non-singular matrix, P, such that $P^{-1}AP = \begin{pmatrix} \lambda_1 & 0 \\ 0 & \lambda_2 \end{pmatrix}$.

Solution

$$\det(tI - A) = (t-1)(t-i) - i = t^2 - (1+i)t$$

implies that $\lambda_1=0$, $\lambda_2=1+i$ are the eigenvalues. Corresponding to this, we find out that $\alpha_1={}^t(1\ i)$, $\alpha_2={}^t(1\ 1)$ are the eigenvectors.

Taking $P=\begin{pmatrix}1 & 1\\ i & 1\end{pmatrix}$, then $P^{-1}=\begin{pmatrix}(1+i)/2 & (-1-i)/2\\ (1-i)/2 & (1+i)/2\end{pmatrix}$

and hence then $P^{-1}AP=\begin{pmatrix}(1+i)/2 & (-1-i)/2\\ (1-i)/2 & (1+i)/2\end{pmatrix}\begin{pmatrix}1 & i\\ 1 & i\end{pmatrix}\begin{pmatrix}1 & 1\\ i & 1\end{pmatrix}$

$$=\begin{pmatrix}0 & 0\\ 0 & 1+i\end{pmatrix}$$

Exercises

1. Find the eigenvalues of A and a non-singular matrix, P, such that

$$P^{-1}AP=\begin{pmatrix}\lambda_1 & 0\\ 0 & \lambda_2\end{pmatrix}$$

a. $A=\begin{pmatrix}1 & 2\\ 3 & 2\end{pmatrix}$ b. $A=\begin{pmatrix}i & 1-i\\ 0 & 1\end{pmatrix}$

2. Find the eigenvalues of $A=\begin{pmatrix}-2 & -1 & 3\\ 6 & 5 & -6\\ 2 & 2 & -1\end{pmatrix}$ and a non-singular matrix, P, such that

$$P^{-1}AP=\begin{pmatrix}\lambda_1 & 0 & 0\\ 0 & \lambda_2 & 0\\ 0 & 0 & \lambda_3\end{pmatrix}$$

3. Show that if A is a square triangular matrix, then the eigenvalues of A are the diagonal elements of A.

4. Let A be a nilpotent matrix. Show that if λ is an eigenvalue of A, then $\lambda=0$.

5. Show that if $A^2=A$ (i.e. A is an idempotent matrix.), and if λ is an eigenvalue of A, then either $\lambda=1$ or $\lambda=0$.

6. Show that A and ${}^tA$ have the same characteristic polynomial.

7. Let X be an eigenvector of A with eigenvalue λ. Show that if P is non-singular, then $P^{-1}X$ is an eigenvector of $P^{-1}AP$ with eigenvalue λ.

8. Let A be non-singular. Show that if λ is an eigenvalue of A, then $\lambda \neq 0$ and $\dfrac{1}{\lambda}$ is an eigenvalue of A^{-1}.

9. Show that if $\lambda > 0$ is an eigenvalue of A^2, then either $\sqrt{\lambda}$ or $-\sqrt{\lambda}$ is an eigenvalue of A.

10. Show that if

$$A = \begin{pmatrix} B & * \\ 0 & D \end{pmatrix}$$

where B and D are square matrices, $*$ is a matrix of certain size, 0 denotes a zero matrix, then the set of eigenvalues of A, $\sigma_p(A)$, is equal to the union of $\sigma_p(B)$ and $\sigma_p(D)$.

§7.2　Diagonalizations and hermitian matrices

依據 Theorem 7.3 以及 7.1 節的 exercise 1 and exercise 2，假設 $\dim V = n$，而且 $\{\lambda_1,...,\lambda_n\}$ 是矩陣 A 的 n 個不同的 eigenvalues 的話，那麼我們可以找到它們相對應的 n 個線性獨立的 eigenvectors，$\alpha_1,...,\alpha_n$。以這 n 個線性獨立的 eigenvectors，$\alpha_1,...,\alpha_n$，為 V 的基底 (我們稱其為 spectral basis) 的話，我們可以求出與矩陣 A 相似的一個對角矩陣，其中對角線的 n 的 elements 剛好是 $\{\lambda_1,...,\lambda_n\}$。也就是說

$$P^{-1}AP = B = \begin{pmatrix} \lambda_1 & 0 & 0 & ... & 0 \\ 0 & \lambda_2 & 0 & ... & 0 \\ 0 & ... & ... & 0 & ... \\ ... & ... & 0 & ... & 0 \\ 0 & 0 & ... & 0 & \lambda_n \end{pmatrix}$$

在這種情形之下，我們稱矩陣 A 為 **diagonalizable**。或者，我們明確的說，

Definition　7.3

> A square matrix, A, is "diagonalizable", if A is similar to a diagonal matrix. I.e. if there exists a non-singular matrix, P, such that $P^{-1}AP$ is a diagonal matrix.

根據這個定義，我們可以說

> An $n \times n$ matrix, A, is diagonalizable, if A possesses a set of n linearly independent eigenvectors.

當然，要是前面的充分條件改為，"if A has n distinct eigenvalues" 的話，上述也是成立的。不過，同學們還是要特別注意，"A possesses a set of n linearly independent eigenvectors" 與 "A has n distinct eigenvalues" 之強弱程度不一樣唷 (Can you see which one is stronger?)。

其實，前面的敘述反過來說也是正確的。

If A is diagonalizable, then A possesses a set of n linearly independent eigenvectors.

因為，若 P 是一個 non-singular matrix，使得

$$P^{-1}AP = D = \begin{pmatrix} d_1 & 0 & \ldots & \ldots & 0 \\ 0 & d_2 & 0 & \ldots & \ldots \\ \ldots & 0 & \ldots & \ldots & \ldots \\ 0 & \ldots & \ldots & \ldots & 0 \\ 0 & \ldots & \ldots & 0 & d_n \end{pmatrix}$$

為一 diagonal matrix 的話，那麼，$AP = PD$。因此，對於 P 每一個 column，$i = 1, 2, \ldots, n$，而言

$$AP^i = d_i P^i$$

由於，P 是 non-singular，所以，$P^1, P^2, \ldots, P^n$ are linearly independent eigenvectors with respect to eigenvalues $d_1, d_2, \ldots, d_n$。(注意，這 n 個數 $d_1, d_2, \ldots, d_n$ 可能有相同的。這就是前面，"which one is stronger?" 的解答。) 這組線性獨立向量，$P^1, P^2, \ldots, P^n$，就是前述所謂的，spectral basis。回頭看看 7.1 節的例 6 與例 8，讓我們再練習三個例題。

Example 1

Show that the matrix, $A = \begin{pmatrix} 1 & 0 \\ 1 & 2 \end{pmatrix}$, is diagonalizable via its eigenvectors.

Proof

Consider for $X = \begin{pmatrix} 1 \\ -1 \end{pmatrix} \in R^2$, $AX = \begin{pmatrix} 1 & 0 \\ 1 & 2 \end{pmatrix}\begin{pmatrix} 1 \\ -1 \end{pmatrix} = \begin{pmatrix} 1 \\ -1 \end{pmatrix}$, so $X = \begin{pmatrix} 1 \\ -1 \end{pmatrix}$ is an eigenvector of A having $\lambda = 1$ as its eigenvalue. And for $X = \begin{pmatrix} 0 \\ 1 \end{pmatrix} \in R^2$, $AX = \begin{pmatrix} 1 & 0 \\ 1 & 2 \end{pmatrix}\begin{pmatrix} 0 \\ 1 \end{pmatrix} = \begin{pmatrix} 0 \\ 2 \end{pmatrix} = 2\begin{pmatrix} 0 \\ 1 \end{pmatrix}$, so $X = \begin{pmatrix} 0 \\ 1 \end{pmatrix}$ is an eigenvector of A having $\lambda = 2$ as its eigenvalue.

However, $\lambda = 1$ and $\lambda = 2$ are distinct, so $\begin{pmatrix} 1 \\ -1 \end{pmatrix}$ and $\begin{pmatrix} 0 \\ 1 \end{pmatrix}$ are linearly independent. And they form a basis for R^2. Therefore, A is diagonalizable. And

$$A \sim \begin{pmatrix} 1 & 0 \\ 0 & 2 \end{pmatrix}$$

Example 2

Given, $A = \begin{pmatrix} 1 & 2 \\ 4 & 3 \end{pmatrix}$. Find a non-singular matrix, P, such that $P^{-1}AP$ is a diagonal matrix.

Solution

$$P_A(t) = \det \begin{pmatrix} t-1 & -2 \\ -4 & t-3 \end{pmatrix} = (t-1)(t-3) - 8 = t^2 - 4t - 5 = (t-5)(t+1)$$

We see that $\lambda = 5$ and $\lambda = -1$ are eigenvalues of A.

Solve the linear system

$$\begin{pmatrix} 1 & 2 \\ 4 & 3 \end{pmatrix} \begin{pmatrix} x \\ y \end{pmatrix} = \lambda \begin{pmatrix} x \\ y \end{pmatrix}$$

we find out that $\alpha_1 = \begin{pmatrix} 1 \\ -1 \end{pmatrix}$ and $\alpha_2 = \begin{pmatrix} 1 \\ 2 \end{pmatrix}$ are eigenvectors of A with respect to $\lambda = -1$ and $\lambda = 5$, respectively.

Taking $\alpha = \{\alpha_1, \alpha_2\}$ as a basis of R^2 and $E = \{E_1, E_2\}$ as the standard basis, then $P = mat(I, \alpha, E) = \begin{pmatrix} 1 & 1 \\ -1 & 2 \end{pmatrix}$, and $P^{-1} = mat(I, E, \alpha) = \begin{pmatrix} 2/3 & -1/3 \\ 1/3 & 1/3 \end{pmatrix}$.

Hence

$$P^{-1}AP = \frac{1}{3} \begin{pmatrix} 2 & -1 \\ 1 & 1 \end{pmatrix} \begin{pmatrix} 1 & 2 \\ 4 & 3 \end{pmatrix} \begin{pmatrix} 1 & 1 \\ -1 & 2 \end{pmatrix}$$

$$= \frac{1}{3} \begin{pmatrix} -3 & 0 \\ 0 & 15 \end{pmatrix} = \begin{pmatrix} -1 & 0 \\ 0 & 5 \end{pmatrix}$$

我們除了說,linearly independent eigenvectors 可以確保 diagonalizability 之外,還有什麼言詞可以取代這個說法呢?我們看看下面一個比較特殊的矩陣。(注意該矩陣是否符合, $A^* = A$?)

Example 3

Show that $A = \begin{pmatrix} 2 & 1+i \\ 1-i & 1 \end{pmatrix}$ is diagonalizable, by finding its linearly independent eigenvectors.

Solution

Since $P_A(t) = \det(tI - A) = t^2 - 3t = t(t-3)$, we see that $\sigma_p(A) = \{0, 3\}$.

Consider that

$$\begin{pmatrix} 2 & 1+i \\ 1-i & 1 \end{pmatrix} \begin{pmatrix} 1+i \\ -2 \end{pmatrix} = \begin{pmatrix} 0 \\ 0 \end{pmatrix}$$

so, $\begin{pmatrix} 1+i \\ -2 \end{pmatrix}$ is an eigenvctor of A with respect to eigenvalue $\lambda_1 = 0$

Also consider that

$$\begin{pmatrix} 2 & 1+i \\ 1-i & 1 \end{pmatrix} \begin{pmatrix} 1+3i \\ 2+i \end{pmatrix} = 3 \begin{pmatrix} 1+3i \\ 2+i \end{pmatrix}$$

we have that $\begin{pmatrix} 1+3i \\ 2+i \end{pmatrix}$ is an eigenvctor of A with respect to eigenvalue $\lambda_2 = 3$

Taking $\alpha = \left\{ \alpha_1 = \dfrac{1}{\sqrt{6}} \begin{pmatrix} 1+i \\ -2 \end{pmatrix}, \alpha_2 = \dfrac{1}{\sqrt{15}} \begin{pmatrix} 1+3i \\ 2+i \end{pmatrix} \right\}$ as a basis of C^2 and $E = \left\{ \begin{pmatrix} 1 \\ 0 \end{pmatrix}, \begin{pmatrix} 0 \\ 1 \end{pmatrix} \right\}$ as the standard basis, then

$$P = mat(I, \alpha, E) = \begin{pmatrix} (1+i)/\sqrt{6} & (1+3i)/\sqrt{15} \\ (-2)/\sqrt{6} & (2+i)/\sqrt{15} \end{pmatrix}$$

and $\quad P^{-1} = mat(I, E, \alpha) = \begin{pmatrix} (1-i)/\sqrt{6} & -2/\sqrt{6} \\ (1-3i)/\sqrt{15} & (2-i)/\sqrt{15} \end{pmatrix}$

Hence
$$P^{-1}AP = \begin{pmatrix} 0 & 0 \\ (3-9i)/\sqrt{15} & (6-3i)/\sqrt{15} \end{pmatrix} \begin{pmatrix} (1+i)/\sqrt{6} & (1+3i)/\sqrt{15} \\ -2/\sqrt{6} & (2+i)/\sqrt{15} \end{pmatrix}$$
$$= \begin{pmatrix} 0 & 0 \\ 0 & 3 \end{pmatrix}$$
∎

我們注意看看，這個例題有什麼特殊結果的地方。首先它的兩個 eigenvectors，α_1 以及 α_2，它們的 standard inner product，$\langle \alpha_1, \alpha_2 \rangle = ({}^t\alpha_1)\overline{\alpha}_2 = 0$；而且，$\langle \alpha_1, \alpha_1 \rangle = 1$，$\langle \alpha_2, \alpha_2 \rangle = 1$。所以我們發現，$P$ 其實是一個 unitary matrix。也就是說，$P^*P = PP^* = I$。此外我們也發現，前面所求出來的 $P^{-1} = mat(I, E, \alpha)$ 剛好也就是 P^*。不錯吧！其實這完全是因為，$A = \begin{pmatrix} 2 & 1+i \\ 1-i & 1 \end{pmatrix}$ 是一個 hermitian matrix 的緣故。沒錯，就是 hermitian matrix。本節後半段，就和大家介紹，一個 hermitian（或 symmetric）matrix 不僅可以被對角化，甚至在它的對角化過程中，所得出來的矩陣 P，是一個 unitary (orthogonal) matrix。這樣的一個過程，在 spectral theorem 中，我們將其稱為 unitary diagonalizable。不過，要研讀這個主題之前，我們得先完成幾個準備工作。

Lemma 7.5

Let $\langle , \rangle$ be the standard inner product on C^n and let A be a complex $n \times n$ matrix. If A is hermitian, then for every $X = {}^t(x_1, x_2, \ldots, x_n) \in C^n$, $\langle AX, X \rangle$ is a real number.

Proof

這個證明不難，我們只消考慮下列式子即可。

$$\langle AX, X\rangle = {}^t(AX)\overline{X}$$
$$= ({}^tX)({}^tA)\overline{X}$$
$$= ({}^tX)(\overline{({}^t\overline{A})X})$$
$$= ({}^tX)(\overline{A^*X})$$
$$= \langle X, A^*X\rangle$$
$$= \langle X, AX\rangle$$

上列最後等式是因為，$A = A^*$。所以，我們發現

$$\langle AX, X\rangle = \langle X, AX\rangle$$

但是由於， $\langle X, AX\rangle = \overline{\langle AX, X\rangle}$
所以， $\langle AX, X\rangle = \overline{\langle AX, X\rangle}$
這說明了， $\langle AX, X\rangle$ is a real number。

其實前述這個 lemma 的逆敘述也是成立的。有關這一點，下面我們馬上會給讀者一個交代。在此之前，我們不如先練習兩個 Exercises 和一個 Proposition。

Exercise

證明：$\langle AX, Y\rangle = \langle X, A^*Y\rangle$，$\forall\ X, Y \in C^n$。(*Hint*：直接計算可得)

這種情形之下我們發現，假如 A 是 hermitian 的話，那麼

$$\langle AX, Y\rangle = \langle X, AY\rangle, \quad \forall\ X, Y \in C^n$$

Exercise

Show that if A is a complex $n \times n$ matrix and $\langle AX, Y\rangle = 0$, $\forall\ X, Y \in C^n$, then A is a zero matrix. (*Hint*：Taking X and Y as standard unit vectors.)

其實，前述 exercise 的條件，可以更為寬鬆一點。

Proposition 7.6

If $\langle AX, X \rangle = 0$, $\forall\ X \in C^n$, then A is a zero matrix.

Proof

Consider for any X and Y in C^n, we have by the hypothesis that

$$\langle A(X-Y), X-Y \rangle = 0$$

This implies that

$$\langle AX, X \rangle - \langle AX, Y \rangle - \langle AY, X \rangle + \langle AY, Y \rangle = 0$$

Hence that $\qquad\qquad \langle AX, Y \rangle + \langle AY, X \rangle = 0 \qquad\qquad (*)$

Substituting X by iX, then equation $(*)$ becomes

$$i\langle AX, Y \rangle - i\langle AY, X \rangle = 0$$

Hence $\qquad\qquad \langle AX, Y \rangle - \langle AY, X \rangle = 0 \qquad\qquad (**)$

Now, $(*)+(**)$ implies,

$$2\langle AX, Y \rangle = 0 \text{ or } \langle AX, Y \rangle = 0$$

The last equation is true for any X and Y. So by the last exercise, A is a zero matrix.

有關這個 proposition 要特別注意。若 C^n 改成 R^n 的話,它就不能成為事實了。我們看看下面矩陣。

$A = \begin{pmatrix} 0 & 1 \\ -1 & 0 \end{pmatrix}$。對於任意,$X = \begin{pmatrix} x \\ y \end{pmatrix} \in R^2$,

$$\langle AX, X \rangle = {}^t(AX)\bar{X} = \begin{pmatrix} y & -x \end{pmatrix}\begin{pmatrix} x \\ y \end{pmatrix} = \begin{pmatrix} 0 \\ 0 \end{pmatrix}$$

可是,$A = \begin{pmatrix} 0 & 1 \\ -1 & 0 \end{pmatrix}$ 並非 zero matrix。

其實就算,$A = \begin{pmatrix} 0 & i \\ -i & 0 \end{pmatrix} = A^*$,對於任意,$X = \begin{pmatrix} x \\ y \end{pmatrix} \in R^2$,相同情況仍然產生

$$\langle AX, X\rangle = {}^t(AX)\overline{X} = \begin{pmatrix} yi & -xi \end{pmatrix}\begin{pmatrix} x \\ y \end{pmatrix} = \begin{pmatrix} 0 \\ 0 \end{pmatrix}$$

所以，我們大家一定要特別注意，在前述 proposition 中的，"$\forall\ X \in C^n$"之條件是不可缺少的。

同樣，下面的 theorem 也將是不成立的，假設 C^n 改成 R^n 的話。(專業一點的說法是，A must be a linear operator on a complex-Hilbert space。)

Theorem 7.7

> If for every $X = {}^t(x_1, x_2, \ldots, x_n) \in C^n$, $\langle AX, X\rangle$ is a real number, then A is a hermitian matrix.

Proof

有了前面的兩個 exercises 和 proposition，這個證明變得非常簡單。要證明 $A = A^*$，只要能夠證明

$$\langle A^*X, X\rangle = \langle AX, X\rangle,\ \forall\ X \in C^n$$

即可。然而，已知 $\langle AX, X\rangle$ 是實數，所以對任意 $X \in C^n$ 而言
$$\langle AX, X\rangle = \overline{\langle AX, X\rangle}$$
或者說 $\quad \langle AX, X\rangle = \langle X, AX\rangle$
又 $\quad\quad\quad \langle X, AX\rangle = \langle A^*X, X\rangle$
所以 $\quad\quad \langle AX, X\rangle = \langle A^*X, X\rangle,\ \forall\ X \in C^n$

Remark

我們當然也可以不透過前述的 proposition，而直接證明
$$\langle A^*X, Y\rangle = \langle AX, Y\rangle,\ \forall\ X, Y \in C^n$$

Consider, for every X, Y in C^n and a complex number, α. By the hypothesis, we see that $\langle A(X + \alpha Y), X + \alpha Y\rangle$ is a real number. Moreover,

$$\langle A(X+\alpha Y), X+\alpha Y\rangle$$
$$=\langle AX,X\rangle+\overline{\alpha}\langle AX,Y\rangle+\alpha\langle AY,X\rangle+|\alpha|^2\langle AY,Y\rangle$$

Since $\langle AX,X\rangle$ and $|\alpha|^2\langle AY,Y\rangle$ are real, $\overline{\alpha}\langle AX,Y\rangle+\alpha\langle AY,X\rangle$ must be real. Hence, $\overline{\alpha}\langle AX,Y\rangle+\alpha\langle AY,X\rangle$ equals its conjugate. That is
$$\overline{\alpha}\langle AX,Y\rangle+\alpha\langle AY,X\rangle=\alpha\overline{\langle AX,Y\rangle}+\overline{\alpha}\,\overline{\langle AY,X\rangle}\ \text{or}$$
$$\overline{\alpha}\langle AX,Y\rangle+\alpha\langle AY,X\rangle=\alpha\langle Y,AX\rangle+\overline{\alpha}\langle X,AY\rangle$$

or, by the previous exercise,
$$\overline{\alpha}\langle AX,Y\rangle+\alpha\langle AY,X\rangle=\alpha\langle A^*Y,X\rangle+\overline{\alpha}\langle A^*X,Y\rangle$$

Now, letting $\alpha=1$ and $\alpha=-i$, we then have the following two equations
$$\langle AX,Y\rangle+\langle AY,X\rangle=\langle A^*Y,X\rangle+\langle A^*X,Y\rangle \tag{!}$$
$$i\langle AX,Y\rangle-i\langle AY,X\rangle=-i\langle A^*Y,X\rangle+i\langle A^*X,Y\rangle \tag{!!}$$

Multiplying both sides of (!) by i, we obtain,
$$i\langle AX,Y\rangle+i\langle AY,X\rangle=i\langle A^*Y,X\rangle+i\langle A^*X,Y\rangle \tag{!!!}$$

Putting (!!) and (!!!) together, we have
$$2i\langle AX,Y\rangle=2i\langle A^*X,Y\rangle$$

This implies, $\quad\langle A^*X,Y\rangle=\langle AX,Y\rangle,\quad \forall\ X,Y\in C^n.$

前面的 Lemma 7.5 (242 頁) 還可以更進一步的告訴我們，有關 hermitian matrix 之 eigenvalue 的一個非常有用的特性。那就是，

Theorem 7.8

If A is an $n\times n$ hermitian matrix and if λ is an eigenvalue of A, then λ is a real number.

Proof

由於，λ is an eigenvalue of A，所以，$\exists\ X\in C^n$, $X\neq 0$，使得 $AX=\lambda X$。利用 Lemma 7.5，得知 $\langle AX,X\rangle$ 是實數。因此
$$\langle AX,X\rangle=\langle \lambda X,X\rangle=\lambda\langle X,X\rangle\ \text{為一實數}。$$

然而，$\lambda \langle X,X \rangle = \lambda \|X\|^2$，且 $\|X\|^2 > 0$，所以 λ 必爲一實數。

Example 4

Find the eigenvalues of $A = \begin{pmatrix} 1 & -1 & 0 \\ -1 & 2 & -1 \\ 0 & -1 & 1 \end{pmatrix}$.

Solution

The characteristic polynomial,

$$P_A(t) = \det(tI - A) = \begin{pmatrix} t-1 & 1 & 0 \\ 1 & t-2 & 1 \\ 0 & 1 & t-1 \end{pmatrix}$$

$$= t^3 - 4t^2 + 3t = t(t-3)(t-1)$$

所以，the eigenvalues of A are 0, 1 and 3。注意，它們全都是實數。只因爲，$A = A^*$。(註：對於 real $n \times n$ matrix, A, 所謂 $A = A^*$，當然就是 $A = {}^t A$ 的意思，或者說，A is symmetric 的意思。) ∎

Example 5

Find the eigenvalues of $A = \begin{pmatrix} 0 & 0 & 1+i \\ 0 & 2 & 0 \\ 1-i & 0 & 1 \end{pmatrix}$.

Solution

The characteristic polynomial,

$$P_A(t) = \det(tI - A) = \begin{pmatrix} t & 0 & -(1+i) \\ 1 & t-2 & 0 \\ -(1-i) & 0 & t-1 \end{pmatrix}$$

$$= t^3 - 3t^2 + 2t - 2(t-2) = t^3 - 3t^2 + 4$$

$$= (t+1)(t-2)^2$$

So, the eigenvalues are -1 and 2.

我們還需要一個預備定理,來完成 hermitian matrix 的對角化工作。

Lemma 7.9

> Given an $n \times n$ hermitian matrix, A, considered as a linear map from C^n into C^n. Let X be an eigenvector of A. If Y is a vector in C^n perpendicular to X, then AY is also perpendicular to X.

Proof

這個證明簡單。假設 $AX = \lambda X$,則
$$\langle AY, X \rangle = \langle Y, A^*X \rangle = \langle Y, AX \rangle = \langle Y, \lambda X \rangle = \lambda \langle Y, X \rangle = 0$$
所以,$AY \perp X$。

其實,上述預備定理可以更進一步的被改寫為

Lemma 7.10

> Let A be an $n \times n$ hermitian matrix. If U is a subspace of C^n such that $A(U) \subseteq U$, then $A(U^\perp) \subseteq U^\perp$.

Proof

Let $v \in U^\perp$. For any $u \in U$
$$\langle Av, u \rangle = \langle v, A^*u \rangle = \langle v, Au \rangle$$
Since $A(U) \subseteq U$, $Au = w \in U$ for some w. Thus,
$$\langle Av, u \rangle = \langle v, A^*u \rangle = \langle v, Au \rangle = 0$$

對於一個 linear map,$A: W \to W$,若 U is a subspace of W and $A(U) \subseteq U$,則我們稱,U is stable under A。現在我們就來看看 hermitian

Chapter Seven　Eigenvalues、Diagonalizations、Idempotent　249

matrix 的對角化問題。這個定理有時被稱為 Spectral theorem。它是矩陣論中的重要定理之一。

Theorem　7.11 (Spectral theorem)

If A is an $n \times n$ hermitian matrix, then A can be diagonalized by a unitary matrix. (i.e. there exists a unitary matrix, P, such that $P^*AP = D$ is a diagonal matrix.)

Proof

首先，令 $X_1 \in C^n$ 為 A 的一個 eigenvector，使得 $AX_1 = \lambda_1 X_1$。此時，若令 $U = SP\{X_1\}$，則 $A(U) \subseteq U$。那麼，前述 Lemma 7.10 告訴我們，$U^\perp$ is also stable under A。而且，$\dim(U^\perp) = n - 1$。

接著，我們把 A 看成一個從 $U^\perp$ 到 $U^\perp$ 的 hermitian linear map。現在，重複前面的動作，我們得出 A 的第二個 eigenvector $X_2 \in U^\perp$，使得 $AX_2 = \lambda_2 X_2$。(注意，$X_2 \perp X_1$)

將這個要領持續重複到第 n 次的時候，我們得到一組由 A 的 eigenvectors 所組成的 C^n 的 orthogonal basis，$\{X_1, X_2, X_3, ..., X_n\}$ (他們的 eigenvalues 分別為 $\{\lambda_1, \lambda_2, ..., \lambda_n\}$)。

現在將這個 orthogonal basis，$\{X_1, X_2, X_3, ..., X_n\}$ 給 normalize，所得出的 orthonormal basis，將其記為

$$\alpha = \{X_1/\|X_1\|,\ X_2/\|X_2\|,\ X_3/\|X_3\|, ..., X_n/\|X_n\|\}。$$

由於，$AX_i = \lambda_i X_i$，$\forall i = 1, 2, ..., n$。所以，$mat(A, \alpha, \alpha) = D$ 是一個對角矩陣，其對角線元素為 $\lambda_1, \lambda_2, ..., \lambda_n$。而，the change of basis matrix，$P = mat(I, \alpha, E)$ 之每一個 column 為，$P^i = X_i/\|X_i\|$，$i = 1, 2, ..., n$。所以，P 是一個 unitary matrix，而且，$P^*AP = D$。定理得證。

Spectral theorem 有時被敘述為，If A is an $n \times n$ hermitian matrix mapping from C^n into C^n, then C^n has an orthogonal basis consisting of eigenvectors of A.

Example 6

Given an $n \times n$ matrix, $A = \begin{pmatrix} 0 & 0 & 1+i \\ 0 & 2 & 0 \\ 1-i & 0 & 1 \end{pmatrix}$. Find a unitary matrix, P, such that $P^*AP = D$ is a diagonal matrix. (with diagonal elements consisting of eigenvalues of A.)

Solution

延續之前的例題，A 的 eigenvalues are -1 and 2。
由於，

$\begin{pmatrix} 0 & 0 & 1+i \\ 0 & 2 & 0 \\ 1-i & 0 & 1 \end{pmatrix} \begin{pmatrix} 0 \\ 1 \\ 0 \end{pmatrix} = 2 \begin{pmatrix} 0 \\ 1 \\ 0 \end{pmatrix}$ $\begin{pmatrix} 0 & 0 & 1+i \\ 0 & 2 & 0 \\ 1-i & 0 & 1 \end{pmatrix} \begin{pmatrix} 1 \\ 0 \\ 1-i \end{pmatrix} = 2 \begin{pmatrix} 1 \\ 0 \\ 1-i \end{pmatrix}$

$\begin{pmatrix} 0 & 0 & 1+i \\ 0 & 2 & 0 \\ 1-i & 0 & 1 \end{pmatrix} \begin{pmatrix} -2 \\ 0 \\ 1-i \end{pmatrix} = -1 \begin{pmatrix} -2 \\ 0 \\ 1-i \end{pmatrix}$

(註：$\lambda = 2$ 是 characteristic polynomial 的重根，而且 $\dim V_{\lambda=2} = 2$)
所以，

$\begin{pmatrix} 0 \\ 1 \\ 0 \end{pmatrix}, \begin{pmatrix} 1 \\ 0 \\ 1-i \end{pmatrix}$ and $\begin{pmatrix} -2 \\ 0 \\ 1-i \end{pmatrix}$ are eigenvectors of A with respect to eigenvalues $2, 2$ and -1, respectively.

現在，我們取其 orthonormal basis 爲

$$\alpha_1 = \begin{pmatrix} 0 \\ 1 \\ 0 \end{pmatrix}, \ \alpha_2 = \frac{1}{\sqrt{3}} \begin{pmatrix} 1 \\ 0 \\ 1-i \end{pmatrix}, \ \alpha_3 = \frac{1}{\sqrt{6}} \begin{pmatrix} -2 \\ 0 \\ 1-i \end{pmatrix}$$

且令 $P = \begin{pmatrix} 0 & 1/\sqrt{3} & -2/\sqrt{6} \\ 1 & 0 & 0 \\ 0 & (1-i)/\sqrt{3} & (1-i)/\sqrt{6} \end{pmatrix}$，則 $P^* = \begin{pmatrix} 0 & 1 & 0 \\ 1/\sqrt{3} & 0 & (1+i)/\sqrt{3} \\ -2/\sqrt{6} & 0 & (1+i)/\sqrt{6} \end{pmatrix}$

此矩陣 P 為一 unitary matrix。而且，

$$P^*AP = \begin{pmatrix} 2 & 0 & 0 \\ 0 & 2 & 0 \\ 0 & 0 & -1 \end{pmatrix}$$

■

前述 Spectral theorem 中的 A 若為單純的 symmetric matrix，並且將其看成一個從 R^n 映至 R^n 的 operator 的話。那麼，由它的 eigenvectors 所組成的矩陣 P 是一個 orthogonal matrix。且 $P^{-1}AP = D$ 是一個對角矩陣，其之對角線元素同樣為它的 eigenvalues。

Hermitian (Symmetric) matrix 的對角化，在求線性微分方程式的解當中，也給我們帶來應用上很大的方便。我們看下面的推論之後，再舉一個例子，大家便可知曉。

Corollary

Given an $n \times n$ real symmetric matrix, A. Let P be the orthogonal matrix consisting of eigenvectors of A such that $P^{-1}AP = D$, where D is a diagonal matrix with diagonal elements consisting of eigenvalues of A. If $X(t) \in R^n$, $Y(t) = P^{-1}X(t)$ and $\dfrac{dX(t)}{dt} = AX(t)$, then $\dfrac{dY(t)}{dt} = DY(t)$.

Proof

我們總喜歡用下列簡圖幫忙理解。這個推論也不例外。圖中的 α 為 A 的 spectral basis，E 為標準基底，$P = mat(I, \alpha, E)$ is the change of basis matrix from α into E。

那麼，本推論的結果可以從下列式子的推導得出。

$$X(t) = PY(t)$$

$$\Rightarrow \frac{dX(t)}{dt} = P\frac{dY(t)}{dt}$$

$$\Rightarrow AX(t) = P\frac{dY(T)}{dt}$$

$$\Rightarrow P^{-1}AX(t) = \frac{dY(t)}{dt}$$

$$\Rightarrow DP^{-1}X(t) = \frac{dY(t)}{dt}$$

$$\Rightarrow DY(t) = \frac{dY(t)}{dt}$$

Example 7

Given, $A = \begin{pmatrix} 1 & -1 & 0 \\ -1 & 2 & -1 \\ 0 & -1 & 1 \end{pmatrix}$. Solve, $\frac{dX(t)}{dt} = AX(t)$.

Solution

The eigenvectors of A with respect to the eigenvalues $\lambda_1=0$, $\lambda_2=1$, $\lambda_3=3$ are ${}^t\alpha_1 = (1,1,1)$, ${}^t\alpha_2 = (1,0,-1)$, ${}^t\alpha_3 = (1,-2,1)$, respectively.

Writing $P = \begin{pmatrix} 1 & 1 & 1 \\ 1 & 0 & -2 \\ 1 & -1 & 1 \end{pmatrix}$, then $P^{-1} = \frac{-1}{6}\begin{pmatrix} -2 & -2 & -2 \\ -3 & 0 & 3 \\ -1 & 2 & -1 \end{pmatrix}$ and

$$P^{-1}AP = \begin{pmatrix} 0 & 0 & 0 \\ 0 & 1 & 0 \\ 0 & 0 & 3 \end{pmatrix} = D$$

利用上述推論，我們先解方程式，$\dfrac{dY(t)}{dt} = DY(t)$，

$$\dfrac{dy_1(t)}{dt} = 0, \quad \dfrac{dy_2(t)}{dt} = y_2, \quad \dfrac{dy_3(t)}{dt} = 3y_3$$

得出 $\quad y_1(t) = c_1, \quad y_2(t) = c_2 e^t, \quad y_3(t) = c_3 e^{3t}$

再利用 $\quad X(t) = PY(t)$

得出 $\quad {}^t X(t) = (c_1 + c_2 e^t + c_3 e^{3t}, \; c_1 - 2c_3 e^{3t}, \; c_1 - c_2 e^t + c_3 e^{3t})$

即為本例題解。

除了前述應用，有關矩陣的**正定則** (positive definite)方面，我們似乎也該研讀一下。記得在第 5.2 節，當我們談論矩陣正定則的時候，我們曾經提及說，

  An $n \times n$ hermitian matrix is positive definite if and only if all its eigenvalues are positive。

沒錯，正是如此。我們看看下面定理。

Theorem 7.12

  Let A be an $n \times n$ hermitian matrix. Then all the eigenvalues of A are positive if and only if $\langle AX, X \rangle > 0$, $\forall\; X \in C^n$.

Proof

($\Rightarrow$)

  從 spectral theorem 的證明過程中，我們發現，存在一組由 A 的 eigenvectors 所組成的 orthonormal basis of C^n，$\{\alpha_1, \alpha_2, \ldots, \alpha_n\}$。

令 $X = \displaystyle\sum_{i=1}^n x_i \alpha_i$，且令 $A\alpha_i = \lambda_i \alpha_i$，則，for any $X \in C^n$，

$$\langle AX, X \rangle = \langle A\sum_{i=1}^n x_i \alpha_i, \sum_{j=1}^n x_j \alpha_j \rangle = \sum_{i=1}^n x_i \langle A\alpha_i, \sum_{j=1}^n x_j \alpha_j \rangle$$

由於，$\langle \alpha_i, \alpha_j \rangle = 0$，$\forall\; i \ne j$，所以，上式變成

$$\langle AX, X \rangle = \sum_{i=1}^{n} x_i \langle A\alpha_i, \sum_{j=1}^{n} x_j \alpha_j \rangle = \sum_{i=1}^{n} x_i \overline{x_i} \langle A\alpha_i, \alpha_i \rangle$$

$$= \sum_{i=1}^{n} |x_i|^2 \langle \lambda_i \alpha_i, \alpha_i \rangle = \sum_{i=1}^{n} \lambda_i |x_i|^2$$

因為，$\lambda_i > 0, \forall i = 1, 2, ..., n$，且有某些個 $x_i \neq 0$，所以

$$\langle AX, X \rangle = \sum_{i=1}^{n} \lambda_i |x_i|^2 > 0$$

($\Leftarrow$)

這個方向應該是比較簡單。

假設，λ is an eigenvalue of A with a non-zero eigenvector, X, such that $AX = \lambda X$。則

$$\langle AX, X \rangle > 0$$
$$\Rightarrow \langle \lambda X, X \rangle > 0$$
$$\Rightarrow \lambda \|X\|^2 > 0$$

所以，λ must be positive.

Example 8

Show that $A = \begin{pmatrix} 2 & 1 \\ 1 & 1 \end{pmatrix}$ is positive definite.

Proof

Consider that

$$\det(tI - A) = t^2 - 3t + 1 = (t - \frac{3+\sqrt{5}}{2})(t - \frac{3-\sqrt{5}}{2})$$

Since $\lambda_1 = \frac{3+\sqrt{5}}{2}$ and $\lambda_2 = \frac{3-\sqrt{5}}{2}$ are both positive, $A = \begin{pmatrix} 2 & 1 \\ 1 & 1 \end{pmatrix}$ is positive definite.

Chapter Seven Eigenvalues · Diagonalizations · Idempotent 255

Indeed, for $\begin{pmatrix} x \\ y \end{pmatrix} \in R^2$,

$$(x \ y)\begin{pmatrix} 2 & 1 \\ 1 & 1 \end{pmatrix}\begin{pmatrix} x \\ y \end{pmatrix} = (2x+y \ \ x+y)\begin{pmatrix} x \\ y \end{pmatrix}$$

$$= 2x^2 + xy + xy + y^2$$

$$= x^2 + (x+y)^2 > 0$$

∎

Exercises

1. Diagonalize each of the following matrices.

   a. $A = \begin{pmatrix} 3 & 2 \\ -1 & 0 \end{pmatrix}$

   b. $A = \begin{pmatrix} -2 & 5 \\ 1 & 2 \end{pmatrix}$

   c. $A = \begin{pmatrix} 2 & 2i \\ -2i & -1 \end{pmatrix}$

   d. $A = \begin{pmatrix} 3 & 2 & 1 \\ 0 & 1 & 2 \\ 0 & 1 & -1 \end{pmatrix}$

   e. $A = \begin{pmatrix} 4 & 0 & 1 \\ -2 & 1 & 0 \\ -2 & 0 & 1 \end{pmatrix}$

2. Diagonalize $A = \begin{pmatrix} 2 & 2i \\ -2i & -1 \end{pmatrix}$ and solve $\dfrac{dX(t)}{dt} = AX(t)$ for $X \in C^2$.

3. Diagonalize $A = \begin{pmatrix} 2 & -1 & 0 \\ -1 & 2 & -1 \\ 0 & -1 & 2 \end{pmatrix}$ and solve $\dfrac{dX(t)}{dt} = AX(t)$ for $X \in R^3$.

4. Determine which of the following matrices are positive definite.

   a. $A = \begin{pmatrix} 1 & 2 \\ 2 & 1 \end{pmatrix}$

   b. $A = \begin{pmatrix} 1 & -1 \\ -1 & 2 \end{pmatrix}$

   c. $A = \begin{pmatrix} 1 & 2 & 3 \\ 2 & 0 & 1 \\ 3 & 1 & 1 \end{pmatrix}$

   d. $A = \begin{pmatrix} 1 & -1 & 0 \\ -1 & 0 & 1 \\ 0 & 1 & 2 \end{pmatrix}$

5. Show that if A is an $n \times n$ hermitian positive definite matrix, then A^{-1} exists and is positive definite. (Hint: $P^{-1} = P^*$ and $P^*AP = D$ is diagonal. D^{-1} is also

diagonal with diagonal elements $1/\lambda_i$)

6. Show that if A is an $n \times n$ hermitian non-singular matrix, then A^2 is positive definite. (*Hint*: A is non-singular, so all eigenvalues of A are non-zero. Besides, $P^* A^2 P = D^2$.)

7. Let A be an $n \times n$ real symmetric matrix and let $f(X) = {}^t XAX$, $\forall\ X \in R^n$ be the associated quadratic form. Show that if $P \in R^n$, $\|P\| = 1$ is such that $f(P)$ is a maximum or minimum for f on the unit sphere, then P is an eigenvector of A. (*Hint*: for any unit vector, $w \in (SP\{P\})^\perp$, constructing a curve, $C(t) = (\cos t)P + (\sin t)w$, on the unit sphere and a function, $g(t) = f(C(t)) = AC(t) \cdot C(t)$, then g attains its max. or min. at $t = 0$. So, $0 = g'(0) = 2w \cdot AP$. This implies, $AP \in SP\{P\}$.)

8. Use exercise 7 to show that if all eigenvalues of an $n \times n$ real symmetric matrix A are positive, then A is positive definite. (*Hint*: $f(P) = {}^t PAP = {}^t P\lambda P = \lambda$)

§7.3　Idempotent matrices

　　第二章我們曾經利用基本列運算,介紹過一般矩陣的上三角化;也研讀過如何利用 eigenvectors 將一些特殊的 $n \times n$ 矩陣,或者 hermitian (symmetric) matrix 給予對角化的過程。下一章,我們則想透過所謂的 Householder 矩陣,和大家共同研讀,如何將一個 $n \times n$ 矩陣給上三角化之後,利用矩陣的上三角化來介紹 normal matrix 的 diagonalization。然而,在這個過程當中,有兩個扮演舉足輕重的矩陣,那就是所謂的 idempotent matrix (冪等矩陣) 以及 Householder matrix。所以,7.3 以及 7.4 節我們將分別和讀者一起研討,這兩個特殊的矩陣以及它們的一些特性。首先,我們稍微回憶一下,以前定義中所謂的 idempotent matrices。依過去所知,

　　A square matrix, A, is called an idempotent matrix (冪等矩陣), if $A^2 = A$. 譬如說,

$$A = \begin{pmatrix} 1 & 0 & 0 \\ 0 & 1 & 0 \\ 0 & 0 & 0 \end{pmatrix} = A^2 \text{ is an idempotent matrix.}$$

　　除了一些基本特性之外,idempotent matrix 還有很多精彩而又應用廣泛的實務上的特性。下面,就讓我們慢慢的一起來欣賞。

Theorem　7.13

　　Let A be an $n \times n$ idempotent matrix. Then
(i) $\,^t A$ and $I - A$ are also idempotent.
(ii) if B is similar to A, then B is also idempotent.
(iii) the eigenvalues of A are either 0 or 1(參看 7.1 節最後的 Exercise 5)。

Proof

(i) $(^t A)(^t A) = \,^t(AA) = \,^t(A^2) = \,^t A$　and
　　$(I - A)(I - A) = I - 2A + A^2 = I - 2A + A = I - A$
(ii) $BB = (P^{-1}AP)(P^{-1}AP) = P^{-1}AAP = P^{-1}AP = B$

(iii) Let λ be an eigenvalue of A such that $AX = \lambda X$ for some nonzero vector X. Then $A^2 X = \lambda AX = \lambda^2 X$, but then $AX = \lambda^2 X$, since A is idempotent. So, $(\lambda^2 - \lambda)X = 0$. Since X is a nonzero vector, $\lambda^2 - \lambda = 0$. And the result follows.

Exercises

1. Show that if A is an $n \times n$ non-singular matrix and if A is idempotent, then A must be the identity matrix. (Hint：$A^{-1}A = AA^{-1}$ and $A^2 = A$)

2. Give an example of a non-zero idempotent matrix which is singular. (Hint：Consider the diagonal matrix.)

此外，在 complex case 當中，假設 A 是 idempotent 的話，那麼，由於 $(A^*)^2 = A^* A^* = (AA)^* = (A^2)^* = A^*$。所以，$A^*$ 也是 idempotent。下面還有更多要陳述的，我們一步一步慢慢的來。

Theorem 7.14

Let A be an $n \times n$ idempotent matrix. Then
(i) $NS(A) = CS(I - A)$ and $CS(A) = NS(I - A) = V_{\lambda = 1}$
(ii) $NS(A) \cap CS(A) = \{0\}$
(iii) $R^n = NS(A) \oplus CS(A) = NS(A) \oplus NS(I - A)$
$\quad = CS(I - A) \oplus NS(I - A) = CS(I - A) \oplus CS(A)$
(iv) $rank(A) = tr(A)$

注意：以 complex case 而言，(iii) 中的 R^n 則為 C^n。

Proof

(i) Let $X \in NS(A)$. Then $AX = 0$ and $X = X - AX = (I - A)X$. This show that $X \in CS(I - A)$. The other hand, suppose that $X \in CS(I - A)$, then $(I - A)Y = X$, for some Y. Hence, $AX = AY - A^2 Y = AY - AY = 0$, and hence $X \in NS(A)$.

The proof of the second part of (i) is quite identical, while knowing that $V_{\lambda=1}$, by definition, is but $NS(I-A)$.

(ii) Let $X \in NS(A) \cap CS(A)$. Then, by (i), $AX = 0$ and $(I-A)X = 0$. And hence then $X = X - AX = (I-A)X = 0$. This Proves $NS(A) \cap CS(A) = \{0\}$.

(iii) 對於這部分，我們先看看下面式子
$$X = AX + (I-A)X$$
這告訴我們，對任意 R^n (或 C^n) 中的 X 而言，$X \in CS(A) + CS(I-A)$。
但是由於，$NS(A) \cap CS(A) = CS(I-A) \cap CS(A) = \{0\}$，所以，
$R^n = CS(A) \oplus CS(I-A)$。其餘的結果，由第 (i) 個公式便可得知。

(iv) To prove this, we let $\{\alpha_1, \alpha_2, ..., \alpha_r\}$ be a basis of $CS(A)$ and $\{\alpha_{r+1}, \alpha_{r+2}, ..., \alpha_n\}$ a basis of $NS(A)$. From the last part, we see that $\alpha = \{\alpha_1, \alpha_2, ..., \alpha_n\}$ is a basis of R^n (or C^n). Defining a map, T, from R^n (or C^n) into itself by
$$TX = AX, \ \forall \ X \in R^n$$
Since $CS(A) = NS(I-A)$, $(I-A)\alpha_i = 0$, $\forall i = 1, ..., r$. So, $T\alpha_i = A\alpha_i = \alpha_i$, $\forall i = 1, ..., r$ and hence, the matrix representation of T from basis α into basis α is
$$mat(T, \alpha, \alpha) = B = \begin{pmatrix} I_r & 0 \\ \hline 0 & 0 \end{pmatrix}$$

Since B is similar to A, A and B have the same characteristic polynomial. That is $P_A(t) = P_B(t)$.

Now, by Theorem 7.2, the coefficient of the term t^{n-1} of the characteristic polynomial is $-tr(A) = -tr(B)$. Therefore,
$$rank(A) = rank(B) = tr(B) = tr(A).$$

上述定理證明的最後過程中，我們得知一個重要的結論，那就是
Every idempotent matrix is similar to a matrix of the form
$$B = \begin{pmatrix} I_r & 0 \\ 0 & 0 \end{pmatrix}$$

或者說，there exists a non-singular matrix, P, such that
$$P^{-1}AP = B = \begin{pmatrix} I_r & 0 \\ 0 & 0 \end{pmatrix} \quad (@)$$

Example 1

Given, $A = \begin{pmatrix} 1 & -1 \\ 0 & 0 \end{pmatrix}$. Find a non-singular matrix, P, such that $P^{-1}AP = B$ is a diagonal matrix as of the form (@).

Solution

We first note that A is idempotent and consider that
$$\begin{pmatrix} 1 & -1 \\ 0 & 0 \end{pmatrix}\begin{pmatrix} a \\ b \end{pmatrix} = \begin{pmatrix} a-b \\ 0 \end{pmatrix} \quad \text{for} \quad \begin{pmatrix} a \\ b \end{pmatrix} \in R^2.$$

So
$$CS(A) = \left\{ x\begin{pmatrix} 1 \\ 0 \end{pmatrix} : x \in R \right\}.$$

And for $\begin{pmatrix} a \\ a \end{pmatrix} \in R^2$,
$$\begin{pmatrix} 1 & -1 \\ 0 & 0 \end{pmatrix}\begin{pmatrix} a \\ a \end{pmatrix} = \begin{pmatrix} 0 \\ 0 \end{pmatrix}.$$

So
$$NS(A) = \left\{ x\begin{pmatrix} 1 \\ 1 \end{pmatrix} : x \in R \right\}$$

Now, taking $\left\{ \alpha_1 = \begin{pmatrix} 1 \\ 0 \end{pmatrix} \right\}$ as a basis of $CS(A)$ and $\left\{ \alpha_2 = \begin{pmatrix} 1 \\ 1 \end{pmatrix} \right\}$ as a basis of $NS(A)$, $\alpha = \{\alpha_1, \alpha_2\}$ is a basis of $R^2 = CS(A) \oplus NS(A)$.

By the way, we see that
$$mat(A, \alpha, \alpha) = B = \begin{pmatrix} 1 & 0 \\ 0 & 0 \end{pmatrix}$$

and the change of basis matrix, $P = mat(I,\alpha,E) = \begin{pmatrix} 1 & 1 \\ 0 & 1 \end{pmatrix}$, $P^{-1} = mat(I,E,\alpha)$
$= \begin{pmatrix} 1 & -1 \\ 0 & 1 \end{pmatrix}$.

$$P^{-1}AP = \begin{pmatrix} 1 & -1 \\ 0 & 1 \end{pmatrix}\begin{pmatrix} 1 & -1 \\ 0 & 0 \end{pmatrix}\begin{pmatrix} 1 & 1 \\ 0 & 1 \end{pmatrix} = \begin{pmatrix} 1 & 0 \\ 0 & 0 \end{pmatrix} = B$$

注意：上述之矩陣 $P = \begin{pmatrix} 1 & 1 \\ 0 & 1 \end{pmatrix}$，它的兩個 columns 不是 orthogonal. ∎

一個 idempotent matrix，A，假如它也是 symmetric (hermitian in the complex case) 矩陣的話，在 7.2 節時我們已經清楚的研讀過，矩陣 P 將是一個 orthonormal (或是 unitary) matrix。不過，這一節我們要換個角度來看，同樣可以說明這個事實。首先，我們確認下面一個基本概念。

For any matrix, A, $NS(A) = (CS({}^t A))^\perp$

這個概念乃是因為，$AX = 0$ 若且唯若 X 垂直於矩陣 A 的每一個 row。也就是說，X 垂直於矩陣 $({}^t A)$ 的每一個 column。以 complex case 而言，前式則應為 $NS(A) = (CS(A^*))^\perp$。這個基本概念告訴我們，

Theorem 7.15

Let A be an $n \times n$ matrix. If A is both idempotent and symmetric (or hermitian), then
(i) $R^n = CS(A) \oplus (CS(A))^\perp$ (or $C^n = CS(A) \oplus (CS(A))^\perp$ in complex case.)
(ii) $AX = X_{CS(A)}$ and $(I - A)X = X_{(CS(A))^\perp}$, $\forall X \in R^n$

Proof

(i) If A is symmetric, then
$$NS(A) = (CS({}^t A))^\perp \quad \text{implies} \quad NS(A) = (CS(A))^\perp$$
Combining this with part(iii) of the last theorem, we have that

$$R^n = CS(A) \oplus (CS(A))^\perp$$

(ii) For this part, we first consider that
$$X = AX + (I - A)X \tag{!}$$
Where we have that
$$(I - A)X \in CS(I - A) = NS(A) = (CS(A))^\perp \text{ and } AX \in CS(A)$$
Because, $R^n = CS(A) \oplus (CS(A))^\perp$, so, for any $X \in R^n$
$$X = X_{CS(A)} + X_{(CS(A))^\perp} \tag{!!}$$
Comparing (!) and (!!), $AX = X_{CS(A)}$ and $(I - A)X = X_{(CS(A))^\perp}$ follow from the direct sum, $R^n = CS(A) \oplus (CS(A))^\perp$.

前述定理告訴我們，一個 symmetric (hermitian) idempotent 矩陣可以被看成，一個從 R^n 映射到 $CS(A)$ 的 orthogonal projection。反過來說，一個 orthogonal projection 同樣的也可以被看成是一個 symmetric (hermitian) idempotent 矩陣。看下面的 proposition。

Proposition 7.16

> Let U be a subspace of V. Defining $T : V \to V$ by
> $$T(X) = X_U, \ \forall \ X = X_U + X_{U^\perp} \in V$$
> Then there exists a basis, α, such that $B = mat(T; \alpha, \alpha)$ is a symmetric idempotent matrix.

Proof

The proof is quite easy by finding an orthonormal basis $\{\alpha_1, \alpha_2, ..., \alpha_r\}$ of U and an orthonormal basis $\{\alpha_{r+1}, \alpha_{r+2}, ..., \alpha_n\}$ of $U^\perp$.

Where $\alpha_1, \alpha_2, ..., \alpha_r, \alpha_{r+1}, ..., \alpha_n$ form an orthonormal basis for V and
$$T(\alpha_i) = \begin{cases} \alpha_i, & \text{if } i \le r \\ 0, & \text{if } i > r \end{cases}$$

Thus the matrix representation, $A = mat(T; \alpha, \alpha)$, is of the form

$$B = \begin{pmatrix} I_r & 0 \\ 0 & 0 \end{pmatrix}$$

which is an idempotent symmetric matrix.

先前的定理說過，一個 symmetric idempotent matrix 可以被視爲一個 orthogonal projection，而剛剛的定理則說，一個 orthogonal projection 其之 matrix representation 的型式可以是 $\begin{pmatrix} I_r & 0 \\ 0 & 0 \end{pmatrix}$。現在，我們將這兩個敘述擺在一起，得出結論爲：

一個 symmetric idempotent matrix，當它被視爲 linear map 的時候，它的 matrix representation 的型式可以是 $\begin{pmatrix} I_r & 0 \\ 0 & 0 \end{pmatrix}$。或者更明確的說，

Theorem 7.17

If A is an idempotent symmetric (hermitian) matrix, then there exists an orthonormal (unitary matrix) matrix, P, such that

$$P^{-1}AP = \begin{pmatrix} I_r & 0 \\ 0 & 0 \end{pmatrix} \qquad (*)$$

Proof

下面這個簡圖再一次的幫助我們，把這一前一後的東西再次搞清楚。

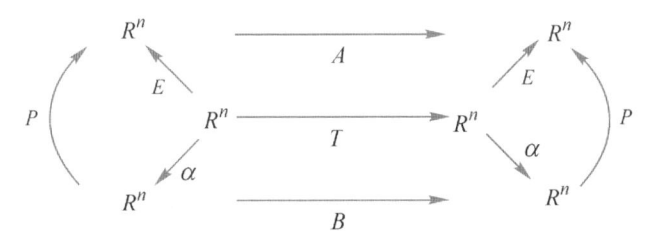

圖中的 E 表示標準基底，α 爲前述定理中的 orthonormal basis，B 爲

$mat(T;\alpha,\alpha) = \begin{pmatrix} I_r & 0 \\ 0 & 0 \end{pmatrix}$,P則為 the change of basis matrix from α into E,也就是說 $P = mat(I,\alpha,E)$。這個簡圖很清楚的告訴我們,$P^{-1}AP = B$。所以,我們僅須證明 $P = mat(I,\alpha,E)$ 是一個 orthonormal matrix 即可。然而,由於 E 是一個標準基底,所以矩陣 P 的每一個 column,$P^i, i = 1,2,...,n$,正是 orthonormal basis α 的第 i 個基底向量 α_i。而由於,α 是一個 orthonormal basis,毫無疑問的 P 當然也就是一個 orthonormal matrix 了。(在 complex case 中,上述的 R^n 改為 C^n,那麼 P^{-1} 改為 P^*)

Example 2

Given, $A = \begin{pmatrix} 1/2 & 1/2 \\ 1/2 & 1/2 \end{pmatrix}$. Find an orthonormal matrix, P, such that $P^{-1}AP = B$ is of the form as (*).

Solution

Also note that A is an idempotent symmetric matrix. Consider for each

$$\begin{pmatrix} x \\ y \end{pmatrix} \in R^2, \begin{pmatrix} 1/2 & 1/2 \\ 1/2 & 1/2 \end{pmatrix}\begin{pmatrix} x \\ y \end{pmatrix} = (\frac{1}{2}x + \frac{1}{2}y)\begin{pmatrix} 1 \\ 1 \end{pmatrix}.$$

So,

$$CS(A) = \left\{ a\begin{pmatrix} 1 \\ 1 \end{pmatrix} : a \in R \right\}$$

For each $\begin{pmatrix} x \\ -x \end{pmatrix} \in R^2, \begin{pmatrix} 1/2 & 1/2 \\ 1/2 & 1/2 \end{pmatrix}\begin{pmatrix} x \\ -x \end{pmatrix} = \begin{pmatrix} 0 \\ 0 \end{pmatrix}.$

So,

$$NS(A) = \left\{ a\begin{pmatrix} 1 \\ -1 \end{pmatrix} : a \in R \right\}.$$

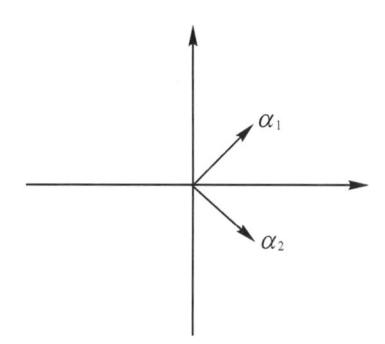

Taking $\left\{\alpha_1 = \begin{pmatrix} 1/\sqrt{2} \\ 1/\sqrt{2} \end{pmatrix}\right\}$ as a basis of $CS(A)$ and $\left\{\alpha_2 = \begin{pmatrix} 1/\sqrt{2} \\ -1/\sqrt{2} \end{pmatrix}\right\}$ as a basis of $NS(A)$, then $\alpha = \{\alpha_1, \alpha_2\}$ is a basis of $R^2 = CS(A) \oplus NS(A)$.

Noting that

$$mat(A,\alpha,\alpha) = B = \begin{pmatrix} 1 & 0 \\ 0 & 0 \end{pmatrix}$$

and that the change of basis matrix,

$$P = mat(I,\alpha,E) = \begin{pmatrix} 1/\sqrt{2} & 1/\sqrt{2} \\ 1/\sqrt{2} & -1/\sqrt{2} \end{pmatrix}, \quad P^{-1} = \begin{pmatrix} 1/\sqrt{2} & 1/\sqrt{2} \\ 1/\sqrt{2} & -1/\sqrt{2} \end{pmatrix}$$

with these we have that

$$P^{-1}AP = \begin{pmatrix} 1/\sqrt{2} & 1/\sqrt{2} \\ 1/\sqrt{2} & -1/\sqrt{2} \end{pmatrix}\begin{pmatrix} 1/2 & 1/2 \\ 1/2 & 1/2 \end{pmatrix}\begin{pmatrix} 1/\sqrt{2} & 1/\sqrt{2} \\ 1/\sqrt{2} & -1/\sqrt{2} \end{pmatrix} = \begin{pmatrix} 1 & 0 \\ 0 & 0 \end{pmatrix} = B$$

注意：此題之矩陣 $P = \begin{pmatrix} 1/\sqrt{2} & 1/\sqrt{2} \\ 1/\sqrt{2} & -1/\sqrt{2} \end{pmatrix}$，它是一個 orthonormal matrix. ∎

本節結束前，我們把頭腦再清理一下。

1. 一個 idempotent matrix is diagonalizable into a matrix of the form $B = \begin{pmatrix} I_r & 0 \\ 0 & 0 \end{pmatrix}$.

2. 一個 symmetric (hermitian) idempotent matrix is an orthogonal projection which is unitary diagonalizable into a matrix as of the form $B = \begin{pmatrix} I_r & 0 \\ 0 & 0 \end{pmatrix}$.

Exercises

1. Two idempotent matrices, A and B, are said to be orthogonal, if $AB = BA = 0$. Show that if $A_1, A_2, ..., A_m$ are mutually orthogonal idempotent matrices, and if for each j, X_j is a non-zero column of A_j, then $X_1, X_2, ..., X_m$ are linearly independent. (*Hint*: For each $j \neq i$, $A_j(X_i) = 0$ and $A_j(X_j) \neq 0$)

2. Given an idempotent matrix, $A = \begin{pmatrix} 1/3 & 1/3 & 1/3 \\ 1/3 & 1/3 & 1/3 \\ 1/3 & 1/3 & 1/3 \end{pmatrix}$. Find an orthonormal matrix, P, such that $P^{-1}AP = B$ is of the form as (∗) of Theorem 7.17. (*Hint*: $CS(A)$ is spanned by $\{{}^t\alpha_1 = \begin{pmatrix} 1/\sqrt{3} & 1/\sqrt{3} & 1/\sqrt{3} \end{pmatrix}\}$, $NS(A)$ is spanned by $\{{}^t\alpha_2 = \begin{pmatrix} 1/\sqrt{2} & 0 & -1/\sqrt{2} \end{pmatrix}$, ${}^t\alpha_3 = \begin{pmatrix} -1/\sqrt{6} & 2/\sqrt{6} & -1/\sqrt{6} \end{pmatrix}\}$. And $\alpha_1, \alpha_2, \alpha_3$ form an orthonormal basis for C^3.)

3. Given that ${}^tX = \begin{pmatrix} -3 & 2 & 0 \end{pmatrix}$. Compute $X({}^tX)$, $({}^tX)X$ and $A = \dfrac{X({}^tX)}{({}^tX)X}$. Show that $A^2 = A$ and $AX = X$.

§7.4 Householder matrices

經過 7.3 節一長串的暖身運動之後，該是我們的主角出場之時。它是一個很特殊的 symmetric (hermitian) idempotent 矩陣。這個矩陣將告訴我們，如何製造出一個 Householder matrix。現在就依 7.3 節最後的 Exercise 3，來做下列之定義。

Definition 7.4

Define, for a non-zero column vector, X, in R^n (or C^n), the $n \times n$ matrix, E_X, by

$$E_X = \frac{X(^tX)}{(^tX)X} \tag{7.1}$$

e.g.

For $^tX = (1, 0, -2)$, $X(^tX) = \begin{pmatrix} 1 & 0 & -2 \\ 0 & 0 & 0 \\ -2 & 0 & 4 \end{pmatrix}$, and $(^tX)X = 1 + 4 = 5$.

So, $E_X = \dfrac{X(^tX)}{(^tX)X} = \begin{pmatrix} 1/5 & 0 & -2/5 \\ 0 & 0 & 0 \\ -2/5 & 0 & 4/5 \end{pmatrix}$

注意，從定義中我們確認了，$X(^tX)$ 為一 $n \times n$ 矩陣，而 $(^tX)X$ 為一純量。當然，in complex case，(7.1) 式中的 (^tX) 則應改為 $X^* = \overline{(^tX)}$。

e.g.

For $X = \begin{pmatrix} 1 \\ i \end{pmatrix}$, $X(X^*) = \begin{pmatrix} 1 & -i \\ i & 1 \end{pmatrix}$ and $(X^*)X = 2$.

So, $E_X = \dfrac{X(X^*)}{(X^*)X} = \begin{pmatrix} 1/2 & -i/2 \\ i/2 & 1/2 \end{pmatrix}$

注意，前面實數例子之 E_X 是 symmetric，複數部分的例子則為 hermitian。

沒錯，除此之外，它還有下面一些特性，讓我們給一併整理出來。

1. $E_X X = X$
2. $E_X^2 = E_X$
3. E_X is symmetric (hermitian)
4. E_X has rank 1
5. Letting $(SP\{X\})^\perp = \{Y \in R^n : Y \perp cX \text{ for any scalar } c\}$, then $E_X Y = 0$, $\forall Y \in \{SP(X)\}^\perp$.

Proof

1. $E_X X = \dfrac{X(^tX)}{(^tX)X} X = \dfrac{X(^tX)X}{(^tX)X} = \dfrac{X(^tXX)}{^tXX} = X$。

2. $E_X^2 = E_X E_X = \dfrac{X(^tX)}{(^tX)X} \dfrac{X(^tX)}{(^tX)X} = \dfrac{X(^tXX)(^tX)}{(^tXX)(^tX)X}$，由於 (^tXX) 是一個純量，分子、分母中的 (^tXX) 可以約去，所以，$\dfrac{X(^tXX)(^tX)}{(^tXX)(^tX)X} = \dfrac{X(^tX)}{(^tX)X} = E_X$ 得證。

3. 由於，(^tXX) 是一個純量，所以，$^t(E_X) = \dfrac{^t(X(^tX))}{(^tX)X} = \dfrac{X(^tX)}{(^tX)X} = E_X$。

4. 由於，$E_X = \dfrac{X(^tX)}{(^tX)X}$ 且 (^tXX) 是一個純量，所以，$rank(E_X) \leq \min\{rank(X), rank(^tX)\} = 1$。因為，$E_X \neq 0$，所以 $rank(E_X) = 1$。

5. 若 $Y \in \{SP(X)\}^\perp$，則 $E_X Y = \dfrac{X(^tX)Y}{(^tX)X} = \dfrac{X(^tXY)}{(^tX)X} = 0$。

Exercise

Defining, $E_X = \dfrac{X(X^*)}{(X^*)X}$, in the complex case, show that $E_X = (E_X)^*$.(*Hint*：證法與實數的 case 相同。可以先計算 $^t(E_X)$，再計算 $\overline{^t(E_X)}$。唯一要注意的是，$(X^*)X$ 是純量，而且 $(^tX)\overline{X}$ 與 $\overline{(^tX)X}$ 相同。)

從上述這一些特性當中，我們發現 E_X 是一個 symmetric (hermitian) idempotent matrix。因此，there exists an orthonormal matrix, P, such that

$$P^{-1}E_X P = \begin{pmatrix} I_1 & 0_{1\times(n-1)} \\ 0_{(n-1)\times 1} & 0_{(n-1)\times(n-1)} \end{pmatrix} \tag{7.2}$$

注意：因為 $rank(E_X)=1$，所以(7.2)式矩陣中的的左上角是 1×1 的 I_1。

除此之外，我們還要特別強調，E_X 是一個從 R^n 映射到 $SP\{X\}$ 的一個 orthogonal projection。現在到了主角出場的時候了，所謂的 Householder matrix, H_X，請看下列定義。

Definition 7.5

Let E_X be as above. The matrix, H_X, defined by
$$H_X = I - 2E_X$$
is called a Householder matrix.

Example 1

Given that ${}^t X = (1, 0, -2)$. Find the Householder matrix, H_X.

Solution

前面例題，我們已經算出

$$E_X = \frac{X({}^t X)}{({}^t X)X} = \begin{pmatrix} 1/5 & 0 & -2/5 \\ 0 & 0 & 0 \\ -2/5 & 0 & 4/5 \end{pmatrix}$$

所以，
$$H_X = I - 2E_X = \begin{pmatrix} 3/5 & 0 & 4/5 \\ 0 & 1 & 0 \\ 4/5 & 0 & -3/5 \end{pmatrix}$$

Example 2

Given that $X = \begin{pmatrix} 1 \\ i \end{pmatrix}$. Find the Householder matrix, H_X.

Solution

前面例題,我們已經算出

$$E_X = \frac{X(X^*)}{(X^*)X} = \begin{pmatrix} 1/2 & -i/2 \\ i/2 & 1/2 \end{pmatrix}$$

所以,

$$H_X = I - 2E_X = \begin{pmatrix} 0 & i \\ -i & 0 \end{pmatrix}$$

這個東西越來越好玩了,首先我們來看看,如此定義的情況之下,它有下面的一些基本特性。

Properties

1. ${}^tH_X = H_X$ (certainly, $(H_X)^* = H_X$, in complex case.)
2. $(H_X)^{-1} = H_X$.(hence, $(H_X)^2 = I$)
3. $H_X X = -X$
4. $H_X Y = Y$, if $Y \perp X$.

Proof

1. Since E_X is symmetric, the symmetry of H_X follows immediately.
2. $H_X = I - 2E_X$ implies that
$$H_X H_X = (I - 2E_X)(I - 2E_X) = I - 2E_X - 2E_X + 4E_X = I$$
也就是說,H_X 為一 orthonormal matrix. (or a unitary matrix in complex case)
3. $H_X X = (I - 2E_X)X = X - 2E_X X = X - 2X = -X$
4. 假若,$Y \perp X$,那麼,$H_X Y = (I - 2E_X)Y = Y - 2E_X Y = Y$。

注意,由特性 3 與 4 的結果來看,我們得知一個很好玩的結果,那就是，"H_X performs a reflection about the subspace, $U = (SP\{X\})^\perp$"。

另外，由 H_X 定義之前的 (7.2) 式，我們更發現，there exists an orthonormal matrix, P, such that $P^{-1}E_XP = \begin{pmatrix} I_1 & 0_{1\times(n-1)} \\ 0_{(n-1)\times 1} & 0_{(n-1)\times(n-1)} \end{pmatrix}$，因此

$$P^{-1}H_XP = P^{-1}(I - 2E_X)P = I - 2P^{-1}E_XP$$

$$= I - 2\begin{pmatrix} 1 & 0_{1\times(n-1)} \\ 0_{(n-1)\times 1} & 0_{(n-1)\times(n-1)} \end{pmatrix} = \begin{pmatrix} -1 & 0 & \cdots & \cdots & 0 \\ 0 & 1 & 0 & \cdots & 0 \\ \cdots & 0 & \cdots & 0 & \cdots \\ \cdots & \cdots & 0 & \cdots & 0 \\ 0 & 0 & \cdots & 0 & 1 \end{pmatrix}$$

這說明了，H_X has trace $n-2$ and has determinant -1.

Example 3

Given, ${}^tX = (1\ 1\ 0)$, ${}^tY = (0\ 1\ 1)$, ${}^tU = (1\ 0\ 1)$, ${}^tV = (0\ 1\ 0)$. Let $Z = X - Y$. Find H_Z and compute H_ZX, H_ZY, H_ZU, H_ZV.

Solution

$$Z = \begin{pmatrix} 1 \\ 0 \\ -1 \end{pmatrix} \qquad E_Z = \begin{pmatrix} 1/2 & 0 & -1/2 \\ 0 & 0 & 0 \\ -1/2 & 0 & 1/2 \end{pmatrix}. \text{ So, } H_Z = \begin{pmatrix} 0 & 0 & 1 \\ 0 & 1 & 0 \\ 1 & 0 & 0 \end{pmatrix} \text{ and}$$

$$H_ZX = \begin{pmatrix} 0 & 0 & 1 \\ 0 & 1 & 0 \\ 1 & 0 & 0 \end{pmatrix}\begin{pmatrix} 1 \\ 1 \\ 0 \end{pmatrix} = \begin{pmatrix} 0 \\ 1 \\ 1 \end{pmatrix} = Y \qquad H_ZY = \begin{pmatrix} 0 & 0 & 1 \\ 0 & 1 & 0 \\ 1 & 0 & 0 \end{pmatrix}\begin{pmatrix} 0 \\ 1 \\ 1 \end{pmatrix} = \begin{pmatrix} 1 \\ 1 \\ 0 \end{pmatrix} = X$$

$$H_ZU = \begin{pmatrix} 0 & 0 & 1 \\ 0 & 1 & 0 \\ 1 & 0 & 0 \end{pmatrix}\begin{pmatrix} 1 \\ 0 \\ 1 \end{pmatrix} = \begin{pmatrix} 1 \\ 0 \\ 1 \end{pmatrix} = U \qquad H_ZV = \begin{pmatrix} 0 & 0 & 1 \\ 0 & 1 & 0 \\ 1 & 0 & 0 \end{pmatrix}\begin{pmatrix} 0 \\ 1 \\ 0 \end{pmatrix} = \begin{pmatrix} 0 \\ 1 \\ 0 \end{pmatrix} = V$$

看到這個例題沒有？好漂亮的 Householder matrix 啊！它把 X 與 Y reflect to each other and remains U and V stable. (注意：U, V are perpendicular to Z) 下面定理，很清楚的說明了 Householder matrix 的這個特性。

Theorem 7.18

> Given two distinct non-zero vectors, X and Y in R^n with $\|X\|=\|Y\|$. Let $Z = X - Y$. Then $H_Z X = Y$ and $H_Z Y = X$.

Proof

We first consider

$$E_Z X = \frac{Z({}^tZ)X}{{}^tZZ} = \frac{(X-Y)({}^tX-{}^tY)X}{({}^tX-{}^tY)(X-Y)} = \frac{(X-Y)({}^tXX-{}^tYX)}{{}^tXX-{}^tXY-{}^tYX+{}^tYY} \tag{7.3}$$

由於，$\|X\|=\|Y\|$，所以，${}^tXX={}^tYY$。又，${}^tXY={}^tYX$，因此 (7.3) 式將變為

$$\frac{(X-Y)({}^tXX-{}^tXY)}{2({}^tXX)-2({}^tXY)} = \frac{(X-Y)({}^tXX-{}^tXY)}{2({}^tXX-{}^tXY)} = \frac{X-Y}{2}$$

(註：上式中之 $({}^tXX-{}^tXY)$ 是一常數，它可以從分數中被消去。)

Now, $H_Z X = (I - 2E_Z)X = X - 2E_Z X = X - 2\dfrac{X-Y}{2} = Y$

The fact that $H_Z Y = X$ follows from the property $(H_Z)^2 = I$.

這個結果讓我們大大開了眼界，那就是"在 R^n 中，存在一個 orthonormal map，它把 X 與 Y 對稱於 $U = (SP\{Z\})^\perp$ 對調。"這種情形，在 R^2 中是比較容易具體化的。我們以下面簡單圖形協助即可明瞭。

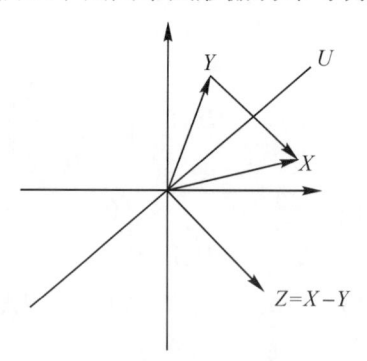

前述定理的證明中，我們使用了，${}^tXY={}^tYX$。而這個式子在 C^n 中而言，是講不通的，因為，$(X^*)Y \neq (Y^*)X$ for $X, Y \in C^n$。正因為如此，所以，在 complex case 中，想要建構這樣一個 unitary (orthonormal) map，H_Z，就有些許不同、需要多費點心思了。不過，無論如何我們還是要試試

看。首先，如同先前一樣，我們以符號，$E = \{e_1, e_2, ..., e_n\}$，表示 C^n 中的標準基底，其中每一個 e_i 代表著一個 column vector，它的 coordinates 除了第 i 個為 1 之外，其餘皆為 0 的意思。

Theorem 7.19

> Given the ith standard unit column vector, e_i, and a unit column vector, X, in C^n such that $X \neq e_i$ and the ith coordinate, x_i, of X is real. If $Z = X - e_i$, then
> $$H_Z X = e_i \quad \text{and} \quad H_Z e_i = X$$

Proof

We first note that $x_i \neq 1$, otherwise, X would be equal to e_i.
Consider that

$$E_Z X = \frac{Z(Z^*)X}{(Z^*)Z} = \frac{(X-e_i)(X^* - e_i^*)X}{(X^* - e_i^*)(X-e_i)} = \frac{(X-e_i)(1-x_i)}{2-2x_i} = \frac{X-e_i}{2}$$

so

$$H_Z X = (I - 2E_Z)X = X - 2E_Z X = X - (X - e_i) = e_i$$

And $H_Z e_i = H_Z H_Z X = (H_Z)^2 X = X$ follows from $(H_Z)^2 = I$.

Example 4

Given, $^t X = (i \ 0 \ 0)$, $^t e_2 = (0 \ 1 \ 0)$. Let $Z = X - e_2$. Find H_Z and compute $H_Z X, H_Z e_2$.

Solution

$$Z = \begin{pmatrix} i \\ -1 \\ 0 \end{pmatrix}, \quad E_Z = \frac{ZZ^*}{Z^*Z} = \frac{1}{2}\begin{pmatrix} i \\ -1 \\ 0 \end{pmatrix}\begin{pmatrix} -i & -1 & 0 \end{pmatrix} = \frac{1}{2}\begin{pmatrix} 1 & -i & 0 \\ i & 1 & 0 \\ 0 & 0 & 0 \end{pmatrix}. \text{ So,}$$

$$H_Z = I - 2E_Z = \begin{pmatrix} 0 & i & 0 \\ -i & 0 & 0 \\ 0 & 0 & 1 \end{pmatrix} \text{ and } H_Z X = \begin{pmatrix} 0 & i & 0 \\ -i & 0 & 0 \\ 0 & 0 & 1 \end{pmatrix}\begin{pmatrix} i \\ 0 \\ 0 \end{pmatrix} = \begin{pmatrix} 0 \\ 1 \\ 0 \end{pmatrix} = e_2 \text{,}$$

$$H_Z e_2 = \begin{pmatrix} 0 & i & 0 \\ -i & 0 & 0 \\ 0 & 0 & 1 \end{pmatrix}\begin{pmatrix} 0 \\ 1 \\ 0 \end{pmatrix} = \begin{pmatrix} i \\ 0 \\ 0 \end{pmatrix} = X$$

一般而言，對任意一個 nonzero vector，$X \in C^n$。假設它的第 i 個座標為，$x_i = r_i e^{i\theta_i}$。今令一 complex number，$c = e^{-i\theta_i}/\|X\|$，則 $cX \in C^n$，$\|cX\| = 1$，且 cX 的第 i 個座標，$r_i/\|X\|$，為一實數。我們將這個事實重新敘述如下：

For any nonzero vector，$X \in C^n$, there exists a complex number c such that the ith coordinate of cX is real and $\|cX\| = 1$.

Example 5

Given a unit vector, $^t X = \begin{pmatrix} 1/\sqrt{2} & 0 & -i/2 & 1/2 \end{pmatrix}$. Find a complex number, c, such that the 3rd coordinate of cX is a real number.

Solution

We first note that $x_3 = \frac{1}{2}e^{i(3\pi/2)}$. Letting $c = e^{-i(3\pi/2)}$, then $^t(cX) = \begin{pmatrix} i/\sqrt{2} & 0 & 1/2 & i/2 \end{pmatrix}$, $\|cX\| = 1$ and the 3rd coordinate of cX is a real.

Exercises

1. Find the Householder matrix, H_X. If X is given by

    a. $X = \begin{pmatrix} i \\ -1 \end{pmatrix}$　　　b. $X = \begin{pmatrix} 2 \\ 0 \\ 1 \end{pmatrix}$

    c. $X = \begin{pmatrix} 0 \\ -1 \\ 2 \end{pmatrix}$　　　d. $X = \begin{pmatrix} 1 \\ 1 \\ 0 \\ 2 \end{pmatrix}$

2. Given, $^tX=(2\ 0\ 1)$, $^tY=(0\ 2\ 1)$, $^tU=(1\ -1\ 0)$, $^tV=(1\ 1\ 3)$. Let $Z = X - Y$. Find H_Z and compute H_ZX, H_ZY, H_ZU, H_ZV. Write a general form of H_ZX for $\forall\ ^tX=(x\ y\ z)\in R^3$.

3. Given, $^tX=(1/\sqrt{2}\ \ i/\sqrt{2}\ \ 0)$, $^te_3=(0\ 0\ 1)$. Let $Z = X - e_3$. Find H_Z and compute H_ZX, H_Ze_3. (Hint: $H_Z = \begin{pmatrix} 1/2 & i/2 & 1/\sqrt{2} \\ -i/2 & 1/2 & i/\sqrt{2} \\ 1/\sqrt{2} & -i/\sqrt{2} & 0 \end{pmatrix}$)

4. Show that the eigenvalue of a Householder matrix, H_X, is either 1 or -1. (Hint: look at the properties of H_X)

Chapter Eight

Spectral Decomposition and Interpolatory Polynomials

8.1 Upper triangulation
8.2 Normal matrices
8.3 Spectral decomposition of a diagonalizable matrix
8.4 Interpolatory polynomials, constituent idempotents and Cayley-Hamilton theorem

　　本章節的主要任務是研讀，可對角化矩陣的"特徵值解構技巧"。除了基本的做法之外，如何利用 interpolatory polynomial 來完成解構工作，是本書的一個特色，它是一個令人激賞的、完美無瑕的藝術臻品。特別是 normal matrix 的解構，更有其極高度的欣賞價值。不過無論如何，我們還是得先從上三角化的工作開始著手。

§8.1　Upper triangulation

　　這一節，我們先舒舒筋骨研究一下，如何利用 7.4 節所學的 Householder matrices 把一個 $n \times n$ 矩陣給上三角化。命題是這樣寫的

Theorem 8.1

Let A be an $n \times n$ complex matrix. Then there exists a unitary matrix, P, such that
$$P^*AP = P^{-1}AP = \begin{pmatrix} \lambda_1 & * & * & * & * \\ 0 & \lambda_2 & * & \cdots & \cdots \\ \cdots & 0 & \cdots & * & \cdots \\ 0 & \cdots & 0 & \cdots & * \\ 0 & \cdots & \cdots & 0 & \lambda_n \end{pmatrix}$$
where $\lambda_1, \lambda_2, \ldots, \lambda_n$ are eigenvalues (they may be repeated) of A.

Note：以 real matrix，A，而言，我們當然得先假設，A 有 n 個 real eigenvalues。那麼，這時所得出來的矩陣 P 就變成是一個 real orthonormal matrix。

Proof

這將是一個很冗長的證明與說明，希望它不會造成讀者太多的困擾。Anyhow, We first let X_1 be a nonzero column eigenvetctor of A with respect to λ_1, and assume without loss of generality that $\|X_1\| = 1$ with its first coordinate x_1 is real. (7.4 節的最後已經交代清楚。)

If it is the case that $X_1 = e_1 = {}^t(1 \ 0 \ \cdots \ \cdots \ 0)$, then $AX_1 = Ae_1 = \lambda_1 e_1$ implies that the first column of the matrix representation of A with respect to the standard basis is ${}^t(\lambda_1 \ 0 \ \cdots \ \cdots \ 0)$.

That means, we may have
$$A = \begin{pmatrix} \lambda_1 & * \\ 0 & A_1 \end{pmatrix}$$
where $*$ is a $1 \times (n-1)$ matrix and A_1 is an $(n-1) \times (n-1)$ matrix.

If it is the case that $X_1 \neq e_1$, we then put $Z_1 = X_1 - e_1$ and let $H_1 = H_{Z_1}$ be the Householder matrix defined in section 7.4.

Consider that
$$H_1 A H_1 e_1 = H_1 A X_1 = H_1 \lambda_1 X_1 = \lambda_1 H_1 X_1 = \lambda_1 e_1$$

We also find out that
$$H_1 A H_1 = \begin{pmatrix} \lambda_1 & * \\ 0 & A_1 \end{pmatrix}$$

上述，是把矩陣 A 上三角化過程的第一步。

接著要看的是，右下的 $(n-1)\times(n-1)$ 矩陣 A_1，我們將 A_1 看成從 C^{n-1} 映至 C^{n-1} 的一個映射。記得 7.1 節的 Exercise10 嗎？從那兒，我們得知，因為 λ_2 是 A 的一個 eigenvalue，所以也是 A_1 的一個 eigenvalue。

第二步，我們令，$Y_2 = {}^t(y_2 \ y_3 \ \ldots \ \ldots \ y_n)$，$y_2$ 為實數，且 Y_2 為一單位向量，使得 $A_1 Y_2 = \lambda_2 Y_2$。今，將 Y_2 換成一個在 C^n 中的一個單位向量，$X_2 = {}^t(0 \ y_2 \ \ldots \ \ldots \ y_n)$。

同樣的，假若 $X_2 = e_2 = {}^t(0 \ 1 \ 0 \ \ldots \ 0)$，則由於 $A_1 Y_2 = \lambda_2 Y_2$，以及

$$\begin{pmatrix} \lambda_1 & * \\ 0 & A_1 \end{pmatrix} X_2 = \lambda_2 X_2 = {}^t(0 \ \lambda_2 \ 0 \ \ldots \ 0)$$

所以，$* = (0 \ *_1)$，其中 $*_1$ 為一個 $1 \times (n-2)$ 矩陣；而且 A_1 為

$$A_1 = \begin{pmatrix} \lambda_2 & *_2 \\ 0 & A_2 \end{pmatrix}$$

也就是說，A 或 $H_1 A H_1$ (第一個步驟的結果) 此時將變成

$$\begin{pmatrix} \lambda_1 & 0 & \vdots & \\ 0 & \lambda_2 & \vdots & * \\ \hdashline & 0 & \vdots & A_2 \end{pmatrix}$$

而此時的 $*$ 為某一，$2 \times (n-2)$ 矩陣，A_2 則為 $(n-2) \times (n-2)$ 矩陣。

當然，假若第二步驟之 $X_2 \neq e_2$，那麼我們要重複第一步驟的做法。令 $Z_2 = X_2 - e_2$ 且令 $H_2 = H_{Z_2}$ 為如上所定義之 Householder 矩陣。那麼，

$$H_2 \begin{pmatrix} \lambda_1 & * \\ 0 & A_1 \end{pmatrix} H_2 e_2 = H_2 \left(\begin{pmatrix} \lambda_1 & 0 \\ 0 & 0 \end{pmatrix} + \begin{pmatrix} 0 & * \\ 0 & A_1 \end{pmatrix} \right) X_2$$

$$= H_2 \begin{pmatrix} \lambda_1 & 0 \\ 0 & 0 \end{pmatrix} \begin{pmatrix} 0 \\ Y_2 \end{pmatrix} + H_2 \begin{pmatrix} 0 & * \\ 0 & A_1 \end{pmatrix} \begin{pmatrix} 0 \\ Y_2 \end{pmatrix}$$

$$= H_2 \begin{pmatrix} 0 & * \\ 0 & A_1 \end{pmatrix} \begin{pmatrix} 0 \\ Y_2 \end{pmatrix}$$

$$= H_2 (\begin{pmatrix} (*)Y_2 \\ 0 \end{pmatrix} + \begin{pmatrix} 0 \\ A_1 Y_2 \end{pmatrix}) = H_2 \begin{pmatrix} (*)Y_2 \\ 0 \end{pmatrix} + H_2 \begin{pmatrix} 0 \\ \lambda_2 Y_2 \end{pmatrix}$$

因為，$\begin{pmatrix} (*)Y_2 \\ 0 \end{pmatrix}$ 垂直於 $Z_2 = X_2 - e_2$，所以，

$$H_2 \begin{pmatrix} (*)Y_2 \\ 0 \end{pmatrix} = \begin{pmatrix} (*)Y_2 \\ 0 \end{pmatrix}$$

因此，

$$H_2 \begin{pmatrix} \lambda_1 & * \\ 0 & A_1 \end{pmatrix} H_2 e_2 = \begin{pmatrix} (*)Y_2 \\ 0 \end{pmatrix} + \lambda_2 H_2 X_2$$

$$= \begin{pmatrix} (*)Y_2 \\ 0 \end{pmatrix} + \lambda_2 e_2$$

$$= \begin{pmatrix} (*)Y_2 \\ 0 \end{pmatrix} + {}^t(0 \quad \lambda_2 \quad 0 \quad \ldots \quad 0) \tag{8.1}$$

我們從另一方面看，$Z_2 = X_2 - e_2$，所以

$$E_{Z_2} = \begin{pmatrix} 0 & 0 \\ \hline 0 & * \end{pmatrix}$$

其中，右上之 0 為一 $1 \times (n-1)$ 之 zero row vector，左下之 0 為一 $(n-1) \times 1$ 之 zero column vector。跟著，

$$H_2 = I - 2E_{Z_2} = \begin{pmatrix} 1 & 0 \\ \hline 0 & V_1 \end{pmatrix}$$

其中，$V_1 = I_{n-1} - 2(*)$ 為一 $(n-1) \times (n-1)$ 矩陣。

所以，

$$H_2 \begin{pmatrix} \lambda_1 & * \\ 0 & A_1 \end{pmatrix} H_2 = \begin{pmatrix} \lambda_1 & *_1 \\ \hline 0 & V_1 A_1 V_1 \end{pmatrix}$$

上述之 $(*_1) = (*)V_1$。

因此，

$$H_2\begin{pmatrix}\lambda_1 & * \\ 0 & A_1\end{pmatrix}H_2 e_2 = \begin{pmatrix}\lambda_1 & *_1 \\ 0 & V_1 A_1 V_1\end{pmatrix}e_2 \qquad (8.2)$$

(8.2) 式乘出來的結果是，矩陣 $\begin{pmatrix}\lambda_1 & * \\ 0 & V_1 A_1 V_1\end{pmatrix}$ 的第二個 column。將此結果與 (8.1) 式比較之後，我們發現

$$V_1 A_1 V_1 = \begin{pmatrix}\lambda_2 & * \\ 0 & A_2\end{pmatrix}$$

或者說

$$H_2\begin{pmatrix}\lambda_1 & * \\ 0 & A_1\end{pmatrix}H_2 = \begin{pmatrix}\lambda_1 & (*)Y_2 & *_2 \\ 0 & \lambda_2 & \\ \hline 0 & & A_2\end{pmatrix}$$

上式最右邊之 $*_2$ 為一 $2\times(n-2)$ 矩陣。

在經過了兩個步驟的 Householder 矩陣的應用之後，原先的矩陣 A 已經可以被改寫成型如，

$$\begin{pmatrix}\lambda_1 & * & * \\ 0 & \lambda_2 & \\ \hline 0 & & A_2\end{pmatrix}$$

按此要領持續下去，我們將可以找到有限個 Householder 矩陣，

$$H_1, H_2, \ldots, H_l$$

使得

$$H_l H_{l-1}\ldots H_1 A H_1 \ldots H_{l-1} H_l = \begin{pmatrix}\lambda_1 & * & \cdots & \cdots & * \\ 0 & \lambda_2 & * & \cdots & * \\ 0 & 0 & \cdots & * & * \\ 0 & \cdots & 0 & \cdots & * \\ 0 & \cdots & \cdots & 0 & \lambda_n\end{pmatrix}$$

最後階段，則令 $P = H_1 H_2 \cdots H_l$，由於每一個 H_i 是 unitary，所以 P 是 unitary，而且 $P^{-1} = P^* = H_l H_{l-1} \cdots H_1$。

結果得證，

$$P^{-1}AP = P^*AP = \begin{pmatrix} \lambda_1 & * & \cdots & \cdots & * \\ 0 & \lambda_2 & * & \cdots & * \\ 0 & 0 & \cdots & * & * \\ 0 & \cdots & 0 & \cdots & * \\ 0 & \cdots & \cdots & 0 & \lambda_n \end{pmatrix}$$

Example 1

Given, $A = \begin{pmatrix} 2 & 1 \\ 2 & 3 \end{pmatrix}$. Find a Householder matrix H such that HAH is an upper triangular matrix.

Solution

Consider that

$$P_A(t) = \det(tI - A) = t^2 - 5t + 4$$

we see that $\lambda_1 = 1$, $\lambda_2 = 4$

Also note that $\begin{pmatrix} 2 & 1 \\ 2 & 3 \end{pmatrix} \begin{pmatrix} 1 \\ -1 \end{pmatrix} = \begin{pmatrix} 1 \\ -1 \end{pmatrix}$

$X = \begin{pmatrix} 1 \\ -1 \end{pmatrix}$ is an eigenvector of A with respect to eigenvalue $\lambda_1 = 1$.

Letting $X_1 = \dfrac{1}{\sqrt{2}}\begin{pmatrix} 1 \\ -1 \end{pmatrix}$ and $Z_1 = X_1 - e_1 = \dfrac{1}{\sqrt{2}}\begin{pmatrix} 1 - \sqrt{2} \\ -1 \end{pmatrix}$, then

$$E_{Z_1} = \dfrac{Z_1({}^tZ_1)}{({}^tZ_1)Z_1} = \dfrac{1}{2-\sqrt{2}}\begin{pmatrix} (3-2\sqrt{2})/2 & (\sqrt{2}-1)/2 \\ (\sqrt{2}-1)/2 & 1/2 \end{pmatrix}$$

$$2E_{Z_1} = \dfrac{1}{2-\sqrt{2}}\begin{pmatrix} 3-2\sqrt{2} & \sqrt{2}-1 \\ \sqrt{2}-1 & 1 \end{pmatrix} \text{ and}$$

$$H_1 = {}^tH_{Z_1} = I - 2E_{Z_1} = \dfrac{1}{2-\sqrt{2}}\begin{pmatrix} \sqrt{2}-1 & 1-\sqrt{2} \\ 1-\sqrt{2} & 1-\sqrt{2} \end{pmatrix}$$

$$= \begin{pmatrix} 1/\sqrt{2} & -1/\sqrt{2} \\ -1/\sqrt{2} & -1/\sqrt{2} \end{pmatrix}$$

So,
$$H_1 A H_1 = H_1 \begin{pmatrix} 1/\sqrt{2} & -3/\sqrt{2} \\ -1/\sqrt{2} & -5/\sqrt{2} \end{pmatrix} = \begin{pmatrix} 1 & 1 \\ 0 & 4 \end{pmatrix}$$
∎

Example 2

Given, $A = \begin{pmatrix} 1 & i \\ 1 & i \end{pmatrix}$. Find a Householder matrix H such that HAH is an upper triangular matrix.

Solution

Consider, $P_A(t) = \det(tI - A) = t^2 - t(1+i) = t(t-(1+i))$, we find out that $\lambda_1 = 1+i$ and $\lambda_2 = 0$ are the eigenvalues of A.

Since $\begin{pmatrix} 1 & i \\ 1 & i \end{pmatrix}\begin{pmatrix} 1 \\ 1 \end{pmatrix} = (1+i)\begin{pmatrix} 1 \\ 1 \end{pmatrix}$, $X = \begin{pmatrix} 1 \\ 1 \end{pmatrix}$ is an eigenvector of A having eigenvalue $\lambda_1 = 1+i$. Normalizing $X = \begin{pmatrix} 1 \\ 1 \end{pmatrix}$ and setting

$$Z_1 = X_1 - e_1 = \frac{1}{\sqrt{2}}\begin{pmatrix} 1 \\ 1 \end{pmatrix} - \begin{pmatrix} 1 \\ 0 \end{pmatrix} = \frac{1}{\sqrt{2}}\begin{pmatrix} 1-\sqrt{2} \\ 1 \end{pmatrix}, \text{ then}$$

$$E_{Z_1} = \frac{Z_1({}^t\overline{Z}_1)}{({}^t\overline{Z}_1)Z_1} = \frac{1}{2-\sqrt{2}}\begin{pmatrix} (3-2\sqrt{2})/2 & (1-\sqrt{2})/2 \\ (1-\sqrt{2})/2 & 1/2 \end{pmatrix},$$

$$H_1 = I - 2E_{Z_1} = \begin{pmatrix} 1/\sqrt{2} & 1/\sqrt{2} \\ 1/\sqrt{2} & -1/\sqrt{2} \end{pmatrix}. \text{ We hence obtain that}$$

$$H_1 A H_1 = \begin{pmatrix} \sqrt{2} & \sqrt{2}i \\ 0 & 0 \end{pmatrix}\begin{pmatrix} 1/\sqrt{2} & 1/\sqrt{2} \\ 1/\sqrt{2} & -1/\sqrt{2} \end{pmatrix} = \begin{pmatrix} 1+i & 1-i \\ 0 & 0 \end{pmatrix}$$
∎

這種題目的計算，用手工做起來真不會太輕鬆。讓我們鼓起勇氣，試試看下面一個 3×3 矩陣的例子。

Example 3

Given, $A = \begin{pmatrix} 1 & 1 & 3 \\ -2 & 0 & 2 \\ 4 & 1 & 0 \end{pmatrix}$. Find Householder matrices, H_1 and H_2, such that $H_2 H_1 A H_1 H_2$ is an upper triangular matrix.

Solution

Consider that

$$\begin{pmatrix} 1 & 1 & 3 \\ -2 & 0 & 2 \\ 4 & 1 & 0 \end{pmatrix} \begin{pmatrix} 1 \\ 2 \\ -2 \end{pmatrix} = -3 \begin{pmatrix} 1 \\ 2 \\ -2 \end{pmatrix}$$

we see that $X = \begin{pmatrix} 1 \\ 2 \\ -2 \end{pmatrix}$ is an eigenvector of A with respect to eigenvalue $\lambda_1 = -3$.

(Actually, the eigenvalues of A are $\lambda_1 = -3, \lambda_2 = 0$ and $\lambda_3 = 4$)

Normalize $\begin{pmatrix} 1 \\ 2 \\ -2 \end{pmatrix}$, and let $X_1 = \frac{1}{3} \begin{pmatrix} 1 \\ 2 \\ -2 \end{pmatrix}$. Then

$$Z_1 = X_1 - e_1 = \frac{-2}{3} {}^t(1 \ -1 \ 1) \text{ and}$$

$$E_{Z_1} = \frac{Z_1({}^tZ_1)}{({}^tZ_1)Z_1} = \frac{1}{3} \begin{pmatrix} 1 & -1 & 1 \\ -1 & 1 & -1 \\ 1 & -1 & 1 \end{pmatrix}, \ H_1 = I - 2E_{Z_1} = \frac{1}{3} \begin{pmatrix} 1 & 2 & -2 \\ 2 & 1 & 2 \\ -2 & 2 & 1 \end{pmatrix}$$

再進一步演算之後，我們得到

$$H_1 A H_1 = \begin{pmatrix} -3 & -1 & 3 \\ 0 & 4 & 0 \\ 0 & -1 & 0 \end{pmatrix}$$

這時候，我們可以先留意一下，$\lambda_1 = -3$已經出現在它應該出現的位置了。接

下來，要考慮的就是矩陣

$$A_1 = \begin{pmatrix} 4 & 0 \\ -1 & 0 \end{pmatrix}$$

由於

$$\begin{pmatrix} 4 & 0 \\ -1 & 0 \end{pmatrix}\begin{pmatrix} 0 \\ 1 \end{pmatrix} = \begin{pmatrix} 0 \\ 0 \end{pmatrix} = 0\begin{pmatrix} 0 \\ 1 \end{pmatrix}$$

所以，我們知道，$Y_2 = \begin{pmatrix} 0 \\ 1 \end{pmatrix}$ is an eigenvector of A_1 with respect to $\lambda_2 = 0$。

現在，令 $X_2 = \begin{pmatrix} 0 \\ 0 \\ 1 \end{pmatrix}$，$Z_2 = X_2 - e_2 = \begin{pmatrix} 0 \\ -1 \\ 1 \end{pmatrix}$，則

$$E_{Z_2} = \frac{1}{2}\begin{pmatrix} 0 & 0 & 0 \\ 0 & 1 & -1 \\ 0 & -1 & 1 \end{pmatrix}, \quad H_2 = I - 2E_{Z_2} = \begin{pmatrix} 1 & 0 & 0 \\ 0 & 0 & 1 \\ 0 & 1 & 0 \end{pmatrix}$$

那麼

$$H_2 H_1 A H_1 H_2 = \begin{pmatrix} 1 & 0 & 0 \\ 0 & 0 & 1 \\ 0 & 1 & 0 \end{pmatrix}\begin{pmatrix} -3 & -1 & 3 \\ 0 & 4 & 0 \\ 0 & -1 & 0 \end{pmatrix}\begin{pmatrix} 1 & 0 & 0 \\ 0 & 0 & 1 \\ 0 & 1 & 0 \end{pmatrix}$$

$$= \begin{pmatrix} 1 & 0 & 0 \\ 0 & 0 & 1 \\ 0 & 1 & 0 \end{pmatrix}\begin{pmatrix} -3 & 3 & -1 \\ 0 & 0 & 4 \\ 0 & 0 & -1 \end{pmatrix} = \begin{pmatrix} -3 & 3 & -1 \\ 0 & 0 & -1 \\ 0 & 0 & 4 \end{pmatrix}$$

即為欲求之上三角矩陣。∎

Exercises

1. Find a Householder matrix H such that HAH is an upper triangular matrix.

a. $A = \begin{pmatrix} 2 & 1 \\ 3 & 4 \end{pmatrix}$ 　　b. $A = \begin{pmatrix} 1 & 0 \\ 1 & 2 \end{pmatrix}$

c. $A = \begin{pmatrix} 2 & 1 \\ 1 & 1 \end{pmatrix}$ 　　d. $A = \begin{pmatrix} 1 & 1 \\ i & 1 \end{pmatrix}$

e. $A = \begin{pmatrix} 2 & 1+i \\ 1-i & 1 \end{pmatrix}$

2. Find Householder matrices, H_1 and H_2, such that $H_2 H_1 A H_1 H_2$ is an upper triangular matrix.

a. $A = \begin{pmatrix} 3 & 1 & 3 \\ -2 & 2 & 2 \\ 4 & 1 & 2 \end{pmatrix}$
b. $A = \begin{pmatrix} 3 & 2 & 1 \\ 0 & 1 & 2 \\ 0 & 1 & -1 \end{pmatrix}$

§8.2 Normal matrices

現在,我們要利用 8.1 節上三角化的技巧,來介紹 normal matrix 的對角化。首先,我們認識一下所謂的 normal matrix。

Definition 8.1

> An $n \times n$ complex matrix, A, is said to be "normal", if it satisfies that
> $$(A^*)A = A(A^*)$$

譬如說,假若

$$A = \begin{pmatrix} i & -i \\ -i & i \end{pmatrix}, \text{ 則 } A^* = \begin{pmatrix} -i & i \\ i & -i \end{pmatrix}, \text{ 而且,}$$

$$AA^* = \begin{pmatrix} i & -i \\ -i & i \end{pmatrix}\begin{pmatrix} -i & i \\ i & -i \end{pmatrix} = \begin{pmatrix} 2 & -2 \\ -2 & 2 \end{pmatrix}$$

$$A^*A = \begin{pmatrix} -i & i \\ i & -i \end{pmatrix}\begin{pmatrix} i & -i \\ -i & i \end{pmatrix} = \begin{pmatrix} 2 & -2 \\ -2 & 2 \end{pmatrix}$$

所以,A is a normal matrix。

當然,要是 A 為一 real matrix 的話,那麼我們說,A is normal 意思是說 $({}^tA)A = A({}^tA)$。很顯然的,假若 A 是一個 hermitian matrix 的話,A 必為 normal($A^*A = AA = AA^*$);又假若 A 是一個 unitary matrix 的話,A 也必為 normal($A^*A = I = AA^*$)。可是,有一點要特別注意,normal matrix 並不一定是 hermitian matrix,更不一定是 unitary matrix。譬如說,

若 $A = \begin{pmatrix} i & -i \\ -i & i \end{pmatrix}$,則 $A^* = \begin{pmatrix} -i & i \\ i & -i \end{pmatrix}$,$A^*A = \begin{pmatrix} 2 & -2 \\ -2 & 2 \end{pmatrix} = AA^*$

但是,$A \neq A^*$。

Exercise

Show that if A is normal and B is unitary similar to A, then B is also normal. (*Hint*: there exists a unitary matrix, P, such that $P^*AP = B$.)

我們把 8.1 節上三角化的結果，再重新敘述一遍，以方便我們清楚的進行 normal matrix 的對角化工作。

對於任意 $n \times n$ 矩陣，A，存在一 unitary matrix，P，使得 P^*AP 是一個上三角矩陣，其之主對角線元素為 A 的 eigenvalues。再清楚一點的說

$$P^*AP = \begin{pmatrix} \lambda_1 & * & \cdots & \cdots & * \\ 0 & \lambda_2 & * & \cdots & * \\ 0 & 0 & \cdots & * & * \\ 0 & \cdots & 0 & \cdots & * \\ 0 & \cdots & \cdots & 0 & \lambda_n \end{pmatrix}$$

其中，$\lambda_1, \lambda_2, \ldots, \lambda_n$ 為 A 的 eigenvalues，它們可能為實數，可能為複數，也可能重複出現。下面就是有關 normal matrix 的對角化工作。

Theorem 8.2

Every normal matrix, A, is unitary diagonalizable with diagonal elements consisting of eigenvalues of A.

Proof

對於一個 $n \times n$ 矩陣，A，there exists a unitary matrix, P, such that $P^*AP = U$ is an upper triangular matrix。令其為

$$U = \begin{pmatrix} u_{11} & u_{12} & \cdots & \cdots & u_{1n} \\ 0 & u_{22} & u_{23} & \cdots & u_{2n} \\ 0 & 0 & \cdots & \cdots & \cdots \\ \cdots & \cdots & 0 & \cdots & \cdots \\ 0 & \cdots & \cdots & 0 & u_{nn} \end{pmatrix}$$

其中，u_{ii}，$i = 1, 2, \ldots, n$ are eigenvalues of A。

Consider that $(P^*AP)^* = U^*$, we have, $P^*A^*P = U^*$, with

$$U^* = \begin{pmatrix} \bar{u}_{11} & 0 & 0 & \cdots & 0 \\ \bar{u}_{12} & \bar{u}_{22} & 0 & \cdots & \cdots \\ \cdots & \bar{u}_{23} & \cdots & 0 & \cdots \\ \cdots & \cdots & \cdots & \cdots & 0 \\ \bar{u}_{1n} & \bar{u}_{2n} & \cdots & \cdots & \bar{u}_{nn} \end{pmatrix}$$

Combining the last two equations, we see that

$$(P^*AP)(P^*A^*P) = UU^* \quad \text{and} \quad (P^*A^*P)(P^*AP) = U^*U$$

or that $(P^*AA^*P) = UU^*$ and $(P^*A^*AP) = U^*U$

Since $A^*A = AA^*$, we find out that

$$U^*U = UU^*$$

Letting $U^*U = D = (d_{ij})_{n\times n} = UU^*$, then we obtain that

$$|u_{11}|^2 = d_{11} = \sum_{j=1}^{n}(|u_{1j}|^2)$$

由上式，我們發現

$$|u_{12}|^2 + |u_{13}|^2 + \cdots + |u_{1n}|^2 = 0$$

從而得知，

$$u_{12} = u_{13} = \cdots = u_{1n} = 0$$

再來，

$$d_{22} = |u_{12}|^2 + |u_{22}|^2 = \sum_{j=2}^{n}(|u_{2j}|^2)$$

前述已知，$u_{12} = 0$，所以

$$d_{22} = |u_{22}|^2 = \sum_{j=2}^{n}(|u_{2j}|^2)$$

這也因此，告訴我們，$u_{23} = u_{24} = \cdots = u_{2n} = 0$.
按此要領，往前做下去的時候，我們將發現

$$u_{34} = u_{35} = \cdots = u_{3n} = 0$$

$$u_{45} = u_{46} = \ldots = u_{4n} = 0$$
$$\ldots\ldots\ldots\ldots\ldots\ldots\ldots\ldots$$
$$u_{(n-1)n} = 0$$

也就是說，$u_{ij} = 0$，$\forall\ i \neq j$。這說明了，$P^*AP = U$ 是一個 diagonal matrix。

其實，這個定理的逆敘述也是成立的。也就是說，假若 A is a unitary diagonalizable matrix, then A is a normal matrix。有關這個事實不難得出。

Suppose that there exists a unitary matrix, P, such that $P^*AP = D$ is a diagonal matrix, then $A = PDP^*$ and $A^* = P\overline{D}P^*$, since D is diagonal. Now,

$$AA^* = (PDP^*)(P\overline{D}P^*) = PD\overline{D}P^* = P\overline{D}DP^*$$
$$= (P\overline{D}P^*)(PDP^*) = A^*A$$

綜合以上，我們可以做出下面一個比較明確的結論。

Theorem 8.3

An $n \times n$ matrix, A, is normal, if and only if A is unitary diagonalizable with diagonal elements consisting of eigenvalues of A.

Example 1

Given, $A = \begin{pmatrix} i & -i \\ -i & i \end{pmatrix}$. Find a unitary matrix, P, such that $P^*AP = \begin{pmatrix} \lambda_1 & 0 \\ 0 & \lambda_2 \end{pmatrix}$, where λ_1, λ_2 are eigenvalues of A.

Solution

We first note that $A = \begin{pmatrix} i & -i \\ -i & i \end{pmatrix}$ is normal and let

$$P_A(t) = \det(tI - A) = t^2 - 2it = 0$$

We find out that $\lambda_1 = 0$ and $\lambda_2 = 2i$.

Chapter Eight Spectral Decomposition and Interpolatory Polynomials 291

Since $\begin{pmatrix} i & -i \\ -i & i \end{pmatrix} \begin{pmatrix} 1/\sqrt{2} \\ 1/\sqrt{2} \end{pmatrix} = \begin{pmatrix} 0 \\ 0 \end{pmatrix}$, $\alpha_1 = \begin{pmatrix} 1/\sqrt{2} \\ 1/\sqrt{2} \end{pmatrix}$ is an eigenvector of A with respect to $\lambda_1 = 0$. Also, $\begin{pmatrix} i & -i \\ -i & i \end{pmatrix} \begin{pmatrix} 1/\sqrt{2} \\ -1/\sqrt{2} \end{pmatrix} = 2i \begin{pmatrix} 1/\sqrt{2} \\ -1/\sqrt{2} \end{pmatrix}$, $\alpha_2 = \begin{pmatrix} 1/\sqrt{2} \\ -1/\sqrt{2} \end{pmatrix}$ is an eigenvector of A with respect to $\lambda_2 = 2i$.

Taking $P = \begin{pmatrix} 1/\sqrt{2} & 1/\sqrt{2} \\ 1/\sqrt{2} & -1/\sqrt{2} \end{pmatrix}$, then $P^* = \begin{pmatrix} 1/\sqrt{2} & 1/\sqrt{2} \\ 1/\sqrt{2} & -1/\sqrt{2} \end{pmatrix}$ and

$$P^*AP = \begin{pmatrix} i & 0 \\ 0 & 2i \end{pmatrix}$$ ∎

Exercises

1. Show that if A is a real normal matrix and all eigenvalues of A are real, then A is symmetric. (*Hint*: there exists an orthonormal matrix, P, such that ${}^t PAP = D$ is a real diagonal matrix.)

2. Show that if A is a normal matrix, then $tI - A$, $t \in C$ is also normal.

3. Let A be a normal matrix. Show that

   a. A^* is normal. (*Note*: $(A^*)^* = ({}^t\overline{A})^* = \overline{{}^t({}^t\overline{A})} = A$)

   b. If λ is an eigenvalue of A, then $\overline{\lambda}$ is an eigenvalue of A^*.
      Hint：A is normal，所以存在一 unitary P，使得
      $P^*AP = D = diag\{\lambda_1, \lambda_2, ..., \lambda_n\}$，其中 λ 是 $\{\lambda_1, \lambda_2, ..., \lambda_n\}$ 集合中的一個元素。
      將上式等號兩邊取 transpose conjugate，得
      $(P^*AP)^* = P^*A^*P = D^* = diag\{\overline{\lambda}_1, \overline{\lambda}_2, ..., \overline{\lambda}_n\}$
      其中，$P^*A^*P = D^*$ implies，$A^*P = PD^*$。而這個式子說明了，矩陣 P 的每一個 column，P^i 是 A^* 的 eigenvector w. r. t. eigenvalue $\overline{\lambda}_i$。
      (參考本節最後的 Exercise 1。)

這一路走來，我們一直惦記著 diagonalizable 比 unitary diagonalizable 來得弱一點。或者說，diagonalizable matrix 不一定是 normal matrix。我們再拿一個例題來說明，以誌永不忘懷。

Example 2

Show that $A = \begin{pmatrix} 3 & 4 \\ 2 & 1 \end{pmatrix}$ is diagonalizable, but it is not normal.

Proof

(參看 7.2 節)

A 的 eigenvalues 為 $\lambda_1 = -1$, $\lambda_2 = 5$，而且 $\alpha_1 = \begin{pmatrix} 1 \\ -1 \end{pmatrix}$, $\alpha_2 = \begin{pmatrix} 2 \\ 1 \end{pmatrix}$ 分別為它們的 eigenvectors。今令，$P = mat(I, \alpha, E) = \begin{pmatrix} 1 & 2 \\ -1 & 1 \end{pmatrix}$，為 the change of basis matrix。則 $P^{-1} = mat(I, E, \alpha) = \begin{pmatrix} 1/3 & -2/3 \\ 1/3 & 1/3 \end{pmatrix}$，而且 $P^{-1}AP = \begin{pmatrix} -1 & 0 \\ 0 & 5 \end{pmatrix}$。所以，$A$ is diagonalizable, but however,

$$A^*A = \begin{pmatrix} 3 & 2 \\ 4 & 1 \end{pmatrix}\begin{pmatrix} 3 & 4 \\ 2 & 1 \end{pmatrix} = \begin{pmatrix} 13 & 14 \\ 14 & 17 \end{pmatrix} \text{ and}$$

$$AA^* = \begin{pmatrix} 3 & 4 \\ 2 & 1 \end{pmatrix}\begin{pmatrix} 3 & 2 \\ 4 & 1 \end{pmatrix} = \begin{pmatrix} 25 & 10 \\ 10 & 5 \end{pmatrix}$$

所以 A 不是 normal matrix。

再重複一遍，一般可對角化的 $n \times n$ 矩陣，A，我們不一定找得到 orthogonal basis 使得它們所形成的 non-singular matrix，P，造成 $P^{-1}AP = D$。然而，在 A 是 normal matrix 的情形之下(與 hermitian matrix 一樣)，我們則保證找得到一組 orthogonal basis 使得它們所形成的 non-singular matrix，P，為一個 unitary matrix 而且 $P^{-1}AP = D$。我們清楚的再給它總結如下。

An $n \times n$ matrix, A, is normal, if and only if C^n has an orthonormal basis consisting of eigenvectors of A.

談到 normal matrix，它還有一個重要而且值得我們注意的特性。我們以下面例題和大家共同來討論。

Example 3

Let A be an $n \times n$ matrix, and let $\langle \, , \, \rangle$ be the standard inner product on C^n. Show that A is normal if and only if $\langle AX, AX \rangle = \langle A^*X, A^*X \rangle$, $\forall \ X \in C^n$.

Proof

($\Rightarrow$)

由第 243 頁的 Exercise
$$\langle AX, Y \rangle = \langle X, A^*Y \rangle, \ \forall \ X, Y \in C^n$$
得知
$$\langle AX, AX \rangle = \langle X, A^*AX \rangle = \langle X, AA^*X \rangle = \langle A^*X, A^*X \rangle, \ \forall \ X \in C^n$$

($\Leftarrow$)

反過來，若 $\langle AX, AX \rangle = \langle A^*X, A^*X \rangle$, $\forall \ X \in C^n$，則
$$\langle A^*AX, X \rangle = \langle AA^*X, X \rangle$$
或者說，
$$\langle A^*AX, X \rangle - \langle AA^*X, X \rangle = 0$$
也就是說，$\langle A^*AX - AA^*X, X \rangle = 0$, $\forall \ X \in C^n$

因此，$\langle (A^*A - AA^*)X, X \rangle = 0$, $\forall \ X \in C^n$

接著，由 7.2 節的 Proposition 7.6，我們得知
$$A^*A - AA^* = 0$$

此即證明，A is normal。 ∎

Exercises

1. Suppose that A is normal. Show that if X is an eigenvector of A with respect to eigenvalue λ, then X is also an eigenvector of A^* with respect to eigenvalue $\overline{\lambda}$.

(*Hint*：
$$0 = \langle (A-\lambda I)X, (A-\lambda I)X \rangle = \langle (A-\lambda I)^* X, (A-\lambda I)^* X \rangle)$$

2. Show that if X_1 and X_2 are eigenvectors of a normal matrix, A, with respect to distinct eigenvalues, λ_1 and λ_2, respectively, then X_1 and X_2 are orthogonal.

(*Hint*: Use 1. and consider that
$$\lambda_1 \langle X_1, X_2 \rangle = \langle \lambda_1 X_1, X_2 \rangle = \langle AX_1, X_2 \rangle$$
$$= \langle X_1, A^* X_2 \rangle = \langle X_1, \overline{\lambda}_2 X_2 \rangle = \lambda_2 \langle X_1, X_2 \rangle)$$

§8.3　Spectral decomposition of a diagonalizable matrix

這節的工作是本章的主要任務之一。我們計畫給讀者介紹，如何用一些**冪等** (idempotent) 矩陣，組織起來構成一個可對角化的矩陣。換句話說，如何將一個可對角化矩陣，表示成冪等矩陣的線性組合，或者所謂的 Spectral decomposition。而，這些冪等矩陣將被稱為原矩陣之**組構冪等矩陣** (constituent idempotents)。

我們從一個可對角化矩陣，A (參閱 7.2 節)，開始談起。假設 $\{\lambda_1, \lambda_2, ..., \lambda_n\}$ 為 A 之 eigenvalue set，P 為一個 non-singular matrix 使得

$$P^{-1}AP = D = diag\{\lambda_1, \lambda_2, ..., \lambda_n\}$$

其中，符號 $diag\{\lambda_1, \lambda_2, ..., \lambda_n\}$ 代表對角矩陣 D，其對角線元素為 $\{\lambda_1, \lambda_2, ..., \lambda_n\}$。

Definition 8.2

Define, for $\forall\ i = 1, 2, ..., n$, an $n \times n$ matrix, E_i, with 1 in the $(i, i)^{th}$ entry and zeros elsewhere.

譬如說

$$E_2 = \begin{pmatrix} 0 & 0 \\ 0 & 1 \end{pmatrix}, \quad E_1 = \begin{pmatrix} 1 & 0 & 0 \\ 0 & 0 & 0 \\ 0 & 0 & 0 \end{pmatrix}, \quad E_3 = \begin{pmatrix} 0 & 0 & 0 & 0 \\ 0 & 0 & 0 & 0 \\ 0 & 0 & 1 & 0 \\ 0 & 0 & 0 & 0 \end{pmatrix}$$

如此定義之下，一個很重要的基本概念，就是 $E_i^2 = E_i$，也就是說，每一個 E_i 皆為 idempotent matrix。而且，$rank(E_i) = 1$。我們再往前看下去，假設 S 是 $\{1, 2, 3, ..., n\}$ 之任意子集合，讓我們來思考一下，$E_S = \sum_{i \in S} E_i$ 這類型矩陣。譬如說，$n = 4$ 而且 $S = \{2, 3\}$，$W = \{1, 2, 4\}$ 的話，那麼

$$E_S = \begin{pmatrix} 0 & 0 & 0 & 0 \\ 0 & 1 & 0 & 0 \\ 0 & 0 & 1 & 0 \\ 0 & 0 & 0 & 0 \end{pmatrix}, \quad E_W = \begin{pmatrix} 1 & 0 & 0 & 0 \\ 0 & 1 & 0 & 0 \\ 0 & 0 & 0 & 0 \\ 0 & 0 & 0 & 1 \end{pmatrix}$$

由此我們發現，E_S，E_W 也仍就是一個冪等矩陣，而且

$rank(E_S) = trace(E_S) = $ the number of elements in S，

$rank(E_W) = trace(E_W) = $ the number of elements in W。

這一節，我們就是想用前述這些 idempotent matrices 加上 eigenvalues 來解構可對角化矩陣 A。

前面所提 A 的 eigenvalues，$\lambda_1, \lambda_2, ..., \lambda_n$，它們可能有重複出現的情形。有關這一點，我們在 7.2 節或是 8.1 節時，曾經一再的提醒過讀者。

現在，我們假設，$d_1, d_2, ..., d_k$，代表矩陣 A 的所有 distinct eigenvalues，而且令 $U_1 = \{j : \lambda_j = d_1\}$，$U_2 = \{j : \lambda_j = d_2\}, ..., U_k = \{j : \lambda_j = d_k\}$。

那麼，此時之 E_{U_j} 就是型如前述所言之冪等矩陣，其對角線不為 0 之數字 1 出現在 eigenvalues 等於 d_j 的位置。而且，$rank(E_{U_j}) = trace(E_{U_j}) = $ the multiplicity of d_j。現在一切準備稍微告一段落，下面我們要看看如何以這些 idempotent matrices 更進一步的來建立矩陣 A 的**組構冪等矩陣** (constituent idempotents)。

Definition 8.3

Let A be the diagonalizable matrix as above with distinct eigenvalues $d_1, d_2, ..., d_k$, and non-singular matrix P such that

$$P^{-1}AP = diag\{\lambda_1, \lambda_2, ..., \lambda_n\}.$$

The constituent idempotents of A are the k matrices $A_1, A_2, ..., A_k$ defined by

$$A_1 = PE_{U_1}P^{-1}, \quad A_2 = PE_{U_2}P^{-1}, \quad ..., \quad A_k = PE_{U_k}P^{-1}$$

(再次的提醒讀者，若矩陣 A 是 hermitian (symmetric) 或是 normal 的話，這裡的 P 會是 unitary matrix，而且 $P^* = P^{-1}$)。下面我們看兩個簡單的例子，先來

研讀一下，這個 constituent idempotents。

Example 1

Given that $A = \begin{pmatrix} 1 & 2 \\ 4 & 3 \end{pmatrix}$, $P = \begin{pmatrix} 1 & 1 \\ -1 & 2 \end{pmatrix}$, $P^{-1} = \begin{pmatrix} 2/3 & -1/3 \\ 1/3 & 1/3 \end{pmatrix}$ with $P^{-1}AP = \begin{pmatrix} -1 & 0 \\ 0 & 5 \end{pmatrix}$. Find the constituent idempotents, A_1 and A_2.

Solution

We first note that $E_{U_1} = \begin{pmatrix} 1 & 0 \\ 0 & 0 \end{pmatrix}$, $E_{U_2} = \begin{pmatrix} 0 & 0 \\ 0 & 1 \end{pmatrix}$, then

$$A_1 = PE_{U_1}P^{-1} = \begin{pmatrix} 1 & 1 \\ -1 & 2 \end{pmatrix}\begin{pmatrix} 1 & 0 \\ 0 & 0 \end{pmatrix}\begin{pmatrix} 2/3 & -1/3 \\ 1/3 & 1/3 \end{pmatrix}$$

$$= \begin{pmatrix} 2/3 & -1/3 \\ -2/3 & 1/3 \end{pmatrix}$$

and

$$A_2 = PE_{U_2}P^{-1} = \begin{pmatrix} 1 & 1 \\ -1 & 2 \end{pmatrix}\begin{pmatrix} 0 & 0 \\ 0 & 1 \end{pmatrix}\begin{pmatrix} 2/3 & -1/3 \\ 1/3 & 1/3 \end{pmatrix}$$

$$= \begin{pmatrix} 1/3 & 1/3 \\ 2/3 & 2/3 \end{pmatrix}$$

讀者不妨自行檢定一下，A_1 與 A_2 是否為 idempotents。也就是說，$A_1^2 = A_1$，$A_2^2 = A_2$。並且更進一步的檢定，$A_1 + A_2 = I_2$。∎

Example 2

Given that $A = \begin{pmatrix} 0 & 0 & 1+i \\ 0 & 2 & 0 \\ 1-i & 0 & 1 \end{pmatrix}$, $P = \begin{pmatrix} 0 & 1/\sqrt{3} & -2/\sqrt{6} \\ 1 & 0 & 0 \\ 0 & (1-i)/\sqrt{3} & (1-i)/\sqrt{6} \end{pmatrix}$, $P^{-1} = P^* = \begin{pmatrix} 0 & 1 & 0 \\ 1/\sqrt{3} & 0 & (1+i)/\sqrt{3} \\ -2/\sqrt{6} & 0 & (1+i)/\sqrt{6} \end{pmatrix}$ with $P^{-1}AP = \begin{pmatrix} 2 & 0 & 0 \\ 0 & 2 & 0 \\ 0 & 0 & -1 \end{pmatrix}$. Find A_1 and A_2.

Solution

首先，$E_{U_1} = \begin{pmatrix} 1 & 0 & 0 \\ 0 & 1 & 0 \\ 0 & 0 & 0 \end{pmatrix}$，$E_{U_2} = \begin{pmatrix} 0 & 0 & 0 \\ 0 & 0 & 0 \\ 0 & 0 & 1 \end{pmatrix}$，此時

$$A_1 = PE_{U_1}P^{-1}$$

$$= \begin{pmatrix} 0 & 1/\sqrt{3} & -2/\sqrt{6} \\ 1 & 0 & 0 \\ 0 & (1-i)/\sqrt{3} & (1-i)/\sqrt{6} \end{pmatrix} \begin{pmatrix} 1 & 0 & 0 \\ 0 & 1 & 0 \\ 0 & 0 & 0 \end{pmatrix} \begin{pmatrix} 0 & 1 & 0 \\ 1/\sqrt{3} & 0 & (1+i)/\sqrt{3} \\ -2/\sqrt{6} & 0 & (1+i)/\sqrt{6} \end{pmatrix}$$

$$= \begin{pmatrix} 1/3 & 0 & (1+i)/3 \\ 0 & 1 & 0 \\ (1-i)/3 & 0 & 2/3 \end{pmatrix}$$

$$A_2 = PE_{U_2}P^{-1}$$

$$= \begin{pmatrix} 0 & 1/\sqrt{3} & -2/\sqrt{6} \\ 1 & 0 & 0 \\ 0 & (1-i)/\sqrt{3} & (1-i)/\sqrt{6} \end{pmatrix} \begin{pmatrix} 0 & 0 & 0 \\ 0 & 0 & 0 \\ 0 & 0 & 1 \end{pmatrix} \begin{pmatrix} 0 & 1 & 0 \\ 1/\sqrt{3} & 0 & (1+i)/\sqrt{3} \\ -2/\sqrt{6} & 0 & (1+i)/\sqrt{6} \end{pmatrix}$$

$$= \begin{pmatrix} 2/3 & 0 & (-1-i)/3 \\ 0 & 0 & 0 \\ (-1+i)/3 & 0 & 1/3 \end{pmatrix}$$

同樣，這裡的 A_1 以及 A_2 也都是冪等矩陣。沒錯，這個道理不難，因為對於每一個 $j = 1, 2, ..., k$ 而言

$$A_j A_j = PE_{U_j}P^{-1}PE_{U_j}P^{-1} = P(E_{U_j})^2 P^{-1} = PE_{U_j}P^{-1} = A_j$$

除此之外，constituent idempotents 還有下面一些重要的特性。

Theorem 8.4

Let $A_1, A_2, ..., A_k$ be the constituent idempotents of the diagonalizable matrix, A. If $d_1, d_2, ..., d_k$ are distinct eigenvalues of A as before, then

Chapter Eight Spectral Decomposition and Interpolatory Polynomials 299

1. $rank(A_j) = trace(A_j) = $ the multiplicity of d_j. And hence, $\sum_{i=1}^{k} rank(A_i) = n$.
2. $A_i A_j = 0$, if $i \neq j$. (pairwise orthogonal)
3. $A_1 + A_2 + \cdots + A_k = I_n$ (a partition of identity)
4. $A = d_1 A_1 + d_2 A_2 + \cdots + d_k A_k$

Proof

1. 按 7.3 節,Theorem 7.14,$rank(A_j) = trace(A_j)$。由於,A_j and E_{U_j} are similar, 而且 $rank(E_{U_j}) = trace(E_{U_j}) = $ the multiplicity of d_j。所以,$trace(A_j) = rank(A_j) = rank(E_j) = $ the multiplicity of d_j。也因此,$\sum_{i=1}^{k} rank(A_i) = n$。

2. If $i \neq j$, then $A_i A_j = P E_{U_i} P^{-1} P E_{U_j} P^{-1} = P E_{U_i} E_{U_j} P^{-1} = P 0 P^{-1} = 0$。

3. $A_1 + A_2 + \cdots + A_k = P(E_{U_1} + E_{U_2} + \cdots + E_{U_k}) P^{-1} = P I_n P^{-1} = I_n$。

4. 對於第 4. 個特性,我們先看看,for each $i = 1, 2, \ldots, k$,
$$AA_i = PDP^{-1} P E_{U_i} P^{-1} = PDE_{U_i} P^{-1}$$
$$= P d_i E_{U_i} P^{-1} = d_i P E_{U_i} P^{-1} \qquad (*)$$
$$= d_i A_i$$
接著將第 3. 個公式之兩邊分別乘上 A,即可得出。

前面這個定理被稱爲,Spectral theorem, the 2nd version。而式子 $A = d_1 A_1 + d_2 A_2 + \cdots + d_k A_k$ 就是所謂的,the Spectral decomposition of A。

注意,前述 (*) 式可以被精確的寫爲,
$$AA_i = d_i A_i = A_i A$$
此外,式子 $AA_i = d_i A_i$,也清楚的告訴我們

Every nonzero column vector of A_i is an eigenvector of A with respect to eigenvalue d_i。

我們先睹為快，用 297 頁的 Example 2 來驗證，這個定理的第 4.個特性。

$d_1 A_1 + d_2 A_2$

$= 2 \begin{pmatrix} 1/3 & 0 & (1+i)/3 \\ 0 & 1 & 0 \\ (1-i)/3 & 0 & 2/3 \end{pmatrix} + (-1) \begin{pmatrix} 2/3 & 0 & (-1-i)/3 \\ 0 & 0 & 0 \\ (-1+i)/3 & 0 & 1/3 \end{pmatrix}$

$= \begin{pmatrix} 0 & 0 & 1+i \\ 0 & 2 & 0 \\ 1-i & 0 & 1 \end{pmatrix} = A$

太好了，真是完美無瑕。我們再看一個例子。

Example 3

Find the spectral decomposition of $A = \begin{pmatrix} -2 & -1 & 3 \\ 6 & 5 & -6 \\ 2 & 2 & -1 \end{pmatrix}$.

Solution

We first search for the eigenvalues of A, and find out that $\lambda_1 = 1$, $\lambda_2 = -1$, $\lambda_3 = 2$ are the distinct eigenvalues. According to these, we also find out that $\alpha_1 = {}^t(1 \ 0 \ 1)$, $\alpha_2 = {}^t(1 \ -1 \ 0)$, and $\alpha_2 = {}^t(1 \ 2 \ 2)$ are eigenvectors of A having $\lambda_1 = 1$, $\lambda_2 = -1$, and $\lambda_3 = 2$ as eigenvalues, respectively. No doubt, they are linearly independent as you may check.

Now, letting $P = \begin{pmatrix} 1 & 1 & 1 \\ 0 & -1 & 2 \\ 1 & 0 & 2 \end{pmatrix}$, then $P^{-1} = \begin{pmatrix} -2 & -2 & 3 \\ 2 & 1 & -2 \\ 1 & 1 & -1 \end{pmatrix}$ and

$P^{-1}AP = \begin{pmatrix} 1 & 0 & 0 \\ 0 & -1 & 0 \\ 0 & 0 & 2 \end{pmatrix}$

To find the constituent idempotents, we first note that $d_1 = 1$, $d_2 = -1$, and

Chapter Eight Spectral Decomposition and Interpolatory Polynomials 301

$d_3 = 2$, they all have multiplicity 1. Hence, $E_{U_1} = \begin{pmatrix} 1 & 0 & 0 \\ 0 & 0 & 0 \\ 0 & 0 & 0 \end{pmatrix}$, $E_{U_2} = \begin{pmatrix} 0 & 0 & 0 \\ 0 & 1 & 0 \\ 0 & 0 & 0 \end{pmatrix}$,

$E_{U_3} = \begin{pmatrix} 0 & 0 & 0 \\ 0 & 0 & 0 \\ 0 & 0 & 1 \end{pmatrix}$, so

$A_1 = PE_{U_1}P^{-1} = \begin{pmatrix} 1 & 1 & 1 \\ 0 & -1 & 2 \\ 1 & 0 & 2 \end{pmatrix}\begin{pmatrix} 1 & 0 & 0 \\ 0 & 0 & 0 \\ 0 & 0 & 0 \end{pmatrix}\begin{pmatrix} -2 & -2 & 3 \\ 2 & 1 & -2 \\ 1 & 1 & -1 \end{pmatrix} = \begin{pmatrix} -2 & -2 & 3 \\ 0 & 0 & 0 \\ -2 & -2 & 3 \end{pmatrix}$

$A_2 = PE_{U_2}P^{-1} = \begin{pmatrix} 1 & 1 & 1 \\ 0 & -1 & 2 \\ 1 & 0 & 2 \end{pmatrix}\begin{pmatrix} 0 & 0 & 0 \\ 0 & 1 & 0 \\ 0 & 0 & 0 \end{pmatrix}\begin{pmatrix} -2 & -2 & 3 \\ 2 & 1 & -2 \\ 1 & 1 & -1 \end{pmatrix} = \begin{pmatrix} 2 & 1 & -2 \\ -2 & -1 & 2 \\ 0 & 0 & 0 \end{pmatrix}$

$A_3 = PE_{U_3}P^{-1} = \begin{pmatrix} 1 & 1 & 1 \\ 0 & -1 & 2 \\ 1 & 0 & 2 \end{pmatrix}\begin{pmatrix} 0 & 0 & 0 \\ 0 & 0 & 0 \\ 0 & 0 & 1 \end{pmatrix}\begin{pmatrix} -2 & -2 & 3 \\ 2 & 1 & -2 \\ 1 & 1 & -1 \end{pmatrix} = \begin{pmatrix} 1 & 1 & -1 \\ 2 & 2 & -2 \\ 2 & 2 & -2 \end{pmatrix}$

Therefore,

$$A = d_1 A_1 + d_2 A_2 + d_3 A_3 = A_1 - A_2 + 2A_3$$

$$= \begin{pmatrix} -2 & -1 & 3 \\ 6 & 5 & -6 \\ 2 & 2 & -1 \end{pmatrix}$$

Exercise

Find the spectral decomposition of

$$A = \begin{pmatrix} 5 & 2 & 2 \\ 2 & 5 & 2 \\ 2 & 2 & 5 \end{pmatrix}$$

(*Hint*: $\lambda_1 = 3$, $\lambda_2 = 3$, $\lambda_3 = 9$ and $P = \begin{pmatrix} 1 & 0 & 1 \\ 0 & 1 & 1 \\ -1 & -1 & 1 \end{pmatrix}$)

看了這個 exercise，我們或許已經發覺，由於矩陣 A 是對稱的，所得出來的 constituent idempotents，A_i，是對稱的，或者正確的說，此時之 A_i 是 hermitian。沒錯，有關這一個觀察，我們作成下列說明。

假若，矩陣 A 是一個 normal matrix 的話，那麼 (根據 8.2 節) there exists a **unitary matrix**，P，使得 $P^*AP = D = diag\{\lambda_1, \lambda_2, ..., \lambda_n\}$。這種情況之下，$A_i = PE_{U_i}P^{-1}$ 指的是 $A_i = PE_{U_i}P^*$。

那麼 $\quad A_i^* = (PE_{U_i}P^*)^* = P(E_{U_i})^*P^* = PE_{U_i}P^* = A_i$

反過來說，假若 A 是一個可對角化矩陣，而且它的每一個 constituent idempotent，A_i，都是 hermitian 的話，則 A 也必是一個 normal matrix。這也不難。假設，$d_1, d_2, ..., d_k$ are distinct eigenvalues of A such that $A = d_1 A_1 + d_2 A_2 + \cdots + d_k A_k$, then $A^* = \overline{d}_1 A_1^* + \overline{d}_2 A_2^* + \cdots + \overline{d}_k A_k^*$. Hence,

$$AA^* = (d_1 A_1 + d_2 A_2 + \cdots + d_k A_k)(\overline{d}_1 A_1^* + \overline{d}_2 A_2^* + \cdots + \overline{d}_k A_k^*)$$
$$= |d_1|^2 A_1 A_1^* + |d_2|^2 A_2 A_2^* + \cdots + |d_k|^2 A_k A_k^*$$
$$= |d_1|^2 A_1^* A_1 + |d_2|^2 A_2^* A_2 + \cdots + |d_k|^2 A_k^* A_k$$
$$= A^* A$$

注意，前面的推導過程中之第二個等式，我們使用了特性，$A_i A_j = 0$, if $i \ne j$。綜合前述所言，我們得出下列推論。

Corollary 1

Let A be a diagonalizable matrix. Then A is normal if and only if the constituent idempotents of A are hermitian.

Example 4

Find the spectral decomposition of $A = \begin{pmatrix} i & -i \\ -i & i \end{pmatrix}$.

Solution

We first note that $A^* = \begin{pmatrix} -i & i \\ i & -i \end{pmatrix}$ and $AA^* = \begin{pmatrix} 2 & -2 \\ -2 & 2 \end{pmatrix} = A^*A$. So, A is normal. Next, we find out that the eigenvalues of A are $\lambda_1 = 0$, $\lambda_2 = 2i$. And that $P = \begin{pmatrix} 1/\sqrt{2} & 1/\sqrt{2} \\ 1/\sqrt{2} & -1/\sqrt{2} \end{pmatrix}$, $P^* = \begin{pmatrix} 1/\sqrt{2} & 1/\sqrt{2} \\ 1/\sqrt{2} & -1/\sqrt{2} \end{pmatrix}$ with these, we have $P^*AP = \begin{pmatrix} i & 0 \\ 0 & 2i \end{pmatrix}$.

Letting $E_{U_1} = \begin{pmatrix} 1 & 0 \\ 0 & 0 \end{pmatrix}$ and $E_{U_2} = \begin{pmatrix} 0 & 0 \\ 0 & 1 \end{pmatrix}$, then

$A_1 = PE_{U_1}P^* = \begin{pmatrix} 1/2 & 1/2 \\ 1/2 & 1/2 \end{pmatrix}$ and $A_2 = PE_{U_2}P^* = \begin{pmatrix} 1/2 & -1/2 \\ -1/2 & 1/2 \end{pmatrix}$.

Therefore, $A = \begin{pmatrix} i & -i \\ -i & i \end{pmatrix} = 0A_1 + 2iA_2$. ∎

這個例題要說明的是，A_1以及A_2都是 hermitian (symmetric)。下面，再看一個本身是 hermitian (當然也就是 normal) 矩陣的例子。

Example 5

Find the spectral decomposition of $A = \begin{pmatrix} 1 & -\sqrt{6}i & 0 \\ \sqrt{6}i & 0 & 0 \\ 0 & 0 & -2 \end{pmatrix}$.

Solution

$$\det(tI - A) = \det\begin{pmatrix} t-1 & \sqrt{6}i & 0 \\ -\sqrt{6}i & t & 0 \\ 0 & 0 & t+2 \end{pmatrix}$$

$$= t(t-1)(t+2) - 6(t+2) = (t+2)^2(t-3)$$

implies that

$\lambda_1 = -2$, $\lambda_2 = -2$ and $\lambda_3 = 3$ are the eigenvalues. $\alpha_1 = {}^t(\sqrt{2/5}\,i \quad \sqrt{3/5} \quad 0)$, $\alpha_2 = {}^t(0 \quad 0 \quad 1)$ and $\alpha_3 = {}^t(\sqrt{3/5} \quad \sqrt{2/5}\,i \quad 0)$ are orthonormal eigenvectors corresponding to $\lambda_1 = -2$, $\lambda_2 = -2$ and $\lambda_3 = 3$, respectively.

Writing $P = \begin{pmatrix} \sqrt{2/5}\,i & 0 & \sqrt{3/5} \\ \sqrt{3/5} & 0 & \sqrt{2/5}\,i \\ 0 & 1 & 0 \end{pmatrix}$, then $P^* = \begin{pmatrix} -\sqrt{2/5}\,i & \sqrt{3/5} & 0 \\ 0 & 0 & 1 \\ \sqrt{3/5} & -\sqrt{2/5}\,i & 0 \end{pmatrix}$ and

$$P^*AP = \begin{pmatrix} -2 & 0 & 0 \\ 0 & -2 & 0 \\ 0 & 0 & 3 \end{pmatrix}$$

Now, $d_1 = -2$ and $d_2 = 3$ are the distinct eigenvalues with

$$E_{U_1} = \begin{pmatrix} 1 & 0 & 0 \\ 0 & 1 & 0 \\ 0 & 0 & 0 \end{pmatrix}, \quad E_{U_2} = \begin{pmatrix} 0 & 0 & 0 \\ 0 & 0 & 0 \\ 0 & 0 & 1 \end{pmatrix}$$

So, $A_1 = PE_{U_1}P^* = \begin{pmatrix} 2/5 & \sqrt{6}\,i/5 & 0 \\ -\sqrt{6}\,i/5 & 3/5 & 0 \\ 0 & 0 & 1 \end{pmatrix}$ and

$A_2 = PE_{U_2}P^* = \begin{pmatrix} 3/5 & -\sqrt{6}\,i/5 & 0 \\ \sqrt{6}\,i/5 & 2/5 & 0 \\ 0 & 0 & 0 \end{pmatrix}$. Therefore, $A = (-2)A_1 + 3A_2$.

(Note：A_1 和 A_2 當然也都是 hermitian。)

我們再看一個比較特別的例子。

Example 6

Find the spectral decomposition of $A = \begin{pmatrix} 1 & -1 \\ 0 & 0 \end{pmatrix}$.

Solution

這有什麼特別的呢？看看，$AA = \begin{pmatrix} 1 & -1 \\ 0 & 0 \end{pmatrix} = A$。它本身是一個 idempotent matrix。這個 matrix 的 eigenvalues 不是 1 就是 0。沒錯！因為

$$P_A(t) = \det(tI - A) = \det\begin{pmatrix} t-1 & 1 \\ 0 & t \end{pmatrix} = t(t-1)$$

Now, $\left\{\alpha_1 = \begin{pmatrix} 1 \\ 0 \end{pmatrix}\right\}$ and $\left\{\alpha_2 = \begin{pmatrix} 1 \\ 1 \end{pmatrix}\right\}$ are eigenvectors of A with respect to $\lambda_1 = 1$ and $\lambda_2 = 0$. And $E_{U_1} = \begin{pmatrix} 1 & 0 \\ 0 & 0 \end{pmatrix}$, $E_{U_2} = \begin{pmatrix} 0 & 0 \\ 0 & 1 \end{pmatrix}$.

Taking $P = \begin{pmatrix} 1 & 1 \\ 0 & 1 \end{pmatrix}$, then $P^{-1} = \begin{pmatrix} 1 & -1 \\ 0 & 1 \end{pmatrix}$ and $P^{-1}AP = \begin{pmatrix} 1 & 0 \\ 0 & 0 \end{pmatrix}$. Hence, $A_1 = PE_{U_1}P^{-1} = \begin{pmatrix} 1 & -1 \\ 0 & 0 \end{pmatrix}$, $A_2 = PE_{U_2}P^{-1} = \begin{pmatrix} 0 & 1 \\ 0 & 1 \end{pmatrix}$.

Therefore, $A = 1A_1 + 0A_2 = A_1$. ∎

這個 matrix 的 spectral decomposition 其實就是它本身。這種現象，一般來說都是如此的。

Corollary 2

If A is an $n \times n$ idempotent matrix, then the spectral decomposition of A is itself.

Proof

$d_1 = 1$, $d_2 = 0$ are the two distinct eigenvalues. So, $E_{U_1} = \begin{pmatrix} I_r & 0 \\ 0 & 0 \end{pmatrix}$ and $E_{U_2} = \begin{pmatrix} 0 & 0 \\ 0 & I_{n-r} \end{pmatrix}$. 7.3 節第 260 頁(@)式提及，存在一 non-singular matrix, P,

such that
$$P^{-1}AP = \begin{pmatrix} I_r & 0 \\ 0 & 0 \end{pmatrix} = E_{U_1}$$

這告訴我們，$A = PE_{U_1}P^{-1} = A_1$。此時，無論 A_2 為何

$$A = 1A_1 + 0A_2 = A_1 \quad 得證。$$

Exercises

Find the spectral decomposition of each of the following matrices.

1. $\begin{pmatrix} 1 & 2 \\ 2 & -2 \end{pmatrix}$

2. $\begin{pmatrix} 1 & i \\ 1 & i \end{pmatrix}$

3. $\begin{pmatrix} 1 & 0 \\ 2 & 0 \end{pmatrix}$

4. $\begin{pmatrix} 1 & 1 & 2 \\ 0 & 5 & -1 \\ 0 & 0 & 7 \end{pmatrix}$

5. $\begin{pmatrix} 3 & 2 & 1 \\ 0 & 1 & 2 \\ 0 & 1 & -1 \end{pmatrix}$

6. $\begin{pmatrix} 1 & 1 & -1 \\ 0 & 1 & 0 \\ 1 & 0 & 1 \end{pmatrix}$

7. $\begin{pmatrix} 0 & 0 & 1+i \\ 0 & 2 & 0 \\ 1-i & 0 & 1 \end{pmatrix}$

§8.4 Interpolatory polynomials、constituent idempotents and Cayley-Hamilton theorem

首先與讀者簡單介紹矩陣的多項式之後，這一節我們打算利用所謂的 interpolatory polynomial，以不同的角度來認識可對角化矩陣的 spectral decomposition。首先，對於矩陣的多項式我們先來個明確的定義。

Definition 8.4

Let A be a square matrix over a scalar field, F (either complex or real). If $f(t) = a_n t^n + a_{n-1} t^{n-1} + \cdots + a_1 t + a_0$ is a polynomial defined on F, then we define,
$$f(A) = a_n A^n + a_{n-1} A^{n-1} + \cdots + a_1 A + a_0 I$$
where I is an identity matrix and A^n denotes, $A \cdot A \cdots A$, n times.

注意，前述定義中的矩陣 A 必得是一個 square matrix，否則無法作自乘運算。舉個簡單例子，假若 $f(t) = t^2 - 3t + 2$ 而且 $A = \begin{pmatrix} 1 & 0 \\ 2 & -1 \end{pmatrix}$，那麼

$$f(A) = A^2 - 3A + 2I = \begin{pmatrix} 1 & 0 \\ 2 & -1 \end{pmatrix}^2 - 3\begin{pmatrix} 1 & 0 \\ 2 & -1 \end{pmatrix} + \begin{pmatrix} 2 & 0 \\ 0 & 2 \end{pmatrix} = \begin{pmatrix} 0 & 0 \\ -6 & 6 \end{pmatrix}$$

再詳細看一下，$f(t) = t^2 - 3t + 2 = (t-2)(t-1)$，所以
$$f(A) = (A - 2I)(A - I) = \begin{pmatrix} -1 & 0 \\ 2 & -3 \end{pmatrix}\begin{pmatrix} 0 & 0 \\ 2 & -2 \end{pmatrix} = \begin{pmatrix} 0 & 0 \\ -6 & 6 \end{pmatrix}$$

有一件事情很好玩，假設有另一矩陣 B 是可逆的話，那麼
$$(B^{-1}AB)^2 = (B^{-1}AB)(B^{-1}AB) = B^{-1}A^2 B$$
或者說 $\quad (B^{-1}AB)^n = (B^{-1}AB)(B^{-1}AB) \cdots (B^{-1}AB) = B^{-1}A^n B$
因此 $\quad f(B^{-1}AB) = B^{-1}f(A)B$
譬如說已知
$$A = \begin{pmatrix} 1 & -1 \\ 0 & 0 \end{pmatrix},\ P = \begin{pmatrix} 1 & 1 \\ 0 & 1 \end{pmatrix},\ P^{-1} = \begin{pmatrix} 1 & -1 \\ 0 & 1 \end{pmatrix} \text{且 } f(t) = t^2 - 3t + 2$$

那麼 $$f(A) = A^2 - 3A + 2I = \begin{pmatrix} 0 & 2 \\ 0 & 2 \end{pmatrix}$$

$$f(P^{-1}AP) = P^{-1}f(A)P = \begin{pmatrix} 1 & -1 \\ 0 & 1 \end{pmatrix}\begin{pmatrix} 0 & 2 \\ 0 & 2 \end{pmatrix}\begin{pmatrix} 1 & 1 \\ 0 & 1 \end{pmatrix} = \begin{pmatrix} 0 & 0 \\ 0 & 2 \end{pmatrix}$$

其實 $$P^{-1}AP = \begin{pmatrix} 1 & -1 \\ 0 & 1 \end{pmatrix}\begin{pmatrix} 1 & -1 \\ 0 & 0 \end{pmatrix}\begin{pmatrix} 1 & 1 \\ 0 & 1 \end{pmatrix} = \begin{pmatrix} 1 & 0 \\ 0 & 0 \end{pmatrix}$$

所以 $$f(P^{-1}AP) = \begin{pmatrix} 1 & 0 \\ 0 & 0 \end{pmatrix}^2 - 3\begin{pmatrix} 1 & 0 \\ 0 & 0 \end{pmatrix} + \begin{pmatrix} 2 & 0 \\ 0 & 2 \end{pmatrix} = \begin{pmatrix} 0 & 0 \\ 0 & 2 \end{pmatrix}$$

觀察前述結果，我們感覺到，可看性比較高的應該是，可對角化矩陣的多項式函數。沒錯，假設 f 是一個多項式，而且 A is diagonalizable such that $P^{-1}AP = D$，那麼依據前述所言，$f(P^{-1}AP) = P^{-1}f(A)P = f(D)$，其中，$D = diag\{d_1, d_2, ..., d_n\}$ 為一對角矩陣，對角線之元素為，$d_1, d_2, ..., d_n$。然而，$f(D) = diag\{f(d_1), f(d_2), ..., f(d_n)\}$，而且，$f(A)P = Pf(D)$，從中我們發現，矩陣 P 的每一個 column，P^i，is an eigenvector of $f(A)$ with respect to $f(d_i)$ as an eigenvalue。我們將這一番說明，寫成下列定理。

Theorem 8.5

If A is a diagonalizable matrix with diagonal elements $d_1, d_2, ..., d_n$, then $f(A)$ is diagonalizable with eigenvalues, $f(d_1), f(d_2), ..., f(d_n)$.

就以前述，$A = \begin{pmatrix} 1 & -1 \\ 0 & 0 \end{pmatrix}$，$f(t) = t^2 - 3t + 2$ 為例。A 的 eigenvalues are 0 and 1。所以，$f(A) = \begin{pmatrix} 0 & 2 \\ 0 & 2 \end{pmatrix}$ 的 eigenvalues 為 $f(0) = 2$ 和 $f(1) = 0$。

Example 1

Given that the spectral decomposition of $A = \begin{pmatrix} 5 & 2 & 2 \\ 2 & 5 & 2 \\ 2 & 2 & 5 \end{pmatrix}$ is

Chapter Eight Spectral Decomposition and Interpolatory Polynomials 309

$$3\begin{pmatrix} 2/3 & -1/3 & -1/3 \\ -1/3 & 2/3 & -1/3 \\ -1/3 & -1/3 & 2/3 \end{pmatrix} + 9\begin{pmatrix} 1/3 & 1/3 & 1/3 \\ 1/3 & 1/3 & 1/3 \\ 1/3 & 1/3 & 1/3 \end{pmatrix}$$ and a polynomial, $f(t) = t^3 - 4t + 1$.

Compute $f(A)$.

Solution

首先看看，$A^2 = (3A_1 + 9A_2)(3A_1 + 9A_2)$。由於，$A_1$, A_2 是 mutually orthogonal idempotent matrices，所以，

$$A^2 = (3A_1 + 9A_2)(3A_1 + 9A_2) = 3^2 A_1 + 9^2 A_2$$

那麼　　　$A^3 = 3^3 A_1 + 9^3 A_2$

或說　　　$A^n = 3^n A_1 + 9^n A_2$

因此　　　$f(A) = f(3)A_1 + f(9)A_2 = 16A_1 + 694A_2$ ∎

最後這個式子說明，從矩陣 A 的 spectral decomposition 可以輕鬆的求出 $f(A)$ 的 spectral decomposition。太棒了，真令人感覺興奮極了。我們當然也要把它整理成定理如下。

Theorem 8.6

> Given a diagonalizable matrix, A, with spectral decomposition $A = d_1 A_1 + d_2 A_2 + ... + d_k A_k$. If $f(t)$ is a polynomial, then
> $$f(A) = f(d_1)A_1 + f(d_2)A_2 + ... + f(d_k)A_k \qquad (xx)$$

Definition 8.5

> Let $d_1, d_2, ..., d_k$ be the distinct eigenvalues of a diagonalizable matrix, A. The interpolatory polynomial, $q_j(t)$, $j = 1, 2, ..., k$, is defined by
> $$q_j(t) = \prod_{i=1, i \neq j}^{k} \frac{t - d_i}{d_j - d_i} = \frac{t - d_1}{d_j - d_1} \cdot \frac{t - d_2}{d_j - d_2} \cdots \frac{t - d_{j-1}}{d_j - d_{j-1}} \cdot \frac{t - d_{j+1}}{d_j - d_{j+1}} \cdots \frac{t - d_k}{d_j - d_k}$$

對於這個陌生的多項式，我們用心詳細的觀察一下。假若，$l \neq j$ 的話，

$$q_j(d_l) = \frac{d_l - d_1}{d_j - d_1} \cdot \frac{d_l - d_2}{d_j - d_2} \cdots \frac{d_l - d_l}{d_j - d_l} \cdot \frac{d_l - d_{l+1}}{d_j - d_{l+1}} \cdots \frac{d_l - d_k}{d_j - d_k} = 0 \text{ 。}$$

又若 $l = j$ 的話，

$$q_j(d_j) = \frac{d_j - d_1}{d_j - d_1} \cdot \frac{d_j - d_2}{d_j - d_2} \cdots \frac{d_j - d_{j-1}}{d_j - d_{j-1}} \cdot \frac{d_j - d_{j+1}}{d_j - d_{j+1}} \cdots \frac{d_j - d_k}{d_j - d_k} = 1 \text{ 。舉個例子看看。}$$

Example 2

Given that the distinct eigenvalues of $A = \begin{pmatrix} -2 & -1 & 3 \\ 6 & 5 & -6 \\ 2 & 2 & -1 \end{pmatrix}$ are $1, -1$ and 2. Find the interpolatory polynomials, q_1, q_2 and q_3. And evaluate each of the values at $1, -1$ and 2.

Solution

$$q_1(t) = \frac{t - d_2}{d_1 - d_2} \cdot \frac{t - d_3}{d_1 - d_3} = \frac{t+1}{1+1} \cdot \frac{t-2}{1-2} \text{ , } q_2(t) = \frac{t - d_1}{d_2 - d_1} \cdot \frac{t - d_3}{d_2 - d_3} = \frac{t-1}{-1-1} \cdot \frac{t-2}{-1-2}$$

$$q_3(t) = \frac{t - d_1}{d_3 - d_1} \cdot \frac{t - d_2}{d_3 - d_2} = \frac{t-1}{2-1} \cdot \frac{t+1}{2+1} \quad \text{and}$$

$q_1(d_1) = 1$, $q_1(d_2) = 0 = q_1(d_3)$, $q_2(d_2) = 1$, $q_2(d_1) = 0 = q_2(d_3)$
$q_3(d_3) = 1$, $q_3(d_1) = 0 = q_3(d_2)$ ◆

我們試試看，將這個新建立起來的多項式代入定理 8.6 的 (xx) 式子中得出，

$$q_1(A) = q_1(d_1)A_1 + q_1(d_2)A_2 + \ldots + q_1(d_k)A_k = A_1$$
$$q_2(A) = q_2(d_1)A_1 + q_2(d_2)A_2 + \ldots + q_2(d_k)A_k = A_2$$
$$\cdots\cdots\cdots\cdots\cdots\cdots\cdots\cdots\cdots\cdots\cdots\cdots\cdots\cdots\cdots\cdots\cdots\cdots$$
$$q_k(A) = q_k(d_1)A_1 + q_k(d_2)A_2 + \ldots + q_k(d_k)A_k = A_k$$

不錯唷，每一個組構冪等矩陣都可以被表示為，矩陣 A 的多項式耶！沒錯，這就是求 constituent idempotent matrices 的另一個方法，這個方法有一個好處就是，我們無需求出矩陣 A 的 eigenvectors，即可求得它的組構冪等矩陣。

Example 3

Use the interpolatory polynomials to find the constituent idempotents of
$$A = \begin{pmatrix} -2 & -1 & 3 \\ 6 & 5 & -6 \\ 2 & 2 & -1 \end{pmatrix}.$$

Solution

From the last example, the distinct eigenvalues are $1, -1, 2$. And

$q_1(t) = \dfrac{t+1}{1+1} \cdot \dfrac{t-2}{1-2}$ this implies, $A_1 = q_1(A) = \dfrac{-1}{2}(A+I)(A-2I) = \begin{pmatrix} -2 & -2 & 3 \\ 0 & 0 & 0 \\ -2 & -2 & 3 \end{pmatrix}$

$q_2(t) = \dfrac{t-1}{-1-1} \cdot \dfrac{t-2}{-1-2} \Rightarrow A_2 = q_2(A) = \dfrac{1}{6}(A-I)(A-2I) = \begin{pmatrix} 2 & 1 & -2 \\ -2 & -1 & 2 \\ 0 & 0 & 0 \end{pmatrix}$

$q_3(t) = \dfrac{t-1}{2-1} \cdot \dfrac{t+1}{2+1} \Rightarrow A_3 = q_3(A) = \dfrac{1}{3}(A-I)(A+I) = \begin{pmatrix} 1 & 1 & -1 \\ 2 & 2 & -2 \\ 2 & 2 & -2 \end{pmatrix}$ ◼

感覺用這個方法做起來，好像比較輕鬆。將這題的結果與第 8.3 節（第 300 頁）的例題 3 比照一下，完全一樣。太棒了！再練習兩個例題吧。

Example 4

Use the interpolatory polynomials to find the constituent idempotents of
$$A = \begin{pmatrix} i & -i \\ -i & i \end{pmatrix}.$$

Solution

The distinct eigenvalues of A are $d_1 = 0$ and $d_2 = 2i$. So, $q_1(t) = \dfrac{t-2i}{-2i}$ and $q_2(t) = \dfrac{t}{2i}$. Hence,

$$A_1 = q_1(A) = \dfrac{1}{-2i}\begin{pmatrix} -i & -i \\ -i & -i \end{pmatrix} = \dfrac{1}{2}\begin{pmatrix} 1 & 1 \\ 1 & 1 \end{pmatrix} \text{ and }$$

$$A_2 = q_2(A) = \dfrac{1}{2i}\begin{pmatrix} i & -i \\ -i & i \end{pmatrix} = \dfrac{1}{2}\begin{pmatrix} 1 & -1 \\ -1 & 1 \end{pmatrix}. \blacksquare$$

Example 5

Use the interpolatory polynomials to find the constituent idempotents of
$A = \begin{pmatrix} 5 & 2 & 2 \\ 2 & 5 & 2 \\ 2 & 2 & 5 \end{pmatrix}$.

Solution

The distinct eigenvalues of A are 3 (multiplicity 2) and 9. So, $q_1(t) = \dfrac{t-9}{3-9}$ and $q_2(t) = \dfrac{t-3}{9-3}$. Therefore,

$$A_1 = q_1(A) = \dfrac{-1}{6}(A-9I) = \dfrac{-1}{6}\begin{pmatrix} -4 & 2 & 2 \\ 2 & -4 & 2 \\ 2 & 2 & -4 \end{pmatrix} = \begin{pmatrix} 2/3 & -1/3 & -1/3 \\ -1/3 & 2/3 & -1/3 \\ -1/3 & -1/3 & 2/3 \end{pmatrix}$$

$$A_2 = q_2(A) = \dfrac{1}{6}(A-3I) = \dfrac{1}{6}\begin{pmatrix} 2 & 2 & 2 \\ 2 & 2 & 2 \\ 2 & 2 & 2 \end{pmatrix} = \begin{pmatrix} 1/3 & 1/3 & 1/3 \\ 1/3 & 1/3 & 1/3 \\ 1/3 & 1/3 & 1/3 \end{pmatrix} \blacksquare$$

事情還沒了，我們回憶一下 8.3 節的部份。上例而言，由於，$\det(A_1) = 0$，而且，A_1 的前兩個 columns are linearly independent，所以，$rank(A_1) = 2$。又由於，$AA_1 = 3A_1$ and $AA_2 = 9A_2$。那麼，A_1 的前兩個 columns are eigenvectors of A having eigenvalue 3。同樣顯然的，$rank(A_2) = 1$，而 A_2 的第一個 column 可以為 A 的第三個 eigenvector with respect to eigenvalue 9。現

在，我們取一個 3×3 non-singular matrix，P，其之三個 columns 為這三個 eigenvectors (將它們分別乘以 3)，即

$$P = \begin{pmatrix} 2 & -1 & 1 \\ -1 & 2 & 1 \\ -1 & -1 & 1 \end{pmatrix}$$。接著，大家算算看 $P^{-1}AP$，

$$P^{-1}AP = \begin{pmatrix} 1/3 & 0 & -1/3 \\ 0 & 1/3 & -1/3 \\ 1/3 & 1/3 & 1/3 \end{pmatrix} \begin{pmatrix} 5 & 2 & 2 \\ 2 & 5 & 2 \\ 2 & 2 & 5 \end{pmatrix} \begin{pmatrix} 2 & -1 & 1 \\ -1 & 2 & 1 \\ -1 & -1 & 1 \end{pmatrix} = \begin{pmatrix} 3 & 0 & 0 \\ 0 & 3 & 0 \\ 0 & 0 & 9 \end{pmatrix}$$

我們發現，利用 interpolatory polynomials，可以輕鬆且快速的從矩陣 A 的 eigenvalues 求得 constituent idempotents，從而觀察出 diagonalizaton 所需的 non-singular matrix，P，而無需事先求知矩陣 A 的 eigenvectors 了。不錯吧。

Example 6

Given that $A = \begin{pmatrix} 2 & 0 & 1 & 0 \\ 0 & 2 & 0 & -1 \\ 1 & 0 & 2 & 0 \\ 0 & -1 & 0 & 2 \end{pmatrix}$

Use the interpolatory polynomials to find the constituent idempotents and non-singular matrix, P, of A.

Solution

$\det(tI - A) = (t-1)^2(t-3)^2$. Writing $d_1 = 1$ and $d_2 = 3$, then $q_1(t) = \dfrac{t-3}{1-3} =$ and $q_2(t) = \dfrac{t-1}{3-1}$. So,

$A_1 = q_1(A) = \dfrac{-1}{2}(A - 3I) = \begin{pmatrix} 1/2 & 0 & -1/2 & 0 \\ 0 & 1/2 & 0 & 1/2 \\ -1/2 & 0 & 1/2 & 0 \\ 0 & 1/2 & 0 & 1/2 \end{pmatrix}$ and

$$A_2 = q_2(A) = \frac{1}{2}(A-I) = \begin{pmatrix} 1/2 & 0 & 1/2 & 0 \\ 0 & 1/2 & 0 & -1/2 \\ 1/2 & 0 & 1/2 & 0 \\ 0 & -1/2 & 0 & 1/2 \end{pmatrix}$$

現在,我們取 A_1 的前面兩個 linearly independent columns (分母 2 約去)為矩陣 P 的 P^1 及 P^2;A_2 的前面兩個 linearly independent columns (同樣分母 2 約去) 為矩陣 P 的 P^3 及 P^4,即

$P^1 = {}^t(1 \ 0 \ -1 \ 0)$,$P^2 = {}^t(0 \ 1 \ 0 \ 1)$,$P^3 = {}^t(1 \ 0 \ 1 \ 0)$,$P^4 = {}^t(0 \ 1 \ 0 \ -1)$

也就是說,$P = \begin{pmatrix} 1 & 0 & 1 & 0 \\ 0 & 1 & 0 & 1 \\ -1 & 0 & 1 & 0 \\ 0 & 1 & 0 & -1 \end{pmatrix}$,$P^{-1} = \begin{pmatrix} 1/2 & 0 & -1/2 & 0 \\ 0 & 1/2 & 0 & 1/2 \\ 1/2 & 0 & 1/2 & 0 \\ 0 & 1/2 & 0 & -1/2 \end{pmatrix}$

那麼, $P^{-1}AP = \begin{pmatrix} 1 & 0 & 0 & 0 \\ 0 & 1 & 0 & 0 \\ 0 & 0 & 3 & 0 \\ 0 & 0 & 0 & 3 \end{pmatrix}$

確實無誤。∎

本節最後,我們簡單介紹認識,可對角化矩陣,A,其之 Cayley-Hamilton Theorem。

Theorem 8.7 (Cayley-Hamilton Theorem)

Let A be a diagonalizable matrix and $P_A(t)$ be the characteristic polynomial of A. Then $P_A(A) = 0$.

Proof

There exists a non-singular matrix, P, such that $P^{-1}AP = D$ is a diagonal matrix with diagonal elements, $\lambda_1, \lambda_2, ..., \lambda_n$, consisting of eigenvalues of A.

Since $A = PDP^{-1}$,$P_A(A) = P_A(PDP^{-1}) = PP_A(D)P^{-1}$. Notice that $P_A(D)$

is also a diagonal matrix with diagonals $P_A(\lambda_i)$, $i=1,2,...,n$. However, $\lambda_1, \lambda_2, ..., \lambda_n$ are zeros of $P_A(t)$, hence $P_A(D) = diag\{P_A(\lambda_1), P_A(\lambda_2), ..., P_A(\lambda_n)\}$ is a zero matrix. Therefore,

$$P_A(A) = P_A(PDP^{-1}) = PP_A(D)P^{-1} = 0$$

下面是 Cayley-Hamilton 的一般型態，在此特別給讀者補充敘述。由於，它不屬於本課程內容的主要焦點，所以我們將不予證明。有興趣的讀者可以參閱 Herstein 編著的 *Topics in Algebra* 第六章第七節，有其相關的細節說明。

Cayley-Hamilton Theorem general form

> Let A be a complex $n \times n$ matrix. If $P_A(t)$ is the characteristic polynomial of A. Then $P_A(A) = 0$.

Exercises

Find the interpolatory polynomial and use it to find the constituent idempotents and non-singular matrix, P, of each of the following diagonalizable matrices.

1. $A = \begin{pmatrix} 1 & 0 \\ 1 & 2 \end{pmatrix}$

2. $A = \begin{pmatrix} 2 & 1+i \\ 1-i & 1 \end{pmatrix}$

3. $A = \begin{pmatrix} 3 & 2 & 1 \\ 0 & 1 & 2 \\ 0 & 1 & -1 \end{pmatrix}$

4. $A = \begin{pmatrix} 1 & -1 & 0 \\ -1 & 2 & -1 \\ 0 & -1 & 1 \end{pmatrix}$

5. $A = \begin{pmatrix} 4 & 0 & 1 \\ -2 & 1 & 0 \\ -2 & 0 & 1 \end{pmatrix}$

6. $A = \begin{pmatrix} 0 & 0 & 1+i \\ 0 & 2 & 0 \\ 1-i & 0 & 1 \end{pmatrix}$

7. $A = \begin{pmatrix} 1 & -\sqrt{6}i & 0 \\ \sqrt{6}i & 0 & 0 \\ 0 & 0 & -2 \end{pmatrix}$

Chapter Nine

Canonical Forms

9.1 Canonical forms for nilpotent matrices
9.2 Jordan canonical forms

　　Canonical forms 是矩陣理論課程中不可缺席的題材。我們以它來當作本書的結尾，有其不能被取代的地位。本章將從如何簡化冪零矩陣成為 canonical form 開始，進而和讀者一起了解，如何簡化一般矩陣成為 Jordan canonical form。

§9.1　Canonical forms for nilpotent matrices

　　下一節所要介紹的 Jordan canonical forms 不會太容易。所以，我們從本節先認識**冪零矩陣** (nilpotent matrix) 的 canonical forms 開始，再慢慢的進入 9.2 節一般矩陣的 Jordan canonical forms。首先回憶一下 3.3 節我們曾經提過，所謂的冪零矩陣：

　　An $n \times n$ matrix, A, is called a nilpotent matrix, if there exists a positive integer, $r \leq n$, such that $A^r = 0$.

Remark

　　Such a number, r, with $A^r = 0$ and $A^{r-1} \neq 0$ is called the **nilpotency (index of nilpotence)** of A.

　　下面我們先來兩個有關冪零矩陣的基本特性，以及一個關鍵定理之後，

317

再和各位探討本節的主題，The canonical form for a nilpotent matrix。

Lemma 9.1

> Given an $n \times n$ nilpotent matrix, A. If a_i, $i = 0, 1, 2, 3, \ldots$ are scalars and $a_0 \neq 0$, $m \in Z^+$, then $a_0 I_n + a_1 A + a_2 A^2 + \cdots + a_m A^m$ is invertible.

Proof

其實 3.3 節的時候，已經有一些類似這個命題的題目（參考第 97 頁的 Theorem 3.13）。現在，我們令

$$S = a_1 A + a_2 A^2 + \cdots + a_m A^m ,$$

則由於 A 是 nilpotent，所以 S 也是 nilpotent。今假設 $S^r = 0$，然後思考下列式子

$$(a_0 I_n + S)(\frac{I_n}{a_0} - \frac{S}{a_0^2} + \frac{S^2}{a_0^3} - \frac{S^3}{a_0^4} + \cdots + (-1)^{r-1} \frac{S^{r-1}}{a_0^r})$$

$$= I_n + (-1)^{r-1} \frac{S^r}{a_0^r} = I_n$$

得知

$$a_0 I_n + S = a_0 I_n + a_1 A + a_2 A^2 + \cdots + a_m A^m \text{ 為可逆的。}$$

Lemma 9.2

> Given an $n \times n$ complex (or real) nilpotent matrix, A, with nilpotency n_1. If $v \in C^n$ (or $v \in R^n$) is such that $A^{n_1-1} v \neq 0$, then $\{v, Av, A^2 v, \ldots, A^{n_1-1} v\}$ is a linearly independent set.

Proof

Let $c_1 v + c_2 Av + c_3 A^2 v + \cdots + c_{n_1} A^{n_1-1} v = 0$, and suppose $1 \leq t \leq n_1$ is the first digit such that c_t is non-zero. Then, $c_t A^{t-1} v + \cdots + c_{n_1} A^{n_1-1} v = 0$. This implies that $(c_t I_n + \cdots + c_{n_1} A^{n_1-t} v) A^{t-1} v = 0$. Since $c_t \neq 0$, by the last lemma,

$(c_t I_n + ... + c_{n_1} A^{n_1-t} v)$ is invertible. Hence $A^{t-1} v = 0$. But this leads to a contradiction, since $t \leq n_1$. So, c_t can not be non-zero and therefore, every $c_i, i = 1, 2, 3, ..., n_1$ has to be zero.

Theorem 9.3

Let $V_1 = SP\{v, Av, A^2 v, ..., A^{n_1-1} v\}$, where $v, Av, A^2 v, ..., A^{n_1-1} v$ are what in the last Lemma.
1. If $u \in V_1$ and $A^{n_1-k} u = 0$ for some $0 < k \leq n_1$, then there exists $u_0 \in V_1$ such that $A^k u_0 = u$.
2. There exists a subspace, W, of C^n invariant under A such that $C^n = V_1 \oplus W$.

Proof
1.
$u \in V_1$
$\Rightarrow u = x_1 v + x_2 Av + \cdots + x_k A^{k-1} v + \cdots + x_{n_1} A^{n_1-1} v$ for some scalars $x_i, i = 1, 2, 3, ..., n_1$. Applying A^{n_1-k} to both sides of the last equation, we obtain,
$0 = A^{n_1-k} u = x_1 A^{n_1-k} v + x_2 A^{n_1-k+1} v + \cdots + x_k A^{n_1-1} v + \cdots + x_{n_1} A^{2n_1-k-1} v$
Since $A^m = 0, \forall m \geq n_1$, the last equation becomes
$0 = x_1 A^{n_1-k} v + x_2 A^{n_1-k+1} v + ... + x_k A^{n_1-1} v$
Because, $A^{n_1-k} v, A^{n_1-k+1} v, ..., A^{n_1-1} v$ are linearly independent. So $x_i = 0, \forall i = 1, ..., k$.
Thus, $u = x_{k+1} A^k v + \cdots + x_{n_1} A^{n_1-1} v = A^k (x_{k+1} v + \cdots + x_{n_1} A^{n_1-k-1} v)$
Letting $u_0 = x_{k+1} v + \cdots + x_{n_1} A^{n_1-k-1} v$, then the result is obtained.

2.
這個證明很冗長，讀者需要一點耐性。首先令 W 為使得

$V_1 \cap W = \{0\}$ and W is invariant under A

之 C^n 中的最大可能維度的子空間。

我們的目標是，要證明 $C^n = V_1 + W$。

我們用反證法：

假設 $V_1 + W \neq C^n$，則 C^n 中存在一向量 z，使得 $z \notin V_1 + W$。

現在，由於 A 的 nilpotency 是 n_1 (亦即 $A^{n_1} = 0$)，所以存在一 $0 < k \leq n_1$ 使得，$A^k z = 0 \in V_1 + W$ 而且 $A^i z \notin V_1 + W$ 當 $0 \leq i < k$ 時。這個結果說明，存在兩個向量 $u \in V_1, w \in W$ 使得，$A^k z = u + w$。因而得出，

$$A^{n_1-k} A^k z = A^{n_1-k} u + A^{n_1-k} w$$
$$\Rightarrow \quad 0 = A^{n_1-k} u + A^{n_1-k} w$$
$$\Rightarrow \quad A^{n_1-k} u = -A^{n_1-k} w$$

由於，V_1 and W are invariant under A，所以，$A^{n_1-k} u \in V_1$，$-A^{n_1-k} w \in W$。因此，$A^{n_1-k} u = -A^{n_1-k} w \in V_1 \cap W = \{0\}$。現在利用 1.的結果得知，存在一 $u_0 \in V_1$ 使得 $A^k u_0 = u$。如此之下，前述 $A^k z = u + w$ 變成 $A^k z = A^k u_0 + w$，或者說，$A^k (z - u_0) = w$。

今令 $z_1 = z - u_0$，則 $A^k z_1 = w \in W$。從而得出，$A^m z_1 \in W$，$\forall m \geq k$。另一方面，若 $i < k$，則 $A^i z_1 = A^i (z - u_0) = A^i z - A^i u_0$。

由於，$A^i u_0 \in V_1$ 且 $A^i z \notin V_1 + W$，所以，$A^i z_1 = A^i z - A^i u_0 \notin V_1 + W, \forall \ i < k$。再來，我們令 W_1 為，W 和 $\{z_1, A z_1, A^2 z_1, ..., A^{k-1} z_1\}$ 所生成的子空間。又由於，$A^k z_1 \in W$ 且 $AW \subseteq W$，所以 $AW_1 \subseteq W_1$。另外，由於 $z \notin V_1 + W$，所以 $z_1 \notin W$。也就是說，W 是 W_1 的真子空間。將此結果與前題假設相比得知，$V_1 \cap W_1 \neq \{0\}$。此式說明，W_1 中存在一非 0 向量，$w_1 = w_0 + x_1 z_1 + x_2 A z_1 + x_3 A^2 z_1 + ... + x_k A^{k-1} z_1$ ($w_0 \in W$)，使得某些個 $x_i \neq 0$ 且 $w_1 = w_0 + x_1 z_1 + x_2 A z_1 + x_3 A^2 z_1 + ... + x_k A^{k-1} z_1 \in V_1$。

(註：若 $x_i = 0$, $\forall \ i = 1, 2, ..., k$，則 $0 \neq w_1 = w_0 \in W \cap V_1$，此與 $V_1 \cap W = \{0\}$ 衝突。)

現在，令 x_t, $t \leq k$ 為第一個不為 0 的係數，則

$$w_1 = w_0 + x_t A^{t-1} z_1 + x_{t+1} A^t z_1 + x_{t+2} A^{t+1} z_1 + \Lambda + x_k A^{k-1} z_1$$
$$= w_0 + (x_t I_n + x_{t+1} A + x_{t+2} A^2 + \Lambda + x_k A^{k-t}) A^{t-1} z_1 \in V$$

前面 Lemma 9.1 提及，$(x_t I_n + x_{t+1} A + x_{t+2} A^2 + \Lambda + x_k A^{k-t})$ 為一可逆矩陣，且

其逆矩陣 B 為一可以用 A 來表示的多項式。因此，$BW \subseteq W$ and $BV_1 \subseteq V_1$。
也因此，$Bw_1 = Bw_0 + B(x_t I_n + x_{t+1} A + x_{t+2} A^2 + \cdots + x_k A^{k-t}) A^{t-1} z_1$
$$= Bw_0 + A^{t-1} z_1 \in V_1$$
然而，$Bw_0 \in W$，所以，$A^{t-1} z_1 = -Bw_0 + Bw_1 \in W + V_1$。此結果與前述 $A^i z_1 = A^i z - A^i u_0 \notin V_1 + W$，$\forall\ i < k$ 相互矛盾。

所以，一開始的大前提假設

$$C^n \text{ 中存在一向量 } z\text{，使得 } z \notin V_1 + W$$

是不可能成立的。也就是說，$C^n = V_1 + W$。得證。

總結這個定理，有一件事情我們先行摘錄下來。前面曾經提及，$\{A^{n_1-1}v, A^{n_1-2}v, ..., Av, v\}$ 為子空間 V_1 之一組基底。今令 A_1 為矩陣 A 局限在子空間 $V_1 = SP\{A^{n_1-1}v, A^{n_1-2}v, ..., Av, v\}$ 上的一個線型映射。亦即，$A_1 = A|_{V_1}$。
那麼，the matrix representation for A_1 with respect to the basis, $\{A^{n_1-1}v, A^{n_1-2}v, ..., Av, v\}$，應為

$$\begin{pmatrix} 0 & 1 & 0 & \cdot & 0 \\ \cdot & 0 & 1 & \cdot & \cdot \\ \cdot & \cdot & 0 & \cdot & 0 \\ \cdot & \cdot & \cdot & \cdot & 1 \\ 0 & 0 & 0 & \cdot & 0 \end{pmatrix}$$

記住這個結果，然後再看看下面的定理。

Theorem 9.4 (Canonical form for nilpotent matrix)

If A is an $n \times n$ nilpotent matrix with nilpotency, n_1, then a basis of C^n can be found such that the matrix representation for A with respect to the basis is of the form,

$$\begin{pmatrix} M_{n_1} & O & \cdot & \cdot & O \\ O & M_{n_2} & O & \cdot & \cdot \\ \cdot & O & \cdot & \cdot & \cdot \\ \cdot & \cdot & \cdot & \cdot & O \\ O & \cdot & \cdot & O & M_{n_r} \end{pmatrix}$$

where $n_1 \geq n_2 \geq \cdots \geq n_r$, $n_1 + n_2 + \cdots + n_r = n$, and for each $t = 1, 2, \ldots, r$,

$$M_{n_t} = \begin{pmatrix} 0 & 1 & 0 & \cdot & 0 \\ 0 & 0 & 1 & 0 & \cdot \\ \cdot & \cdot & 0 & \cdot & 0 \\ \cdot & \cdot & \cdot & \cdot & 1 \\ 0 & \cdot & \cdot & 0 & 0 \end{pmatrix}_{n_t \times n_t}$$

Proof

依據前述之 Lemma 以及定理，$C^n = V_1 \oplus W$，其中 $V_1 = SP\{A^{n_1-1}v, A^{n_1-2}v, \ldots, Av, v\}$，而且 V_1 and W are invariant under A。現在，找出 W 的一組基底加到 V_1 的基底 $\{A^{n_1-1}v, A^{n_1-2}v, \ldots, Av, v\}$ 上擴充而為 C^n 的基底。然後，根據定理之前的補充說明，可以得知，以這個基底所表示出來之 A 的 matrix representation 應為如下型式

$$\begin{pmatrix} M_{n_1} & O_{n_1 \times (n-n_1)} \\ O_{(n-n_1) \times n_1} & A_2 \end{pmatrix}$$

其中，矩陣 M_{n_1} 如上所示；矩陣 A_2 則為一 $(n-n_1) \times (n-n_1)$ 矩陣，而矩陣 $B = \begin{pmatrix} O_{n_1 \times n_1} & O_{n_1 \times (n-n_1)} \\ O_{(n-n_1) \times n_1} & A_2 \end{pmatrix}$ 是 A 局限在子空間 W 的 matrix representation with respect to the above basis。

接著由於，$A^{n_1} = 0$，所以存在一整數 $0 < n_2 \leq n_1$ 使得 $B^{n_2} = 0$。把前面得出矩陣 M_{n_1} 的程序應用在矩陣 B 上，則空間 W 也可以被分解為 $W = V_2 \oplus W_1$，而且也能夠找出 V_2 的一組基底 $\{B^{n_2-1}u, B^{n_2-2}u, \ldots, Bu, u\}$，使得矩陣 A 的 matrix

representation with respect to the basis, $\{A^{n_1-1}v, A^{n_1-2}v, ..., Av, v\} \cup \{B^{n_2-1}u, B^{n_2-2}u,$
$..., Bu, u\} \cup \{$a basis for $W_1\}$之型式為

$$\begin{pmatrix} M_{n_1} & O & O \\ O & M_{n_2} & O \\ O & O & A_3 \end{pmatrix}$$

其中矩陣 A_3 則為一 $(n-(n_1+n_2)) \times (n-(n_1+n_2))$ 矩陣，而 $\begin{pmatrix} O_{n_1 \times n_1} & O & O \\ O & O_{n_2 \times n_2} & O \\ O & O & A_3 \end{pmatrix}$

是 A 局限在子空間 W_1 的 matrix representation with respect to the above basis。

持續前述步驟，我們最終將得出 C^n 的 r 個子空間，$V_1, V_2, ..., V_r$，以及 C^n 的一組基底，使得 $C^n = V_1 \oplus V_2 \oplus \cdots \oplus V_r$；其中對於每一個 $i = 1, 2, 3, ..., r$ 而言，$\dim V_i = n_i$，$n_1 \geq n_2 \geq \cdots \geq n_r$，$n_1 + n_2 + \cdots + n_r = n$，而且 the matrix representation for A is of the form

$$\begin{pmatrix} M_{n_1} & O & \cdot & \cdot & O \\ O & M_{n_2} & O & \cdot & \cdot \\ \cdot & O & \cdot & \cdot & \cdot \\ \cdot & \cdot & \cdot & \cdot & O \\ O & \cdot & \cdot & O & M_{n_r} \end{pmatrix}.$$

下面有兩個新的名詞，在此順便與讀者補充說明一下。

1. The integers, $n_1 \geq n_2 \geq \cdots \geq n_r$, are called the invariants of A.
2. Let A be a nilpotent matrix. An invariant subspace, W, of C^n with $\dim W = m$ is said to "A-cyclic", if
   a. $A^m w = 0$, $\forall w \in W$ and $A^{m-1} u \neq 0$ for some $u \in W$.
   b. $\exists v \in W$ such that $v, Av, A^2 v, ..., A^{m-1} v$ form a basis for W.

Example 1

Given a 3×3 real nilpotent matrix, $A = \begin{pmatrix} 0 & 1 & 1 \\ 0 & 0 & 0 \\ 0 & 0 & 0 \end{pmatrix}$.

Define $A: R^3 \to R^3$ by

$$A\begin{pmatrix} x \\ y \\ z \end{pmatrix} = \begin{pmatrix} 0 & 1 & 1 \\ 0 & 0 & 0 \\ 0 & 0 & 0 \end{pmatrix}\begin{pmatrix} x \\ y \\ z \end{pmatrix} = \begin{pmatrix} y+z \\ 0 \\ 0 \end{pmatrix}, \quad \forall \begin{pmatrix} x \\ y \\ z \end{pmatrix} \in R^3.$$

Find the canonical form for A.

Solution

首先確認，$A^2 = 0$，A是 nilpotent，它的 nilpotency 為 2。

令 $v = \begin{pmatrix} 0 \\ 0 \\ 1 \end{pmatrix}$，則 $Av = \begin{pmatrix} 1 \\ 0 \\ 0 \end{pmatrix} \neq 0$。所以，$v, Av$ 為線性獨立的，而且它們形成 $V_1 = Sp\{Av, v\}$ 的一組基底。用這個基底得出，$M_{n_1} = M_2 = \begin{pmatrix} 0 & 1 \\ 0 & 0 \end{pmatrix}$。接著令 $w = \begin{pmatrix} 0 \\ 1 \\ -1 \end{pmatrix}$，$W = Sp\{w\}$；我們發現 $V_1 \cap W = \{0\}$，$R^3 = V_1 \oplus W$ 以及 W is invariant under A (註：$Aw = 0$，$n_2 = 1$)。而且，$M_{n_2} = M_1 = (0)_{1 \times 1}$。

最後，得出在 $\{Av, v, w\}$ 為 R^3 之基底之下，矩陣 A 的 canonical form 為

$$B = \begin{pmatrix} \begin{pmatrix} 0 & 1 \\ 0 & 0 \end{pmatrix} & 0 \\ 0 & 0 & (0) \end{pmatrix}$$

我們總喜歡利用下列簡圖，回味一下矩陣的相似性質。

Chapter Nine Canonical Forms 325

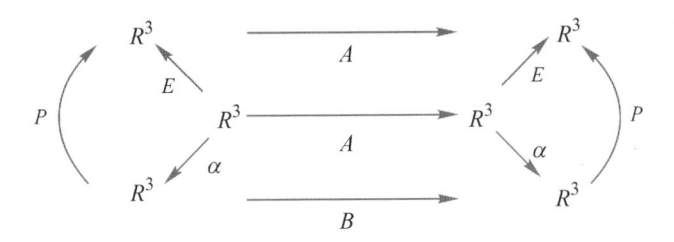

Let $E = \left\{ \begin{pmatrix} 1 \\ 0 \\ 0 \end{pmatrix}, \begin{pmatrix} 0 \\ 1 \\ 0 \end{pmatrix}, \begin{pmatrix} 0 \\ 0 \\ 1 \end{pmatrix} \right\}$ be the standard basis, and

$$\alpha = \{Av, v, w\} = \left\{ \begin{pmatrix} 1 \\ 0 \\ 0 \end{pmatrix}, \begin{pmatrix} 0 \\ 0 \\ 1 \end{pmatrix}, \begin{pmatrix} 0 \\ 1 \\ -1 \end{pmatrix} \right\}.$$

Write, $P = mat(I, \alpha, E)$, the change of basis matrix from α into E, then

$$P = \begin{pmatrix} 1 & 0 & 0 \\ 0 & 0 & 1 \\ 0 & 1 & -1 \end{pmatrix}, \text{ and } P^{-1} = mat(I, E, \alpha) = \begin{pmatrix} 1 & 0 & 0 \\ 0 & 1 & 1 \\ 0 & 1 & 0 \end{pmatrix}.$$

Furthermore, $P^{-1}AP = \begin{pmatrix} 1 & 0 & 0 \\ 0 & 1 & 1 \\ 0 & 1 & 0 \end{pmatrix} \begin{pmatrix} 0 & 1 & 1 \\ 0 & 0 & 0 \\ 0 & 0 & 0 \end{pmatrix} \begin{pmatrix} 1 & 0 & 0 \\ 0 & 0 & 1 \\ 0 & 1 & -1 \end{pmatrix} = \begin{pmatrix} 0 & 1 & 0 \\ 0 & 0 & 0 \\ 0 & 0 & 0 \end{pmatrix} = B$

which is the canonical form for A.

Example 2

Given a 3×3 real nilpotent matrix, $A = \begin{pmatrix} 0 & 1 & 2 \\ 0 & 0 & 3 \\ 0 & 0 & 0 \end{pmatrix}$. Define $A: R^3 \to R^3$

by

$$A \begin{pmatrix} x \\ y \\ z \end{pmatrix} = \begin{pmatrix} 0 & 1 & 2 \\ 0 & 0 & 3 \\ 0 & 0 & 0 \end{pmatrix} \begin{pmatrix} x \\ y \\ z \end{pmatrix} = \begin{pmatrix} y + 2z \\ 3z \\ 0 \end{pmatrix}, \quad \forall \begin{pmatrix} x \\ y \\ z \end{pmatrix} \in R^3.$$

Find the canonical form for A.

Solution

首先，$A^2 = \begin{pmatrix} 0 & 0 & 3 \\ 0 & 0 & 0 \\ 0 & 0 & 0 \end{pmatrix}$，$A^3 = 0$。$A$ 是一個 nilpotent matrix with nilpotency 3。令 $v = \begin{pmatrix} 0 \\ 0 \\ 1 \end{pmatrix}$，則 $Av = \begin{pmatrix} 2 \\ 3 \\ 0 \end{pmatrix}$，$A^2v = \begin{pmatrix} 3 \\ 0 \\ 0 \end{pmatrix}$。$\{A^2v, Av, v\}$ 是 $V_1 = R^3$ 的一組基底。以此基底得出，A 的矩陣表示為 $\begin{pmatrix} 0 & 1 & 0 \\ 0 & 0 & 1 \\ 0 & 0 & 0 \end{pmatrix}$。由於，$V_1 = R^3$，所以，$\begin{pmatrix} 0 & 1 & 0 \\ 0 & 0 & 1 \\ 0 & 0 & 0 \end{pmatrix}$ 就是 A 的 canonical form。∎

Example 3

Given a 4×4 complex nilpotent matrix, $A = \begin{pmatrix} i & -i & 0 & 0 \\ i & -i & 0 & 0 \\ 0 & 0 & 1 & -1 \\ 0 & 0 & 1 & -1 \end{pmatrix}$. Find the canonical form for A.

Solution

We first note that $A^2 = 0$，$n_1 = 2$。令 $v = \begin{pmatrix} 1 \\ 0 \\ 0 \\ 0 \end{pmatrix}$，則 $Av = \begin{pmatrix} i \\ i \\ 0 \\ 0 \end{pmatrix}$。令 $V_1 = SP\{Av, v\}$，且令 $W = \left\{ \begin{pmatrix} 0 \\ 0 \\ x \\ y \end{pmatrix} : x, y \in C \right\}$，那麼 $C^4 = V_1 \oplus W$。令

$w_1 = \begin{pmatrix} 0 \\ 0 \\ 1 \\ 0 \end{pmatrix}$, $w_2 = \begin{pmatrix} 0 \\ 0 \\ 0 \\ 1 \end{pmatrix} \in W$，$\{Av, v, w_1, w_2\}$ 為 C^4 之一組基底。以此基底，求出

矩陣 A 的 representation 為 $\begin{pmatrix} 0 & 1 & 0 & 0 \\ 0 & 0 & 0 & 0 \\ 0 & 0 & 1 & -1 \\ 0 & 0 & 1 & -1 \end{pmatrix}$。亦即 $M_{n_1} = \begin{pmatrix} 0 & 1 \\ 0 & 0 \end{pmatrix}$，

$A_2 = \begin{pmatrix} 1 & -1 \\ 1 & -1 \end{pmatrix}$。再如定理所示，令 $B = \begin{pmatrix} 0 & 0 & 0 & 0 \\ 0 & 0 & 0 & 0 \\ 0 & 0 & 1 & -1 \\ 0 & 0 & 1 & -1 \end{pmatrix}$，則 $B = A|_W$，$B^2 = 0$，

且 $n_2 = 2$。今令 $u = \begin{pmatrix} 0 \\ 0 \\ 1 \\ 0 \end{pmatrix}$，$Bu = \begin{pmatrix} 0 \\ 0 \\ 1 \\ 1 \end{pmatrix}$，$V_2 = SP\{Bu, u\} = W$；then the matrix

representation for A with respect to the basis, $\{Av, v\} \cup \{Bu, u\}$, is of the form

$\begin{pmatrix} \begin{pmatrix} 0 & 1 \\ 0 & 0 \end{pmatrix} & O_{2 \times 2} \\ O_{2 \times 2} & \begin{pmatrix} 0 & 1 \\ 0 & 0 \end{pmatrix} \end{pmatrix}$。而這就是矩陣 A 的 canonical form。 ∎

Exercises

1. 本節 317 頁一開始以及 321 頁的冪零矩陣之 canonical form 定理中，都說明了，"假設一個 $n \times n$ nilpotent matrix A 的 nilpotency 是 k 的話，則 $k \leq n$。" 現在，以不用上述這兩個結果的情況之下，請讀者證明前列敘述的正確性。

2. Find the canonical form for $A = \begin{pmatrix} 1 & 1 & 1 \\ -1 & -1 & -1 \\ 1 & 1 & 0 \end{pmatrix}$.

3. Find the canonical form for $A = \begin{pmatrix} 0 & 0 & 0 & 0 \\ 1 & 0 & 0 & 0 \\ 1 & 0 & 0 & 0 \\ 0 & 0 & 1 & 0 \end{pmatrix}$.

§9.2 Jordan canonical forms

首先回憶第三章的一件事情。那就是，若 $\{\alpha_1,...,\alpha_n\}$ 是 $V = C^n$ 的一組 column basis 而且 A 是一個複數體系的 $n \times n$ 矩陣的話，那麼 $\{A\alpha_1,...,A\alpha_n\}$ 生成 A 的值域，$A(V)$。也就是說，for any $X \in C^n$, there exist scalars, $x_1, x_2,...,x_n$, such that $AX = \sum_{i=1}^{n} x_i A\alpha_i$。

接著，在介紹 Jordan canonical forms 之前，我們需要先來一些準備工作。

Lemma 9.5

> Given a polynomial $P(t)$. Let $P(t) = f(t)g(t)$ with deg.f, deg.$g \geq 1$ and let f and g be relatively prime. If A is an $n \times n$ matrix such that $P(A) = 0$, and if $W_1 = Kerf(A)$, $W_2 = Kerg(A)$, then $C^n = W_1 \oplus W_2$.

Proof

由於，f and g are prime to each other，存在兩多項式，$c(t), d(t)$，使得 $c(t)f(t) + d(t)g(t) = 1$。這說明，$c(A)f(A) + d(A)g(A) = I_n$。今令，$v \in C^n$，則 $(c(A)f(A))v + (d(A)g(A))v = v$。由於，
$g(A)(c(A)f(A))v = c(A)f(A)g(A)v = c(A)P(A)v = 0$，
$f(A)(d(A)g(A))v = d(A)f(A)g(A)v = d(A)P(A)v = 0$，
所以，$(c(A)f(A))v \in W_2$，$(d(A)g(A))v \in W_1$。這證明了，$C^n = W_1 + W_2$。又令，$v \in W_1 \cap W_2$，則 $f(A)v = 0 = g(A)v$。因而，
$$c(A)f(A)v + d(A)g(A)v = 0 = v。$$
所以，$C^n = W_1 \oplus W_2$。

Corollary

Let $P(t)$ be a polynomial such that $P(t) = f_1(t)f_2(t)\cdots f_k(t)$ with each deg.$f_i \geq 1$. Let $f_1, f_2, \ldots, f_k$ be relatively prime. If A is an $n \times n$ matrix such that $P(A) = 0$, and if $W_i = \mathrm{Ker} f_i$, $i = 1, 2, 3, \ldots, k$, then $C^n = W_1 \oplus W_2 \oplus \cdots \oplus W_k$.

Proof

利用前述 Lemma 加上歸納法可以得證本推論。

Example 1

Given a 3×3 complex matrix, $A = \begin{pmatrix} 1 & -\sqrt{6}i & 0 \\ \sqrt{6}i & 0 & 0 \\ 0 & 0 & -2 \end{pmatrix}$. Find its characteristic polynomial, $P_A(t)$. Decompose $P_A(t)$ into relatively prime factors, $f(t), g(t)$ such that $C^3 = W_1 \oplus W_2$, if $W_1 = \mathrm{Ker} f(A)$, $W_2 = \mathrm{Ker} g(A)$.

Solution

$$P_A(t) = \det(tI - A) = \det \begin{pmatrix} t-1 & \sqrt{6}i & 0 \\ -\sqrt{6}i & t & 0 \\ 0 & 0 & t+2 \end{pmatrix}$$

$$= t(t-1)(t+2) - 6(t+2) = (t+2)^2(t-3)$$

$f(t) = (t+2)^2$ and $g(t) = (t-3)$

Replacing t by A, we have that

$$P_A(A) = (A + 2I_3)^2 (A - 3I_3) = \begin{pmatrix} 15 & -5\sqrt{6}i & 0 \\ 5\sqrt{6}i & 10 & 0 \\ 0 & 0 & 0 \end{pmatrix} \begin{pmatrix} -2 & -\sqrt{6}i & 0 \\ \sqrt{6}i & -3 & 0 \\ 0 & 0 & -5 \end{pmatrix} = O_{3 \times 3}$$

This satisfies the hypothesis of the previous theorem.

Now,

$$W_1 = Kerf(A) = \left\{ \begin{pmatrix} x \\ y \\ z \end{pmatrix} \in C^3 : \begin{pmatrix} 15 & -5\sqrt{6}i & 0 \\ 5\sqrt{6}i & 10 & 0 \\ 0 & 0 & 0 \end{pmatrix} \begin{pmatrix} x \\ y \\ z \end{pmatrix} = 0 \right\}$$

$$= \left\{ c \begin{pmatrix} 1 \\ (-\sqrt{6}i)/2 \\ 0 \end{pmatrix} + d \begin{pmatrix} 0 \\ 0 \\ 1 \end{pmatrix} : c, d \in C \right\}$$

$$W_2 = Kerg(A) = \left\{ \begin{pmatrix} x \\ y \\ z \end{pmatrix} \in C^3 : \begin{pmatrix} -2 - \sqrt{6}i & 0 \\ \sqrt{6}i & -3 & 0 \\ 0 & 0 & -5 \end{pmatrix} \begin{pmatrix} x \\ y \\ z \end{pmatrix} = 0 \right\}$$

$$= \left\{ s \begin{pmatrix} 1 \\ (\sqrt{6}i)/3 \\ 0 \end{pmatrix} : s \in C \right\}$$

Since $\begin{pmatrix} 1 \\ (-\sqrt{6}i)/2 \\ 0 \end{pmatrix}, \begin{pmatrix} 0 \\ 0 \\ 1 \end{pmatrix}$ and $\begin{pmatrix} 1 \\ (\sqrt{6}i)/3 \\ 0 \end{pmatrix}$ are linearly independent, $C^3 = W_1 \oplus W_2$.

Lemma 9.6

Let A be an $n \times n$ nilpotent matrix and let W be A-cyclic with $\dim W = m$. Then for each $k \leq m$, the dimension of $A^k W$ is $m - k$.

Proof

依據 A-cyclic 之定義，我們令 $\{A^{m-1}v, A^{m-2}v, ..., Av, v\}$ 為 W 之一基底。接著 Apply A^k to the basis, $\{A^{m-1}v, A^{m-2}v, ..., Av, v\}$，我們得到，$\{A^{m+k-1}v, A^{m+k-2}v, ..., A^{k+1}v, A^k v\} = \{0, ..., 0, A^{m-1}v, ..., A^{k+1}v, A^k v\}$。由本節 329 頁一開始的說明，$\{0, ..., 0, A^{m-1}v, ..., A^{k+1}v, A^k v\}$ 生成 the image of W under A^k。亦即，$SP\{A^{m-1}v, ..., A^{k+1}v, A^k v\} = A^k W$。注意，$\{A^{m-1}v, ..., A^{k+1}v, A^k v\}$ 是 $\{A^{m-1}v, A^{m-2}v, ..., Av, v\}$ 的一部分，所以，$A^{m-1}v, ..., A^{k+1}v, A^k v$ are linearly independent，也因此，它們形成了 $A^k W$ 的一個基底。這個結果告訴我們，

$$\dim(A^k W) = m - 1 - k + 1 = m - k \text{ 。}$$

在 9.1 節中,我們曾經提及

"對於任何一個 nilpotent matrix, A, 存在正整數, $n_1 \geq n_2 \geq ... \geq n_r$, 以及 A-cyclic subspaces, $V_1, V_2, ..., V_r$, 使得 $\dim V_i = n_i$, $i = 1, 2, ..., r$, $C^n = V_1 \oplus V_2 \oplus ... \oplus V_r$。"

現在問題來了,我們想要問的是,這些 invariants, $n_1 \geq n_2 \geq ... \geq n_r$, 以及 A-cyclic subspaces, $V_1, V_2, ..., V_r$ 是否為唯一的?也就是說,我們是否能夠找到另一組 invariants, $m_1 \geq m_2 \geq ... \geq m_s$, 以及 A-cyclic subspaces, $U_1, U_2, ..., U_s$, 使得 $\dim U_j = m_j$, $j = 1, 2, ..., s$, $C^n = U_1 \oplus U_2 \oplus \cdots \oplus U_s$? We now answer the problem by the following theorem.

Theorem 9.7

The invariants of a nilpotent matrix, A, are uniquely determined.

Proof

Suppose that $m_1 \geq m_2 \geq ... \geq m_s$ is the set of other invariants of A and suppose that m_i is the first integer such that $m_i \neq n_i$. Say, $m_i < n_i$, and apply A^{m_i} to both sides of $C^n = V_1 \oplus V_2 \oplus ... \oplus V_r$, then

$$A^{m_i}(V) = A^{m_i}(V_1) \oplus A^{m_i}(V_2) \oplus \cdots \oplus A^{m_i}(V_r)$$

(why?). By the last lemma, we see that

$$\dim A^{m_i}(V_j) = n_j - m_i, \quad \forall j = 1, ..., i . \ (\textit{Note: } n_j \geq m_i, \forall j \leq i), \text{ so}$$

$$\dim A^{m_i}(C^n) \geq \sum_{j=1}^{i} \dim A^{m_i}(V_j) = \sum_{j=1}^{i}(n_j - m_i) .$$

On the other hand, $C^n = U_1 \oplus U_2 \oplus \cdots \oplus U_s$, thus in the same manner, $A^{m_i}(C^n) = A^{m_i}(U_1) \oplus A^{m_i}(U_2) \oplus \cdots \oplus A^{m_i}(U_s)$.

Since $A^{m_i}(U_j) = 0$, $\forall j \geq i$, $A^{m_i}(C^n) = A^{m_i}(U_1) \oplus A^{m_i}(U_2) \oplus \cdots \oplus A^{m_i}(U_{i-1})$.

Hence, $\dim A^{m_i}(C^n) = \sum_{j=1}^{i-1} \dim A^{m_i}(U_j) = \sum_{j=1}^{i-1} (m_j - m_i)$.

By our choice of i, $n_1 = m_1, n_2 = m_2, ..., n_{i-1} = m_{i-1}$,

so $\dim A^{m_i}(C^n) = \sum_{j=1}^{i-1} (m_j - m_i)$ becomes $\dim A^{m_i}(C^n) = \sum_{j=1}^{i-1} (n_j - m_i)$. This leads to a contradiction to $\dim A^{m_i}(C^n) \geq \sum_{j=1}^{i} (n_j - m_i)$.

We therefore conclude that $n_i = m_i$, $\forall i$. And since $n_1 + n_2 + \cdots + n_r = n = m_1 + m_2 + \cdots + m_s$, r and s must be equal.

And the proof is completed.

一件事情特別補充說明一下。假設 A 與 B 為兩個 $n \times n$ nilpotent matrices，而且假設，它們有相同的 invariants，$n_1 \geq n_2 \geq ... \geq n_r$。則由 9.1 節定理 9.4，我們發現分別存在兩個 C^n 的 column bases，$\alpha = \{v_1, v_2, ..., v_n\}$ 以及 $\beta = \{w_1, w_2, ..., w_n\}$，使得矩陣 A 與 B 的 matrix representations，$mat(A, \alpha, \alpha)$ 與 $mat(B, \beta, \beta)$ 都是

$$M = \begin{pmatrix} M_{n_1} & O & \cdot & \cdot & O \\ O & M_{n_2} & O & \cdot & \cdot \\ \cdot & O & \cdot & \cdot & \cdot \\ \cdot & \cdot & \cdot & \cdot & O \\ O & \cdot & \cdot & O & M_{n_r} \end{pmatrix}$$

把這個概念記起來，再往下看。假設，L 是另外一個 $n \times n$ 矩陣使得 $Lv_i = w_i$，$\forall i = 1, 2, ..., n$，(參看，Theorem 4.1, 102 頁。)

$$\begin{array}{ccccccc} C^n & \xrightarrow{P_1} & C^n & \xrightarrow{L} & C^n & \xleftarrow{P_2} & C^n \\ A\uparrow & & M\uparrow & & M\uparrow & & B\uparrow \\ C^n & \xrightarrow{P_1} & C^n & \xrightarrow{L} & C^n & \xleftarrow{P_2} & C^n \end{array}$$

上圖中之 $P_1 = mat(I, E, \alpha)$，$P_2 = mat(I, E, \beta)$，且 L 是可逆的，再則由上面

簡圖可以得知，$A = P_1^{-1}L^{-1}P_2BP_2^{-1}LP_1$。由於，$P_1$，$P_2$ 以及 L 都是可逆的，所以，$P_2^{-1}LP_1$ 是可逆的，而且 $(P_2^{-1}LP_1)^{-1} = P_1^{-1}L^{-1}P_2$。這說明，$A$ and B are similar。所以，總歸一句話，兩個有相同 invariants 的 nilpotent matrices 是相似的。

現在到了談論，我們的主要目標，**Jordan canonical form**，的時候了。Jordan canonical form 是 matrix representations 的最高境界。它的重要性、它的可看性，自然不言可喻。

Theorem 9.8

Let $\lambda_1, \lambda_2, ..., \lambda_k$ be distinct eigenvalues of an $n \times n$ complex matrix, A. Then a basis of C^n can be found in which the matrix representation for A according to the basis is of the form,

$$\begin{pmatrix} J_1 & & & \\ & J_2 & & \\ & & \cdots & \\ & & & J_k \end{pmatrix}, \text{where each } J_i = \begin{pmatrix} B_{i1} & & & \\ & B_{i2} & & \\ & & \cdots & \\ & & & B_{ir_i} \end{pmatrix}, \quad i = 1, 2, ..., k$$

and each $B_{ij} = \begin{pmatrix} \lambda_i & 1 & 0 & \cdot & 0 \\ 0 & \lambda_i & 1 & \cdot & \cdot \\ \cdot & 0 & \cdot & \cdot & 0 \\ \cdot & \cdot & \cdot & \cdot & 1 \\ 0 & \cdot & 0 & 0 & \lambda_i \end{pmatrix}$, $j = 1, 2, ..., r_i$.

(註：the square matrix, B_{ij}, with λ on the main diagonal, 1 on the 2nd upper diagonal, and 0 elsewhere is called a **basic Jordan block belonging to** λ. If B_{ij} is an $m \times m$ square matrix, then $B_{ij} = \lambda I_m + M_m$, where M_m is the $m \times m$ square matrix, defined in the last section, with 1 on the 2nd upper diagonal and 0 elsewhere.)

Proof

首先令 $P_A(t) = (t-\lambda_1)^{m_1}(t-\lambda_2)^{m_2}\cdots(t-\lambda_k)^{m_k}$ 爲矩陣 A 的 characteristic polynomial。那麼，8.4 節的 Caley-Hamilton theorem 告訴我們，$P_A(A) = (A-\lambda_1 I)^{m_1}(A-\lambda_2 I)^{m_2}\cdots(A-\lambda_k I)^{m_k} = 0$。現在，假設 $W_i = Ker(A-\lambda_i I)^{m_i}$, $\forall i=1,2,\cdots,k$，那麼由 330 頁的推論得知 $C^n = W_1 \oplus W_2 \oplus \cdots \oplus W_k$。接著，我們的做法是，把矩陣 A 限制在每一個子空間 W_i 上運算。也就是說，令 $A_i = A|_{W_i}$, $i=1,2,...,k$，則得知 $(A_i - \lambda_i I)^{m_i} = 0$, $\forall i=1,2,...,k$。這告訴我們，每一個矩陣 $(A_i - \lambda_i I)$ 都是 nilpotent matrix。此時，由 9.1 節定理 9.4 的結果以及本定理之前的論述得知，存在一組 W_i 中的 basis，使得 $(A_i - \lambda_i I)$ 的 matrix representation according to the basis 唯一被決定爲如下型式

$$\begin{pmatrix} M_{i1} & O & \cdot & \cdot & O \\ O & M_{i2} & O & \cdot & \cdot \\ \cdot & O & \cdot & \cdot & \cdot \\ \cdot & \cdot & \cdot & \cdot & O \\ O & \cdot & \cdot & O & M_{ir_i} \end{pmatrix}$$

或者說，A_i 的 matrix representation according to the basis 唯一被決定爲如下型式

$$\begin{pmatrix} \lambda_i & & & & \\ & \lambda_i & & & \\ & & \cdot & & \\ & & & \cdot & \\ & & & & \lambda_i \end{pmatrix} + \begin{pmatrix} M_{i1} & O & \cdot & \cdot & O \\ O & M_{i2} & O & \cdot & \cdot \\ \cdot & O & \cdot & \cdot & \cdot \\ \cdot & \cdot & \cdot & \cdot & O \\ O & \cdot & \cdot & O & M_{ir_i} \end{pmatrix} = \begin{pmatrix} B_{i1} & & & & \\ & B_{i2} & & & \\ & & \cdot & & \\ & & & \cdot & \\ & & & & B_{ir_i} \end{pmatrix} = J_i$$

這正說明了，所謂的 Jordan canonical form。

Example 2

Given a 4×4 matrix, $A = \begin{pmatrix} 3 & 1 & -1 & 2 \\ 0 & 3 & 0 & 1 \\ 0 & 0 & 3 & 0 \\ 0 & 0 & 0 & 2 \end{pmatrix}$. Find its Jordan canonical form.

Solution

首先看看它的 characteristic polynomial,
$$P_A(t) = \det(tI - A) = (t-3)^3(t-2)$$

今令 $f(t) = (t-3)^3$, $g(t) = (t-2)$，則

$$f(A) = (A - 3I_4)^3 = \begin{pmatrix} 0 & 1 & -1 & 2 \\ 0 & 0 & 0 & 1 \\ 0 & 0 & 0 & 0 \\ 0 & 0 & 0 & -1 \end{pmatrix}^3 = \begin{pmatrix} 0 & 0 & 0 & 1 \\ 0 & 0 & 0 & 1 \\ 0 & 0 & 0 & 0 \\ 0 & 0 & 0 & -1 \end{pmatrix}$$

$$g(A) = (A - 2I_4) = \begin{pmatrix} 1 & 1 & -1 & 2 \\ 0 & 1 & 0 & 1 \\ 0 & 0 & 1 & 0 \\ 0 & 0 & 0 & 0 \end{pmatrix}$$

$$W_1 = \mathrm{Ker} f(A) = \left\{ \begin{pmatrix} x \\ y \\ z \\ w \end{pmatrix} : \begin{pmatrix} 0 & 0 & 0 & 1 \\ 0 & 0 & 0 & 1 \\ 0 & 0 & 0 & 0 \\ 0 & 0 & 0 & -1 \end{pmatrix} \begin{pmatrix} x \\ y \\ z \\ w \end{pmatrix} = \begin{pmatrix} 0 \\ 0 \\ 0 \\ 0 \end{pmatrix} \right\}$$

$$= \left\{ s \begin{pmatrix} 1 \\ 0 \\ 0 \\ 0 \end{pmatrix} + t \begin{pmatrix} 0 \\ 1 \\ 0 \\ 0 \end{pmatrix} + w \begin{pmatrix} 0 \\ 0 \\ 1 \\ 0 \end{pmatrix} : s, t, w \text{ are scalars} \right\}$$

注意，$(A - 3I_4)$ is hence a nilpotent matrix restricting on W_1。

令 $v = \begin{pmatrix} 0 \\ 1 \\ 0 \\ 0 \end{pmatrix}$，則 $(A - 3I_4)v = \begin{pmatrix} 1 \\ 0 \\ 0 \\ 0 \end{pmatrix}$ 且 $(A - 3I_4)^2 v = \begin{pmatrix} 0 \\ 0 \\ 0 \\ 0 \end{pmatrix}$。現在，以 $\{(A - 3I_4)v, v\}$

為基底，所得出矩陣 $(A - 3I_4)$ 之 matrix representation is of the form,
$$M = \begin{pmatrix} 0 & 1 \\ 0 & 0 \end{pmatrix}。$$

跟著，再令 $w = \begin{pmatrix} 0 \\ 1 \\ 1 \\ 0 \end{pmatrix}$，得出 $(A-3I_4)w = 0$。此時，以 $\{(A-3I_4)v, v, w\}$ 為基底，$(A-3I_4)$ 之 matrix representation 則為 $M_1 = \begin{pmatrix} \begin{pmatrix} 0 & 1 \\ 0 & 0 \end{pmatrix} & 0 \\ 0 & 0 & (0) \end{pmatrix}$。或者說，$A$ 在 W_1 上的 canonical form 是 $J_1 = 3I + M_1 = \begin{pmatrix} \begin{pmatrix} 3 & 1 \\ 0 & 3 \end{pmatrix} & 0 \\ 0 & 0 & (3) \end{pmatrix}$。接著，

$$W_2 = Kerg(A) = \left\{ \begin{pmatrix} x \\ y \\ z \\ w \end{pmatrix} : \begin{pmatrix} 1 & 1 & -1 & 2 \\ 0 & 1 & 0 & 1 \\ 0 & 0 & 1 & 0 \\ 0 & 0 & 0 & 0 \end{pmatrix} \begin{pmatrix} x \\ y \\ z \\ w \end{pmatrix} = \begin{pmatrix} 0 \\ 0 \\ 0 \\ 0 \end{pmatrix} \right\}$$

$$= \left\{ s \begin{pmatrix} -1 \\ -1 \\ 0 \\ 1 \end{pmatrix} : s \text{ is a scalar} \right\}$$

令 $u = \begin{pmatrix} -1 \\ -1 \\ 0 \\ 1 \end{pmatrix}$，$(A-2I_4)u = 0$。Restricting $(A-2I_4)$ on W_2, its matrix representation according to the single basis, u, is $M_2 = (0)$。也因此，$J_2 = 2I + M_2 = (2)$。

最後得出，矩陣 A 的 Jordan canonical form 為

$$\begin{pmatrix} J_1 & \\ & J_2 \end{pmatrix} = \begin{pmatrix} \begin{pmatrix} 3 & 1 \\ 0 & 3 \end{pmatrix} & 0 & 0 \\ 0 & 0 & (3) & 0 \\ 0 & 0 & 0 & (2) \end{pmatrix}$$

Exercises

1. Given a 3×3 matrix, $A = \begin{pmatrix} 3 & 2 & 1 \\ 0 & 1 & 2 \\ 0 & 1 & -1 \end{pmatrix}$. Find its characteristic polynomial, $P_A(t)$.

   Decompose $P_A(t)$ into relatively prime factors, $f(t)$, $g(t)$ and $h(t)$ such that $C^3 = W_1 \oplus W_2 \oplus W_3$, where $W_1 = Kerf(A)$, $W_2 = Kerg(A)$ and $W_3 = Kerh(t)$. (Hint: $P_A(t) = (t - \sqrt{3})(t + \sqrt{3})(t - 3)$)

2. Given a 3×3 matrix, $A = \begin{pmatrix} 0 & 0 & 1+i \\ 0 & 2 & 0 \\ 1-i & 0 & 1 \end{pmatrix}$. Find its characteristic polynomial, $P_A(t)$.

   Decompose $P_A(t)$ into relatively prime factors, $f(t)$, $g(t)$ such that $C^3 = W_1 \oplus W_2$, where $W_1 = Kerf(A)$, $W_2 = Kerg(A)$. (Hint: $P_A(t) = (t+1)(t-2)^2$)

3. Given a 4×4 matrix, $A = \begin{pmatrix} 2 & 0 & 1 & 0 \\ 0 & 2 & 0 & -1 \\ 1 & 0 & 2 & 0 \\ 0 & -1 & 0 & 2 \end{pmatrix}$. Find its characteristic polynomial, $P_A(t)$. Decompose $P_A(t)$ into relatively prime factors, $f(t)$, $g(t)$ such that $C^4 = W_1 \oplus W_2$, where $W_1 = Kerf(A)$, $W_2 = Kerg(A)$. (Hint: $P_A(t) = (t-1)^2(t-3)^2$)

Index

A

A-cyclic   323
adjoint   142
alternating   185
associative law   35
augmented matrix   47

B

backward substitution   82
bases   13
basic Jordan block   334
basis   15
bijective   103
bilinear form   131
block matrix   209

C

canonical form   321
Cayley-Hamilton theorem   314
C-B-S inequality   154
change of basis matrix   122
change of coordinates matrix   123
characteristic polynomial   226
characteristic value   225
characteristic vector   225
cofactor   186

column space   46
column vector   36
complex conjugate   40
complex field   2
composite mapping theorem   121
composition   110
conjugate linear   132
constituent idempotent   296
coordinates   17
Cramer's rule   216
$CS(A)$ 44
column operations   190
commutative law   35

D

determinant   174
determinant function   186
diagonal matrix   78
diagonalizable   238
dimension   19
direct sum   25
distinguished column   50
distributive law   36

E

eigenspace   224

eigenvalue 224
eigenvector 224
elementary matrix 61
epimorphism 104
$\varepsilon(\sigma)$ 198
existence 188
expansion according to columns 188

F
field 2
fixed line 225
Fourier coefficient 158
Fourier series 161

G
Gauss-Jordan elimination 56
Gram matrix 136
Gram-Schmidt orthogonalization 159

H
hermitian bilinear form 139
hermitian matrix 41
homogeneous linear system 47
Householder matrix 267
hyperplane 107

I
idempotent matrix 257
identity map 99

identity matrix 32
injective 104
inner product 150
inner product space 151
interpolatory polynomial 309
inverse map 111
inverse matrix 37
invertible matrix 37
isomorphism 104

J
Jordan canonical form 334

K
kernel 103
$KerT$ 103

L
least error problem 165
LDU-factorization 209
Legendre polynomials 161
linear combination 10
linear dependence 13
linear independence 13
linearly dependent 13
linear map 100
linear mapping 100
linear operator 111
linear space 2

linear system   43
linear transformation   99
lower triangular matrix   78
LU-factorization   82
$L(V, W)$ 120

M
matrix   31
matrix representation   115
$mat(T, \alpha, \beta)$ 118
maximal set of linear independence   20
method of block matrices   208
Minkowski's inequality   154
monomorphism   104
multiplication of matrices   34

N
negative definite   145
negative semi-definite   145
nilpotency (index of nilpotence)   317
nilpotent   94
n-linear   184
non-definite   145
non-singular   38
non-trivial solution   47
norm   152
normal form   163
normal matrix   287
$NS(A)$ 67

null space   67

O
orthogonal basis   156
orthogonal complement   161
orthogonal vectors   155
orthonormal basis   156
orthonormal column   172
orthonormal matrix   172
orthonormal row   172
ortho-triangular factorization   166

P
parallelogram law   154
Parseval's identity   162
permutation   194
pivotal condensation   213
point spectrum, $\sigma_p(T)$ 224
positive definite   145
positive semi-definite   145
projection   164
Pythagoras law   157

Q
quadratic form   144

R
range   103
rank   67

ranks of matrices   71
RanT   103
real field   2
row-echelon matrix   50
row equivalent   63
row operation   54
row space   45
row vector   36
RS(A)   46

S

scalar   2
scalar product   150
self-adjoint   142
set of eigenvalues   224
similar matrices   127
singular   38
smallest subspace   9
spanning set   10
spectral basis   238
spectral decomposition   299
spectral theorem   249
SP(S)   10
square matrix   32
standard basis   15
standard inner product   152
strictly upper triangular matrix   94
subspace   5
sum   25

surjective   103
symmetric   40
symmetric bilinear form   140

T

trace of a matrix   143
transpose   39
transpose conjugate   40
transposition   195
trivial solution   47
trivial subspace   6

U

unique   38
uniqueness   188
unitary diagonalizable   242
unitary matrix   172
upper row-echelon matrix   82
upper triangular matrix   78
upper unit triangular matrix   79

V

vector   5
vector space   2

Z

zero column vector   47
zero map   100
zero matrix   32

Bibliography

Richard E. Phillips
Lecture note on a course in matrix theory, 1982, Michigan State University.

John B. Conway
A course in functional analysis, Springer-Verlag, 1985.

Serge Lang
Linear Algebra, 3rd edition, Springer-Verlag, 1987.

Otto Bretscher
Linear Algebra with applications, Prentice-Hall, 1977.

Steven J. Leon
Linear Algebra with application, 5th edition, Prentice-Hall, 1998.

Hal G. Moore
A first course in linear algebra, 2rd edition, Harper Collins, 1992.

W. Keith Nicholson
Linear Algebra with application, 4th edition, McGraw-Hill, 2002.

Lee W. Johnson
Introduction to Linear Algebra, 5th edition, Pearson, 2002.

Richard C. Penney
Linear Algebra, ideas and applications, John Wiley & Sons, 1998.

Tom M. Apostol
Linear Algebra, A first course with applications to differential equations, John Wiley & Sons, 1997.

Stephen H. Friedberg
Linear Algebra, 3rd edition, Prentice-Hall, 1977.

I. N. Herstein
Topics in Algebra, 2nd edition, the Southeast book company, 1987.

Nathan Jacobson
Basic Algebra I, Freeman, 1974.